Low Emission Power Generation Technologies and Energy Management

Low Emission Power Generation Technologies and Energy Management

Edited by
Jean-Claude Sabonnadière

iSTE

WILEY

First published in France in 2007 by Hermes Science/Lavoisier entitled: *Nouvelles technologies de l'énergie volumes 2 et 4* © LAVOISIER, 2007
First published in Great Britain and the United States in 2009 by ISTE Ltd and John Wiley & Sons, Inc.
Translated from the French by Professor Albert Foggia and Ms Florence Martin.

ISTE Ltd
27-37 St George's Road
London SW19 4EU
UK

www.iste.co.uk

John Wiley & Sons, Inc.
111 River Street
Hoboken, NJ 07030
USA

www.wiley.com

Library of Congress Cataloging-in-Publication Data

Stockage et technologies à émission réduite. English
Low emission power generation technologies and energy management / edited by Jean-Claude Sabonnadière.
p. cm.
Originally published: Paris : Hermes science, 2006: under title: Nouvelles technologies de l'énergie, Pt. 2. Stockage et technologies à émission réduite.
Includes bibliographical references and index.
ISBN 978-1-84821-136-0
1. Power resources--Management. I. Sabonnadière, Jean-Claude II. Title.
TJ163.3.S74713 2009
621.042--dc22

2009017411

British Library Cataloguing-in-Publication Data
A CIP record for this book is available from the British Library
ISBN: 978-1-84821-136-0

Printed and bound in Great Britain by CPI Antony Rowe, Chippenham and Eastbourne.

FSC
Mixed Sources
Product group from well-managed forests and other controlled sources
Cert no. SGS-COC-2953
www.fsc.org
© 1996 Forest Stewardship Council

Table of Contents

Preface

1.1. Electric energy

Energy, which was the basis of the industrial revolution, has had an exponential growth since the birth of industrial electricity at the end of the 19th century. The discovery of the rotating field by Nikola Tesla and the invention of the transformer allowed the expansion of three phases: alternating currents for generation, transmission, and delivery and uses of electric energy in the cheapest way.

At the beginning of the 20th century the development of large electric power grids enabled many countries throughout the world to bring the benefits of electric energy to their citizens, while intensively developing the industrial and tertiary applications of electricity. This growth led to generalization of the use of electric energy in domestic applications and all sectors of industry.

Generation, transportation, transmission and distribution of electric energy were considered to be such strategic operations by most countries that they decided to build them as monopolistic state companies in order to control their development. These decisions and the heavily capitalistic nature of generating and transmission systems led to a vertical integration of electric energy utilities for economic reasons. This is the classic paradigm that for more than a century has allowed the creation of an industry that reached the heights of its power in a slow but constant improvement of the reliability of equipment, the main objective being to ensure the supply of electricity to domestic and industrial customers connected to the grid. The tremendous growth of electric energy consumption during the middle of the last century led to the construction of very large and complex electric power systems (for instance, in the French grid, more than 66,000 MW are flowing at any time in about 1,300,000 km of lines and cables).

Figure 1 shows a schematic representation of these systems.

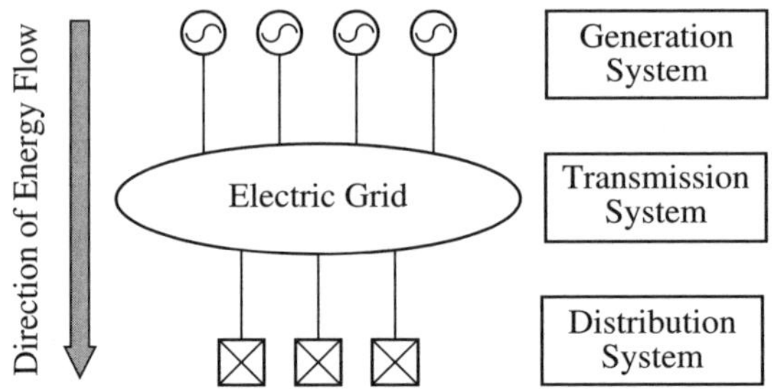

Figure 1. *The historic paradigm*

All electric power systems have been built and operated according to this diagram from the beginning, and for almost all, of the 20th century. Their organization and their operation from generation to the consumer were integrated inside only one, generally monopolistic, private or public company.

The deregulation that started at the beginning of the 1980s introduced a tremendous change by imposing the unbundling of the functions of generation, transmission and distribution. This change introduced a new mode of organization according to a broken up model we shall describe in the next paragraph as the new paradigm.

1.2. The new paradigm

The goal aimed at by the instigators of the deregulation of electric systems has always been to promote a new organization of the electric systems in order to create the conditions of commercial competition among all the stakeholders: the aim is to lower the price of electric energy supply for consumers.

The setting up of new regulations took place in a context where geographic constraints were prominent due to the territorial settlement of power grids that were in fact naturally monopolistic.

However, the new system instituted in the UK at the beginning of the 1990s, and then in the USA, is now installed in almost all developed countries, although with some difficulties of adaptation. As a matter of fact, the systems designed and built on to be operated as an integrated company in a well-defined territory must today be operated on a continental scale without prior modification of the infrastructures of transmission and interconnection.

1.2.1. *The energy system in a deregulated context*

The major functions which must be carried out in order to satisfy consumer demand are the same as previously described but with a different mode of interaction, and they are operated by different actors according to the five essential links described in Figure 2.

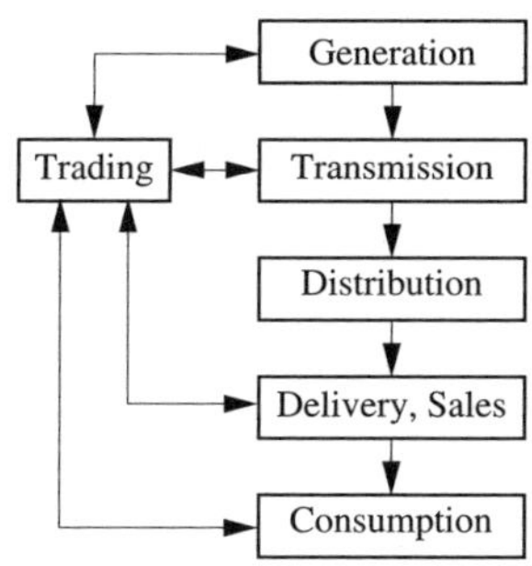

Figure 2. *The new organization of electric systems*

This new organization creates a total hierarchical independence between the producers and the other members of the system. The producers of course are in charge of the generation. They sell to consumers according to several kinds of contracts for short or medium term delivery. The traders are the link between generation and consumption, taking into account the power transmission capacity through transmission and distribution grids.

The power grids must ensure the transmission of energy to the consumers according to the trading exchanges they had with the producers, insuring equitable treatment for all producers. Independent system operators for transmission and distribution grids are in charge of this duty.

This new set of concepts generates not only drastic changes in the economic conditions of the electric system's operation, but also significant technical changes that favor the new technologies with dispersed generation, especially renewable energy technologies that will continue to be the foundation of consumer power.

1.3. Dispersed generation

The economic issues arising from the new regulations incite the big consumers to buy internal generation units in order to smooth the price fluctuations coming from deregulation. They are of course small generation units of limited power, generally connected to distribution grids. These units will be able to control the power imported by the consumer, who will be able to sell his energy on the market by injecting it into the grid if price conditions are favorable. The development of a large number of small generation units from solar, wind, small hydraulic or thermal units, with combined heat and power, alongside classical large generation machines, will superimpose a new phenomenon on the normal operation of the distribution grids: bidirectional energy flow, generally intermittent and random according to wind and sun production.

This phenomenon will create new difficult problems such as the management of these energies while keeping the grid security at the required level.

The new paradigm, which integrates dispersed generation with the economic environment, will lead to a new operation scheme for electric power systems, one that will increasingly be a substitute for the present scheme as described in Figure 3.

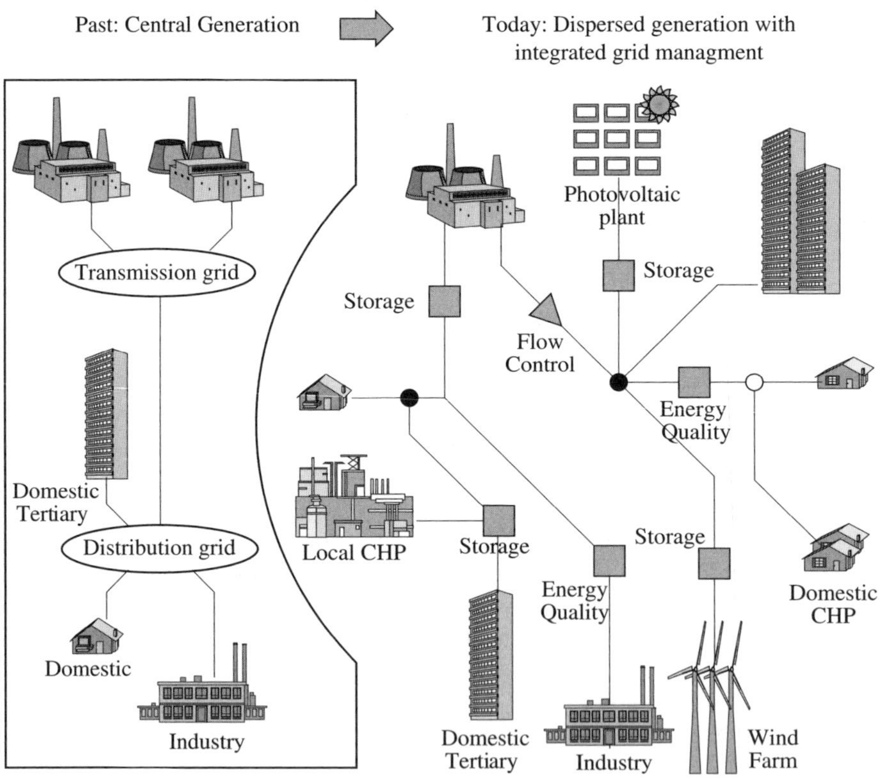

Figure 3. *The new electric power system*

1.4. The new energy generation and management

Independently of these new schemes of grid operation, in the following chapters we shall describe the main characteristics of the emerging generation technologies that will be installed on the grid of the next decades.

In the first volume, *Renewable Energies*, we reviewed the new energy technologies in terms of the types of renewable energy. This second volume, *Low Emission Technologies and Energy Management*, will deal with reduced emission technologies and energy saving.

Volume 1 includes various forms of solar energy such as photovoltaic, thermal and thermodynamic energy conversion.

Wind technologies are in full development today. The chapters dedicated to them describe the state of the art, taking into consideration the question of insertion into the grids of a large quantity of this intermittent energy.

Energy from the sea completes this general view, with a chapter discussing very small hydraulic plants that become of interest when fossil energies become more and more expensive.

We then provide an analysis of geothermal energy, followed by energy from biomass. This entails a full description of biofuels and biogas, and especially energy from wood as a substitute to fossil energy heating.

Volume 2 is dedicated to energy storage, low emission technologies and energy management.

We start with the new generation of nuclear energy, which is presently at a crossroads of its history with these future new generations. It then analyzes combined heat and power generation, by which it is possible to produce heat and electricity jointly as a complement to heat generation in order to improve the efficiency of heat plants. A very careful analysis of the economic conditions of the operation of combined heat and power leads to a description of the economic conditions for which this technology is advantageous.

Energy storage describes the various means and methods of storage in association with intermittent energies like photovoltaic and wind energy plants. Their efficiency and cost are strongly dependent upon their operational facilities and investment costs.

A projection of the use of hydrogen as an energy vector similar to electricity will give readers a comprehensive view of the way to create, store and transport this gas, which is generally improperly considered as very dangerous. Its future using fuel cells as a conversion facility allows us to foresee significant development of this energy vector with a large set of applications.

Finally, we take up the very important subject of energy management, control of energy demand, and energy saving. We describe positive energy houses, low consumption public and domestic lighting, and power moderation by control of the load from the grid.

Jean-Claude SABONNADIÈRE

Chapter 1

Energy Storage: Applications to the Electricity Vector

Energy is a fundamental need for human beings. We can extract it from nature and use it in the way best adapted to our needs, either directly (heat, mechanical energy, etc.) or by transforming it into a more polyvalent form, such as electricity.

The human body is itself an energy source, and also an efficient converter of biochemical energy.

Exploitable energy is stored in nature in various forms and densities:

– fossil storage, in the terrestrial geological layers;

– gravitational storage, e.g. using water;

– biochemical storage, for biomass or the human body (glycogen);

– thermal storage for the sun, the effects of which also influence wind and water energy;

– electromagnetic storage, in electric and magnetic fields.

1.1. Energy density

The level of energy stored in materials is very different to that stored in storage components. Figure 1.1 shows the great advantage of fossil sources over secondary

---

Chapter written by Yves BRUNET.

storage sources. Nuclear sources are even more concentrated since we obtain 10^8 Wh/kg from the natural fission of uranium.

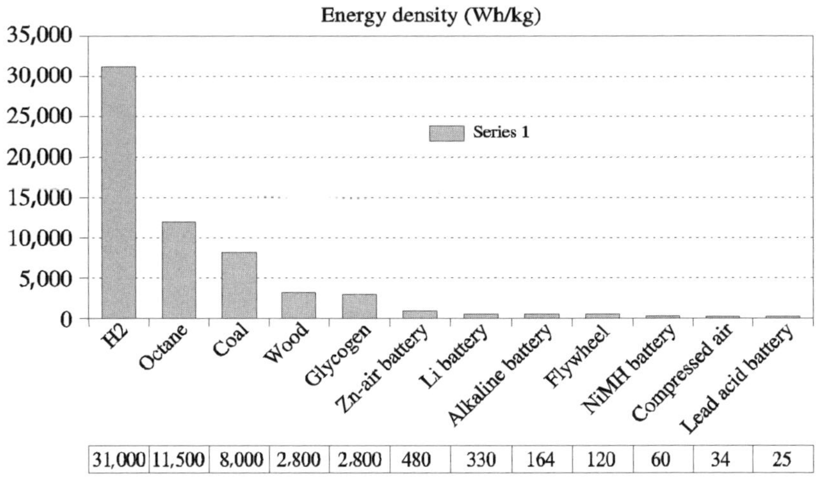

Figure 1.1. *Energy density*

Once the energy has been extracted and generated in an easily transportable form by an energy vector, it is brought to the consumer in the form best suited to the needs of the consumer. Electricity is one of the most versatile vectors known to date (Figure 1.2).

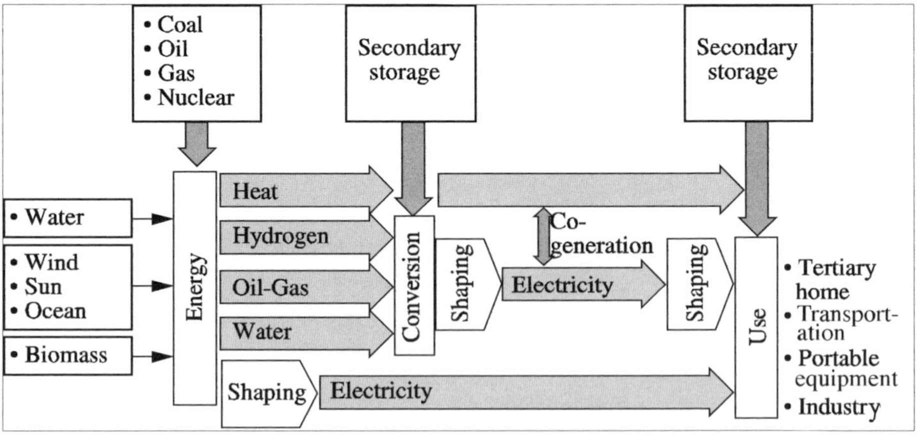

Figure 1.2. *Main energy vectors*

In the energy chain, storage may be implemented at each step. Storage is particularly important for electricity networks since the balance between delivered supply and demand must be assured. The most suitable choice of storage technology must be made taking into account the intended application and economic parameters such as cost of investment, energy or power densities, life cycle, and impact on the environment (Figure 1.3).

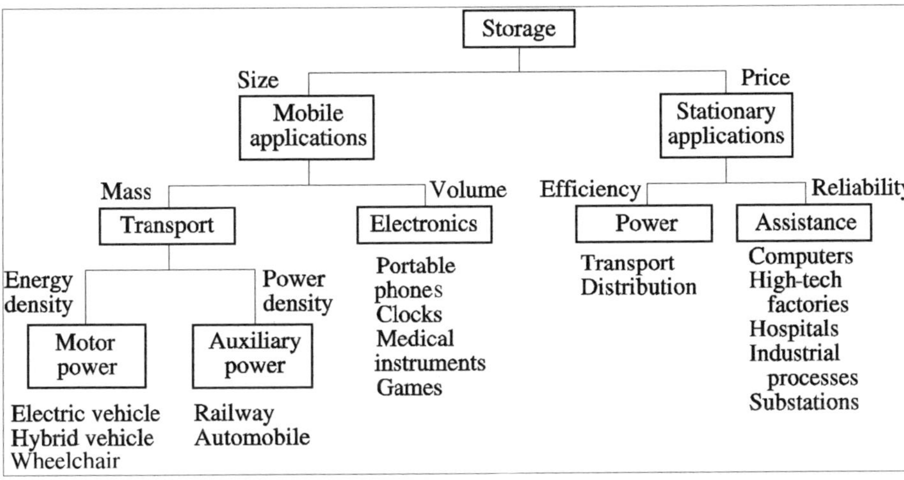

Figure 1.3. *Constraints and criteria for storage choices versus applications*

There are so many application ranges that it is better to start with some simple applications. What can we do with 1 kWh?

– drive 1 km in a car that uses 8 liters of fuel for 100 km;

– run a refrigerator for one day;

– light a house for one evening;

– manufacture 200 g of steel or 100 g of plastic.

On average, the total energy used in France, per capita and taking in to account all types of energy, corresponds to 40 MWh/year, which means 4.5 kWh/hour per person.

1.2. Storage problem

All energy sources are stored, either on a geological scale, or in the case of the sun, for even longer, and then released as they are needed. The notion of renewable energy makes sense only in terms of human timescales.

We can distinguish the primary storage of fossil sources that exist naturally, and for which we only have to pay for extraction, and the secondary storage created by man, for which we have to pay for the storage and extraction. This chapter focuses on secondary storage.

Sources	Units of time
Biomass	Years
Thermal gradient of oceans	Hundreds of years
Fossil fuels	Millions of years
Tides, waves	Hours
Geothermal	Days or years
Thermal masses	Hours
Batteries	Minutes
Super conducting magnetic energy storage (SMES)	Seconds
Capacitors	Seconds
Hydraulic pumping	Hours

Table 1.1. *Rechargeable time constants of the sources (source: [HER 05])*

1.2.1. *Electrical networks*

Storage is essential in mobile applications (vehicles, onboard systems, portable equipment) in order to carry the necessary energy for their operation.

In the case of electrical networks, electricity cannot be stored efficiently and in a sufficient quantity so supply must always be able to meet demand, taking into account the daily and seasonal changes in consumer requirements (Figure 1.4).

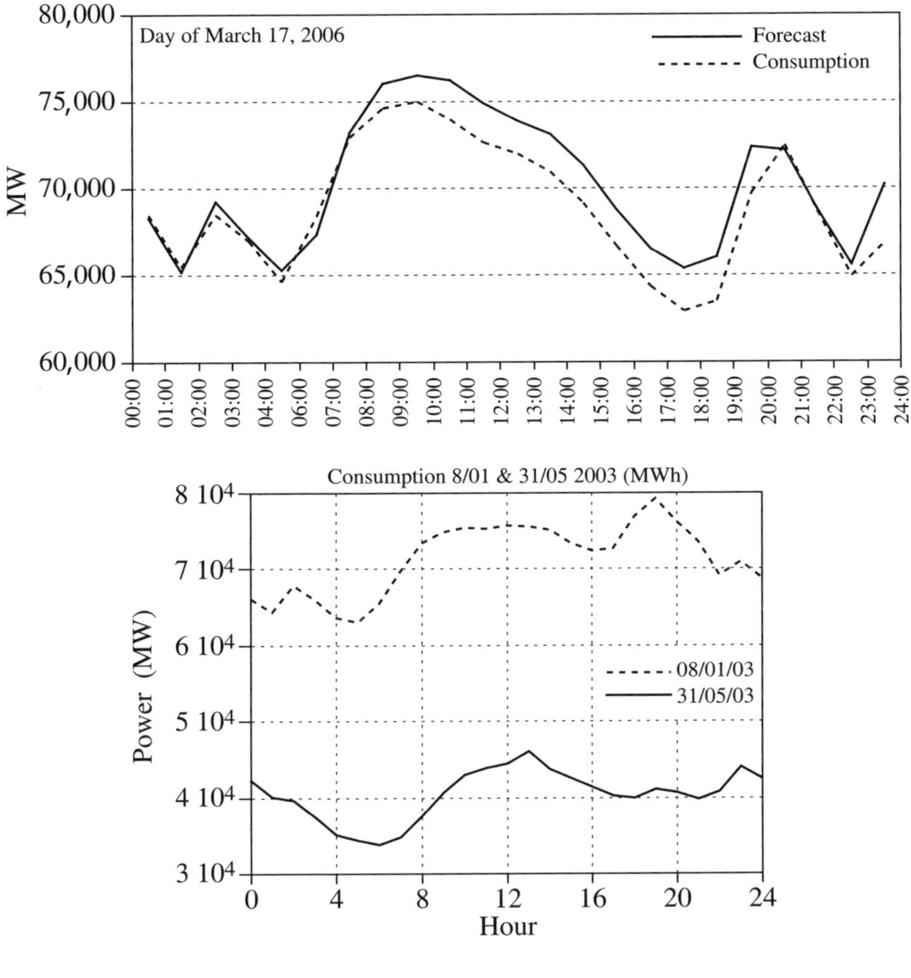

Figure 1.4. *Forecast curves of electric energy consumption in France*

It is therefore necessary to store energy in an auxiliary physical form (mechanical, thermal, chemical) and to convert this stored energy into electricity (battery, generator). This can be done using converters and energy adaptors, which have an efficiency rating of 80 to 90%, but it results in both energy and financial costs.

1.2.2. *Electric energy in France: forecast and consumption curves*

To provide the differential energy in relation to the average value of daily consumption would lead to a potential storage requirement of dozens of GWh.

The financial cost resulting from momentary or prolonged interruptions (blackouts) of the electricity network is very important because it concerns all the economic sectors (such financial cost is estimated at many tens of billions of dollars per year in the USA) and this cost must be compared to that of energy storage that is capable of reducing the risk of interruptions.

Power and energy including distributed generation sources must also be globally managed at the network-operator level, by means of network operation systems like ICT (information and communication technology).

ICT, in addition to its classic control-command functions, is also able to manage production, storage, and load in the form of virtual plants.

Storage increases the flexibility and the reliability of electricity networks at each level of the electricity chain (Figure 1.5):

– at the generation level, by reducing the investment and operating costs: power reserve, frequency regulation, deferral of necessary new investments (lines, plants, etc. (see Figure 1.6);

– by enabling supply during momentary failure of classic generation sources;

– as electricity prices vary at different times of the day and year (due to the law of supply and demand in an increasingly open market), by enabling the cost of energy to be evened out, but it also provides the opportunity to speculate on its cost (although this operation has its own limits, see Figure 1.11);

– by enabling the reduction of losses in the transmission lines (see Figure 1.7);

– by giving flexibility and compensating for the intermittence of renewable energies, facilitates their integration in the networks (because of the intermittence of wind energy and its low unitary production capability, the penetration of wind energy is limited to 30% (Figure 1.8));

– by increasing the use rate of transport and distribution lines, it allows for delay in purchasing new equipment, and participates in voltage regulation;

– by guaranteeing the quality of energy at consumer levels (UPS: uninterruptible power supply) and reducing costs by taking care of peak power demands;

– by permiting the supply to isolated zones that are not connected to the grid.

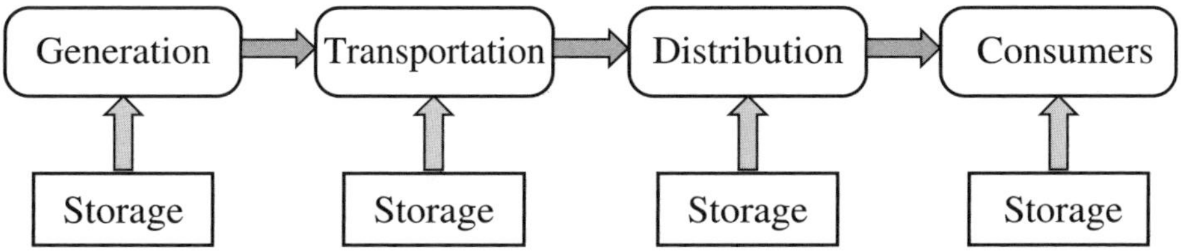

Figure 1.5. *Insertion points of storage means in a grid*

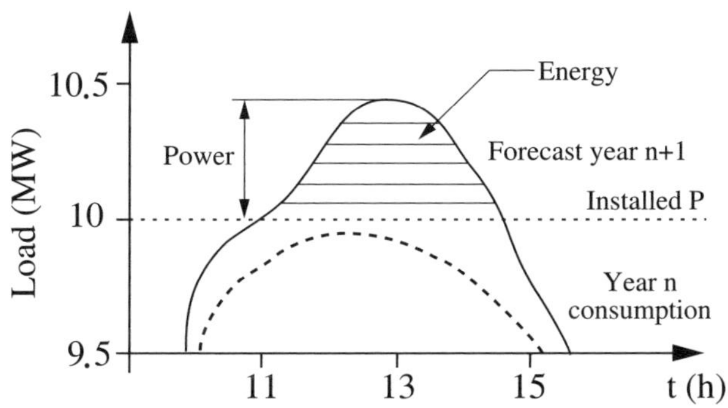

Figure 1.6. *Storage. One of the systems facilitating investment reports: a storage source of 1 MWh, 0.5 MW can avoid the implementation of a new production source*

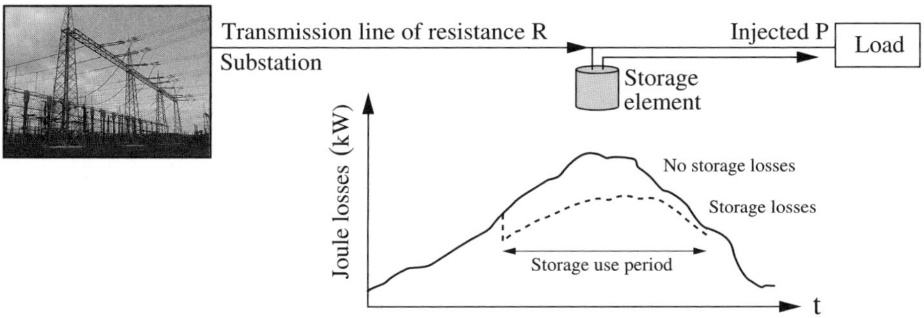

Figure 1.7. *Line losses reduction by insertion of a storage unit near the charge*

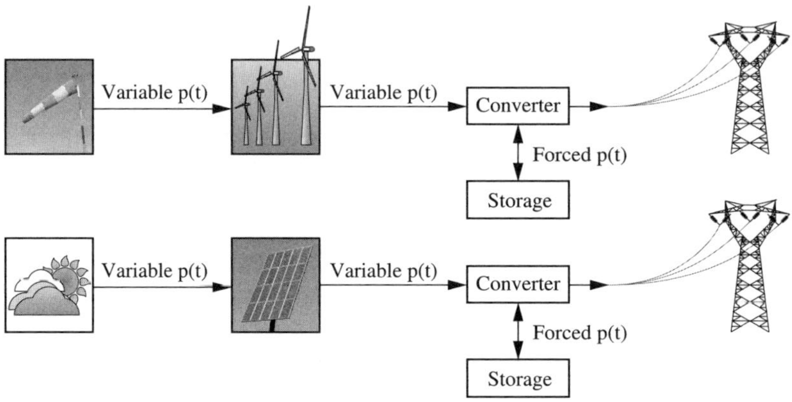

Figure 1.8. *Intermittent renewable sources and storage association*

At a consumer level, the problem is slightly different since the energy and power quantities are far less important; it might therefore be interesting to consider storage solutions closer to the point of use[1] (Figure 1.9).

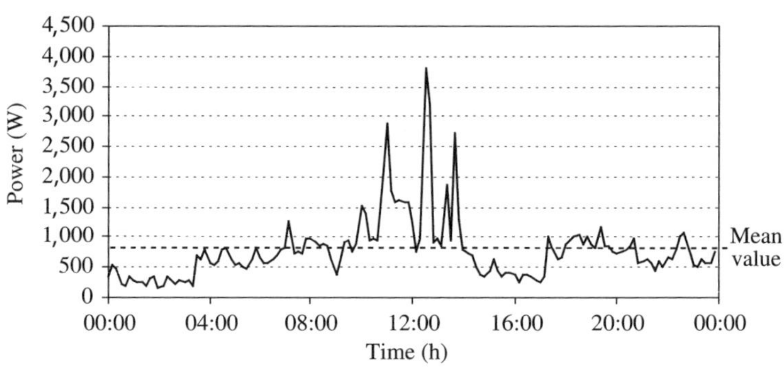

Figure 1.9. *Daily consumption in a house (November 2003). Instead of a 6 kW power installation, corresponding to a standard installation already above the necessary maximum, a 1 kW installation could be connected to the grid (mean value = 780 kW) with a 4 kW, 4 kWh storage facility, capable of smoothing the peaks*

In electricity networks, the types of applications can be distinguished depending upon their energy and power ranges. For each type of application, the preferred choice of storage technology depends on the quantity of power and energy required.

1. EPRI 2.4 kW, 15 kWh, Salt River Residential Photovoltaic-Battery Energy Storage System Project 1997 [EPR 03].

Type of application	Discharge Power Range	Discharge Time Range	Stored Energy Range	Applications
Energy storage	10–1,000 MW	1–10 h	10–10,000 MWh	Power reserve, generation smoothing, congestion management, investment deferment.
Distributed generation	0.1–2 MW	0.5–4 h	0.05–10 MWh	Investment deferment, power peak support, control.
Energy quality	0.1–2 MW	1–30 s	0.03–20 kWh	Power quality and reliability, PQ regulation, voltage support, transients, operation continuity.

Table 1.2. *Storage needs according to application types*

Energy Storage	Distributed Generation	Energy Quality
Pb acid batteries	Pb acid batteries	Pb acid batteries
NaS batteries	NaS batteries	Li-ion batteries
Redox flow	Redox flow	Flywheel
Zn-Br batteries	Li-ion batteries	Super capacitors
Ni-Cd batteries	Zn-Br batteries	SMES
Hydroelectric storage	Ni-Cd batteries	
Hydraulic pumping	Flywheel	
Super capacitors	Super capacitors	
Compressed air	Compressed air	
	Fuel cell	

Table 1.3. *Storage types according to applications*

Any storage problem must integrate the technical-economic aspects. We must pay for energy storage and restorage operations, with each operation having its own efficiency.

For each application it is important to optimize the choice of the storage type according to the primary fuel prices and the investment and operating costs. It is also important to manage the storage/restorage strategy (Figure 1.10).

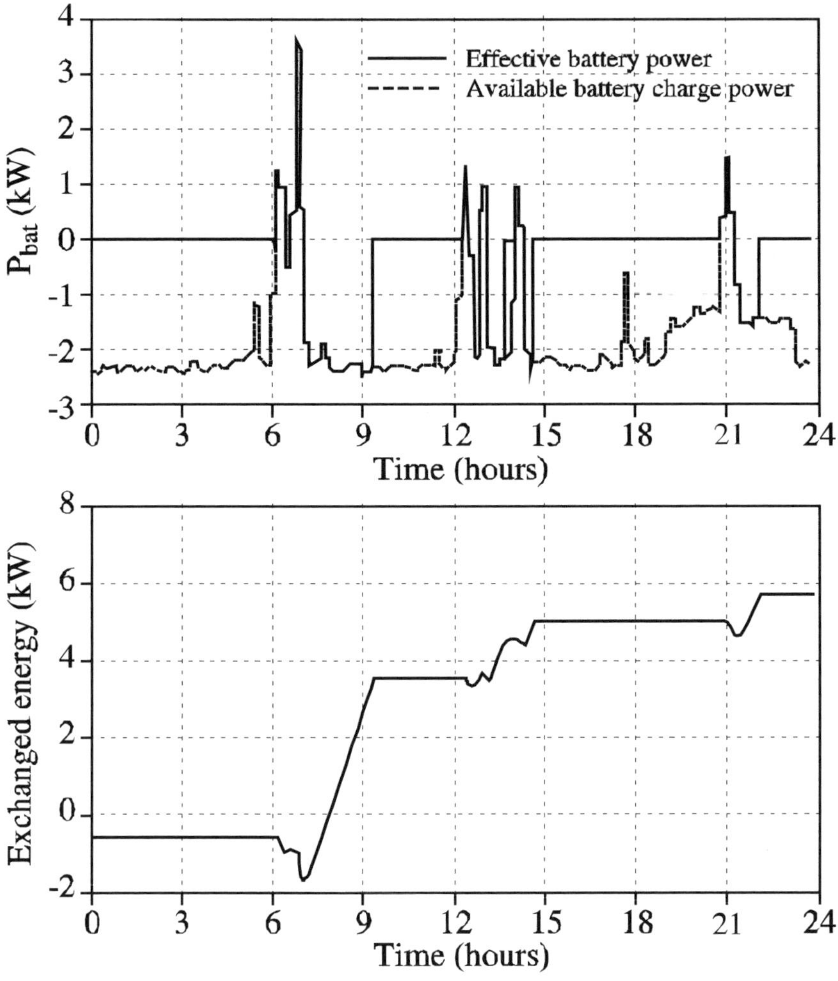

Figure 1.10. *Battery recharge strategy by consumption smoothing, [GER 02] with and without recharge power limitation*

Hydraulic pumping or compressed air storage (CAES) is, for example, more interesting when the price difference of electricity between off-peak hours and peak hours is important. A generalized development of local storage (batteries, fuel cells, etc.) may even lead to a smoothing of load curves, which could reduce the economic interest of some types of storage (Figure 1.11).

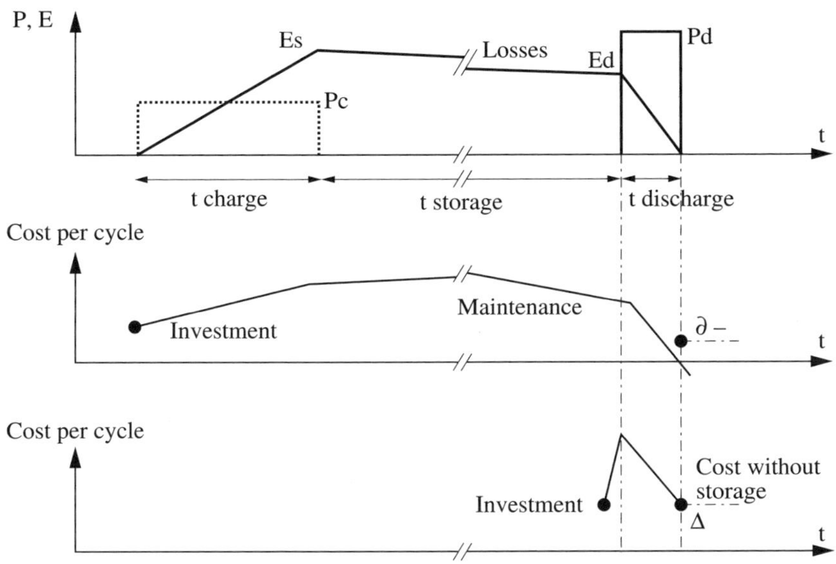

Figure 1.11. *Relevant limits of storage using the differential off-peak hours/peak hours*

1.2.3. *Relevant limits of storage using the differential off-peak hours/peak hours (Figure 1.11)*

During the loading period, low-cost electricity is used (off-peak hours). This energy (storing time may be large compared to charging time and discharge time). When it is used, the stored energy, minus losses (leakages), Ed < Es, is converted into electricity and used during peak hours, when electricity costs are high.

If we consider the economics of such a cycle, ∂ represents the producer cost with for a system with storage and Δ represents the same cost for a system without storage which requires specific equipment, e.g. gas turbine, with both systems giving the same service to the consumer. As long as $\partial < \Delta$, the system with storage is economically profitable.

Once energy is stored, it is restored according to the target applications, either in its power form (W), or in its energy form[2] (J or Wh), e.g. power for a certain period of time. Sometimes we will use storage sources that associate a quantity of stored energy with instantly available power. Different storage strategies may lead to different choices of solutions (Figure 1.12).

2. 3600 J = 1 Wh, 1 MWh = 0.0857 ton oil equivalent (toe), 1 toe = 11.7 MWh.

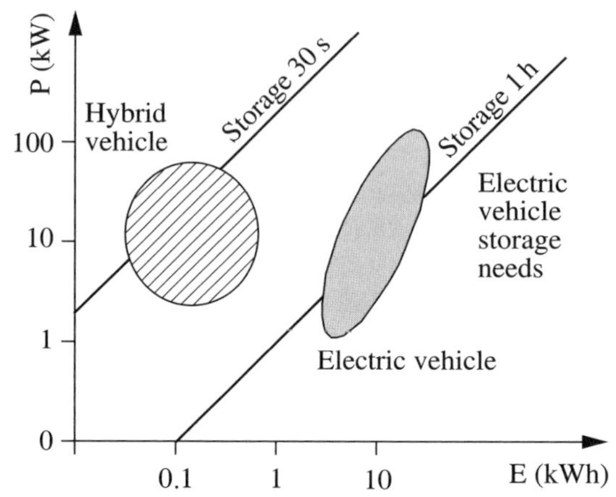

Figure 1.12. *Needs of an electric vehicle: a hybrid vehicle needs to store power whereas an electric vehicle will need both power and energy. For a hybrid vehicle, the objectives to be targeted are around 12 Wh/kg and 500 W/kg, with an available energy of 300 Wh and available power of 10 kW during 2 s and a 15-year life*

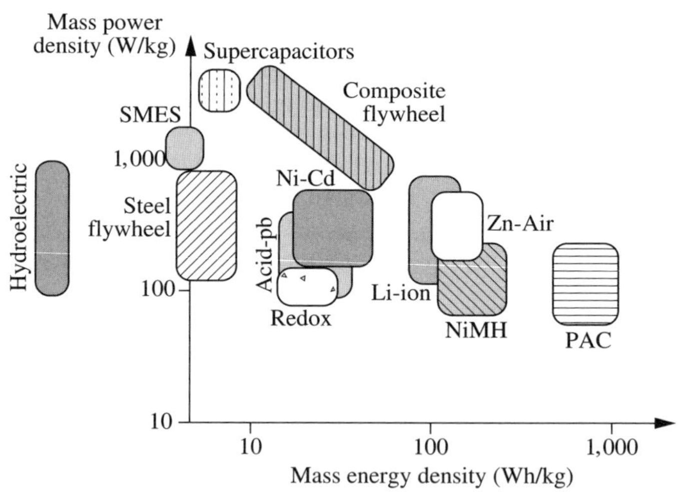

Figure 1.13a. *Rangone diagrams of power and mass energy densities of a selection of storage sources*

The different storage sources can thus be characterized by their power densities and energy per mass or per volume (Figures 1.13a and b).

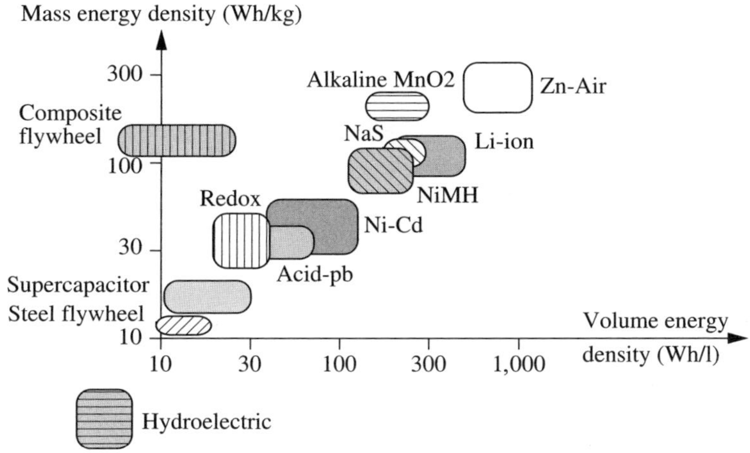

Figure 1.13b. *Rangone diagrams of power and mass energy densities of some storage sources*

Storage technologies offer ranges of power and energy extending over many orders of magnitude and adapted to very different needs; the efficiencies of storing and restoring operations and the number of cycles used are also relatively different (Figures 1.14 and 1.15).

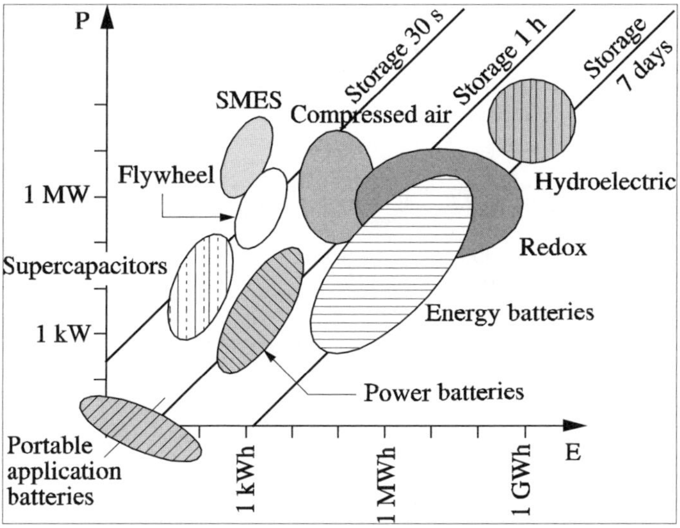

Figure 1.14. *Use range of storage sources*

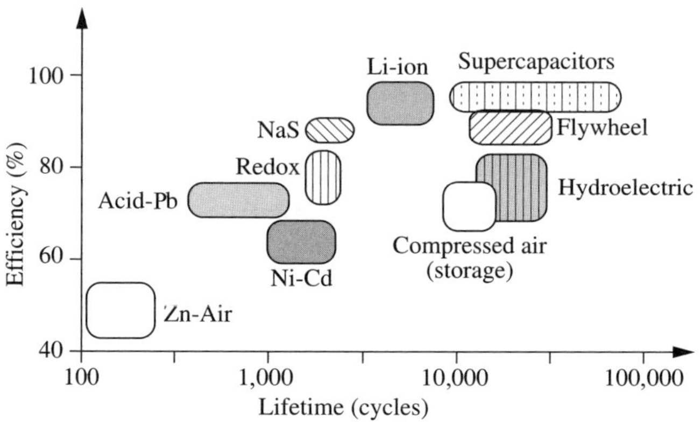

Figure 1.15. *Efficiency and lifetime of storage sources*

Moreover, in order to define the quality of a type of energy storage, the concept of exergy, which measures the absolute energy efficiency, must be considered.

Whilst energy efficiency is clearly defined for the main energies (mechanical, hydraulic, electric), usable thermal energy depends on its temperature; we are richer if we have a 2,200° source than a 80° source because when a thermal motor is used, more main energy can be extracted from the high temperature source than from the low temperature source.

The exergy concept, which takes into account the quality of energy, quantifies this property. The choice of a storage mode, e.g. heat, will take into account the required quality (Figure 1.16).

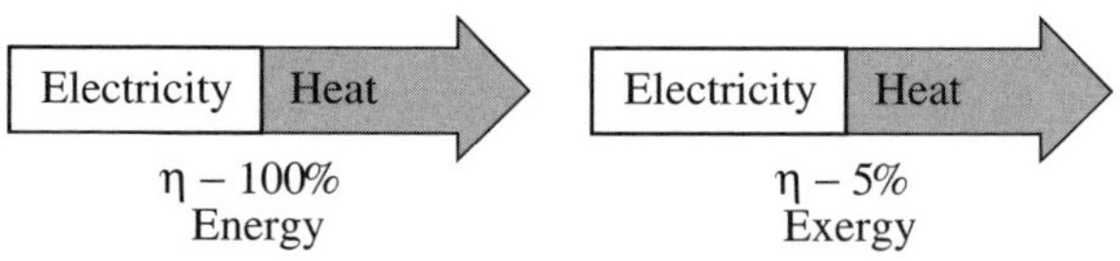

Figure 1.16. *Electrical heat efficiencies*

1.3. Types of storage

Storage technologies use different physical or chemical principles. These technologies can be classified according to physical-chemical phenomena they are based on.

Origin of stored energy	Technologies
Mechanical	Hydraulic Flywheel Compressed air Spring
Electrochemical	Batteries Hydrogen/fuel cell
Electromagnetic	Capacitor/supercapacitor Inductance/magnetic storage (SMES)
Thermal	Sensitive energy ($C\Delta T$): water, thermal masses Latent heat (L): melted salts with phase change

Table 1.4. *Physical principles and storage technologies*

These technologies are characterized by:

– storable quantity of energy;

– available power (discharge capacity);

– time constants of charge and discharge;

– lifetime;

– mass and volume densities;

– charge and discharge efficiency;

– ease of conversion to electricity – thermo-dynamic machines, e.g. gas turbines coupled to synchronous generators, and direct conversion technologies (batteries, fuel cells) are used;

– investment and maintenance costs;

– availability and local feasibility (especially mass storage sites);

– social and environmental considerations.

At very low scales, we can also consider micro-energy storage sources for personal applications of microelectronic technologies in the power range less than 100 mW. The range of potentially usable sources may then be enlarged with the sources using ambient radiation, a few temperature differences, mechanical distortion energy, etc.

The advantages and disadvantages of each storage system are outlined in the following table. The concept of power source is linked to the speed with which the energy is available. Thus, if the hydraulic pumping is able to supply large amounts of power, it will take a few minutes to start up.

Technology	Advantages	Drawbacks	Power source	Energy source
Hydraulic pumping	Important storage capacity Low cost	Specific sites Construction delays		λ
Compressed air CAES	Important storage capacity Low cost	Specific sites Cast with natural gas		λ
Redox flow battery	Important storage capacity	Low energy density	μ λ	λ
NaS battery	Large energy and power density	Cost, operation security	λ	λ
Metal-air battery	Large energy density	Difficult recharging		λ
Li-ion battery	Large energy and power density Good efficiency	Cost Special charge circuit	λ	μ
Ni-Cd battery	Large energy and power density Efficiency		λ	μ λ
Pb-acid battery	Low cost	Lifetime	λ	μ
Flywheel	High power	Low energy density	λ	μ
Magnetic storage SMES	High power	Low energy density Cost	λ	
Super capacitor	Lifetime Good efficiency	Low energy density	λ	μ λ

Table 1.5. *Comparison of storage technologies (source: ESA Electricity Storage Association)*
λ *= pertinent solution,* μ λ *= envisaged solution,* μ *= non-mature solution*

The main storage technologies are detailed in the following sections.

1.3.1. *Gravity storage*

The use of gravity leads to low-energy densities: with a difference in height of 100 m for 3.6 t of water we can store 1 kWh.

The stored energy and power are:

$$W = mgh \text{ and } P = \dot{m}gh$$

where $\dot{m}$ is the mass flow and h the height difference, which gives mass and volume densities equal to g h, i.e. respectively in the order of 0.2 to 2 Wh/kg and of 10^2 to 10^3 W/kg.

This type of storage is, thus, an excellent source of continuous power (starting times are around one minute) which requires large volumes in order to store significant energy, and strong flows in order to increase the power.

Some water reserves are mainly primary reserves (there is no storage cost apart from the investment for building the dam): they fill naturally. A dam that prevents water from flowing leads to a differential in water height. This water height differential, possibly increased by a pipe, produces energy converted into electricity by means of a turbine generator. Considerable amounts of power are thus obtained:

– the world hydroelectric production is of 2,600 TWh/year;

– the Churchill Falls (Canada) site has a reservoir of 6,988 km^2, a fall height of 312 m and an installed power of 5.4 GW for a flow of 1,700 m^3/s; this power is available for more than 50 days if a level reduction of 1 m is accepted;

– the Three Gorges plant (China) will have an installed power of 18 GW for a 85 TWh/year production.

Pumping station plants (STEP) are secondary storage sources. They use the cost differential between peak and off-peak hours to make storage profitable. The efficiency is about 70–80%, the cost is small, but this type of storage needs adapted sites (Figure 1.17).

For example, the Grand-Maison site stores 137,000,000 m^3 of water for a lake surface of 2.19 km^2, with a fall height of 955 m. It generates 1,800 MW with a flow of 217 m^3/s, and could store 1.3 TWh if the water reserve is completely emptied.

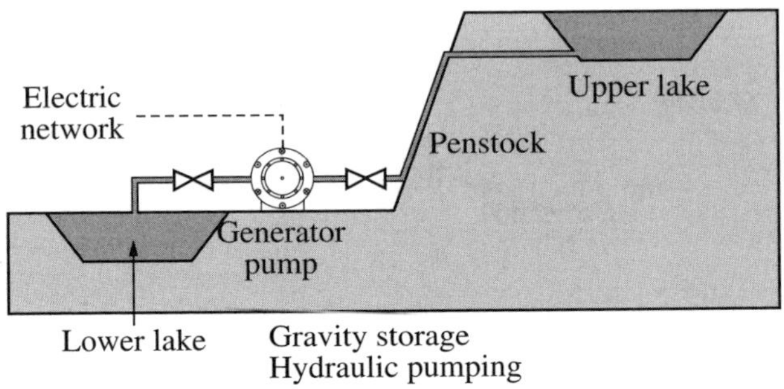

Figure 1.17. *Gravity storage STEP*

The hydraulic potential remains considerable – around 12,000 TWh/year (more than 50% of this is in Asia) – even if about half of this equipment is not considered economically profitable.

1.3.2. *Inertial storage*

In rotating devices potential energy is replaced by kinetic energy, which transfers to a wheel of inertia momentum J rotating at a speed Ω, an energy $W = 1/2\, J\Omega^2$, with $J = KMR^2$. The centrifugal force cannot exceed the traction stress limits of the materials, which leads to a limitation of the energy density, proportional to σ / ρ,

$$\frac{W}{M} = \frac{1}{2} K \frac{\sigma}{\rho} \text{ and } \frac{W}{V} = \frac{1}{2} K\sigma \qquad [1.1]$$

where σ is the traction elastic limit and ρ the volume mass. The optimum geometry is a thin-wall cylinder for which $K = 1$.

As shown in Table 1.6 the best materials are therefore carbon-fiber composite materials, which can reach more than 300 Wh/kg.

Material	ρ (kg/m3)	σ (MPa)	σ / ρ (Wh/kg)
Steel	8,000	600	20
Al 2017 (Dural)	2,800	180	18
Titanium 6.4	4,400	1,000	63
Kevlar	1,440	2,760	530
Fiberglass epoxy composite	1,150	330	79
Carbon-fiber epoxy composite	1,500	3,000	555

Table 1.6. *Materials characteristics for inertial storage*

In order to limit friction losses, a flywheel is placed in a vacuum enclosure using contactless magnetic bearings and, possibly, superconductors. A motor generator directly coupled to the shaft enables the wheel to rotate at its own speed and recover energy (Figure 1.18).

The pulse power can reach 2 kW/kg and an efficiency rate of 90%.

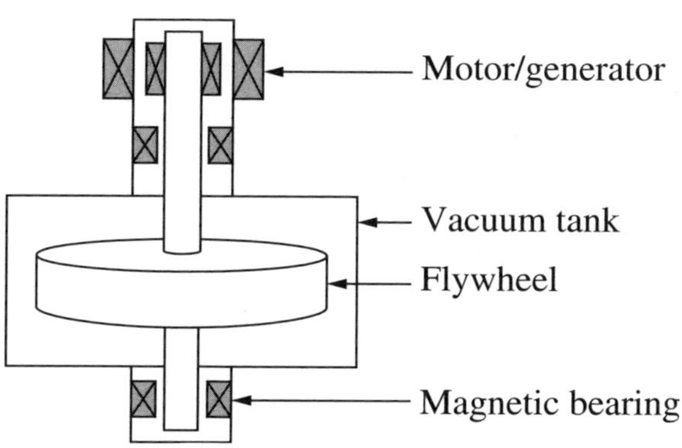

Figure 1.18. *Principle of a flywheel*

Many industrial installations, with single flywheels[3] (vehicles, UPS) or wheel stocks for network stability[4] have been set up. Rotating speeds may reach 100,000 rpm.

Low-speed flywheels (<10,000 rpm) use standard bearings and steel rotors, and therefore have densities limited to about 10 Wh/kg. Up to 20,000 rpm, standard bearings are used with composite material rotors to reach densities of about 150 Wh/kg.

For higher speeds, magnetic bearings (superconductors) must be used and the rotating parts are in a vacuum.

1.3.3. *Compressed air energy storage (CAES)*

This technique consists of storing energy in the form of compressed air, either in large geological caverns, e.g. salt mines (Figure 1.19), or, on a simpler scale, in pressurized gas bottles.

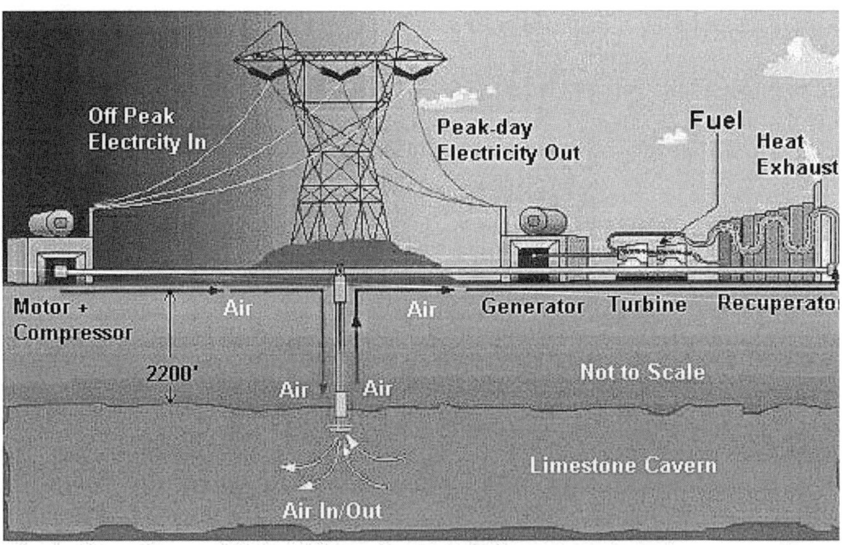

Figure 1.19. *Principle of a compressed air storage plant (photo courtesy of CAES Development Company)*

3. Beacon Power System 7 kWh, 2 kW, *Powerbridge*.
4. Project NEDO 10 kWh with superconductors bearings, Beacon Power 105 kW = 7X15 kW flywheels.

The air is compressed at about 100 bars during the periods of low-cost/off-peak electricity. This pneumatic energy, improved while heated in a combustion chamber of a gas turbine, is restored in its electric form by a synchronous generator coupled to the gas turbine during peak hours.

Because a standard gas turbine uses around two-thirds of the available power on the shaft to compress the combustion air, by separating the two processes it is possible to generate three times more power using the same amount of gas.

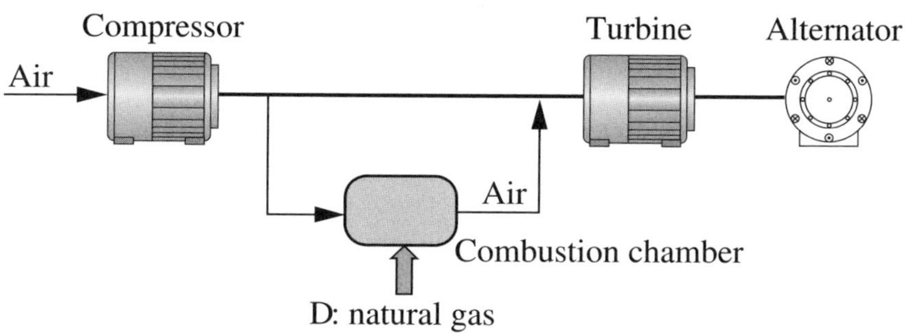

Figure 1.20. *Principle of a single cycle gas turbine operation*

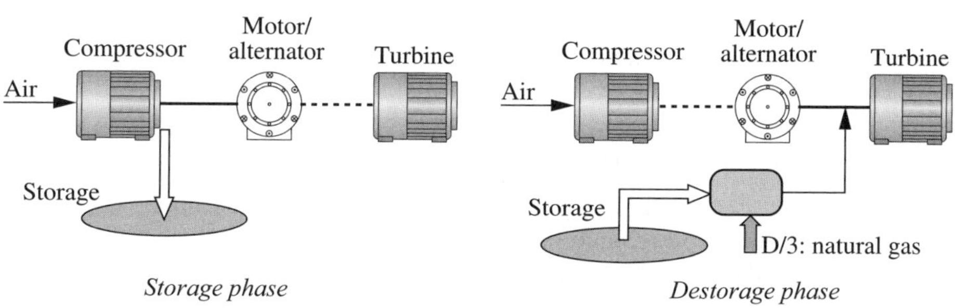

Figure 1.21. *Principle of a compressed air storage system coupled to a gas turbine operation*

The problems with this system are the high energy costs of compression which have an impact on cost effectiveness, the slow start up time of around 15 min, and the high cost of investment. Therefore it is necessary to have a significant Δ between the peak rates and off-peak rates, and to find nearby underground caverns.

Gas consumption is also significant, even if it is lower (30–40%) than in the case of a standard gas turbine.[5]

This storage system is promising for the future for electric systems of about 10 MW where it is possible to store compressed air and heat at the same time in order to reduce the gas turbine's fuel consumption. This system is also useful for the production of many GWhs of energy. To produce 100 MW for 8 hours with an 80% efficiency of a gas turbine, 1 GWh must be stored. If the air is heated up to 1,000°C from a 20°C source, the stored air mass must be 3,670 tons (Cp air = 1 kJ/kgK), which gives an energy density of the gas turbine and storage set of 270 Wh/kg.

For example, a CAES source can be connected to a wind farm in order to create a flexible plant. The latter can be used in a quasi-permanent manner, thus smoothing the intermittent problems and taking benefit from the low operating cost of the wind energy (Figure 1.22a).

Coupling CAES and heat (thermal and compressed air storage (TACAS)) is also an alternative – for a few minutes – to batteries and flywheels for power of some MW (Figure 1.22b).

1.3.4. *Electrochemical storage: batteries*

A battery generates electricity from an oxidation/reduction chemical reaction. A battery contains one or more cells in series that are made of three basic components:

– Anode and cathode. An anode is a negative reductive electrode, which supplies the electrons to the external circuit; it is generally a metal component. A cathode is a positive oxidant electrode, which recovers the electron from the external circuit; it is generally a metallic oxide component.

– Electrolyte. The electrolyte is where the electrochemical reactions take place; this is an ionic conductor with no conductivity for electrons; the current flow is due to the motion of the ions affected by the electric field between the electrodes; this is generally an aqueous solution.

– Membrane. The membrane which is permeable to the electrolyte may sometimes separate the electrodes.

5. AEC 110 MW, 3 GWh, 540,000 m^3 at 70 bars, McIntosh plant, Alabama; 290 MW, 580 MWh Huntdorf Germany.

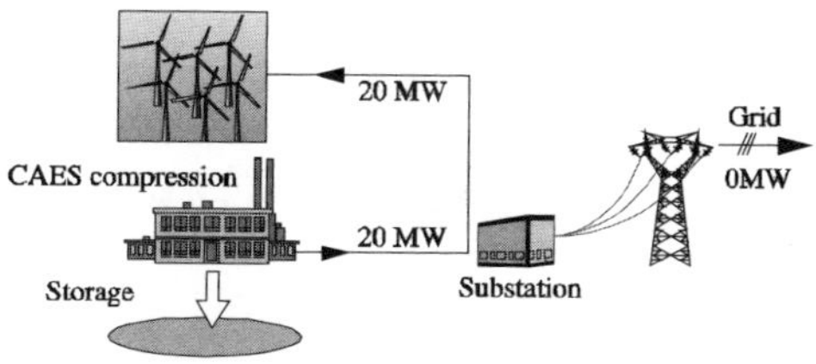

a) Compression: wind turbines used to compress air without using the grid

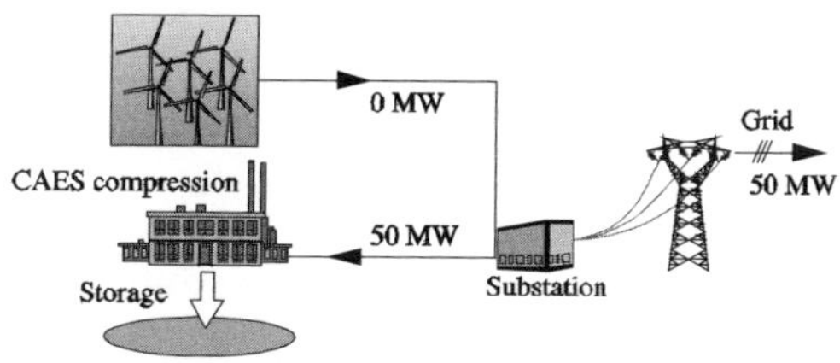

b) Compression: grid used to compress air without using wind turbines (peak hours, absence of wind)

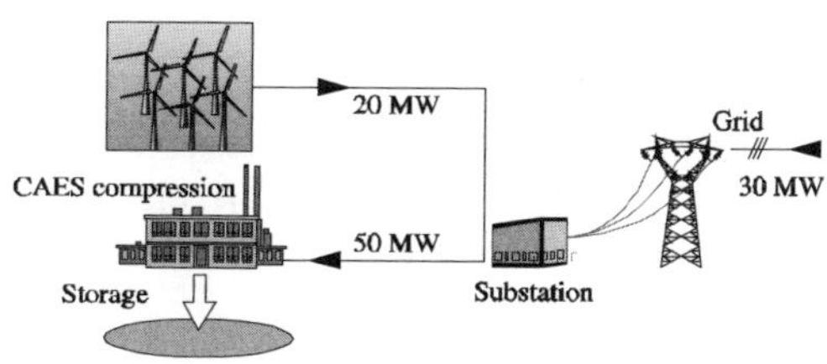

c) Compression: combination of wind turbines and grid to compress air

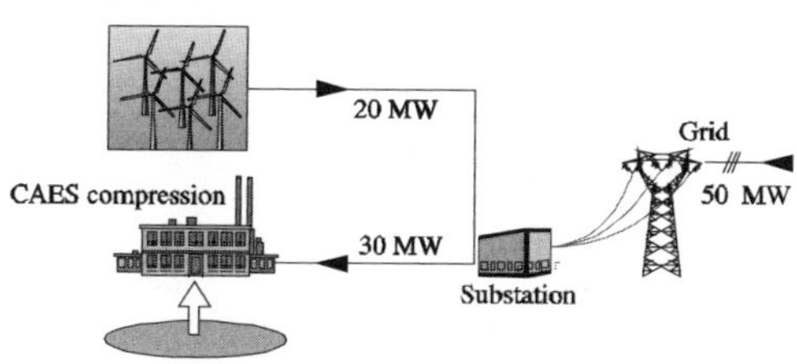

d) Generation: wind turbines and CAES supplying grid (peak hours, sufficient wind)

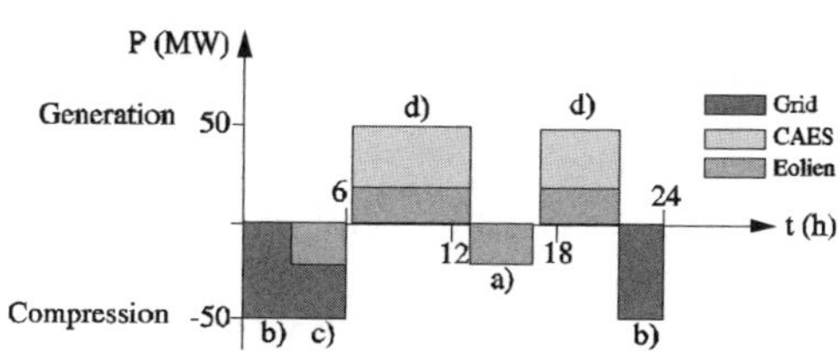

Figure 1.22a. *Integration wind and CAES*

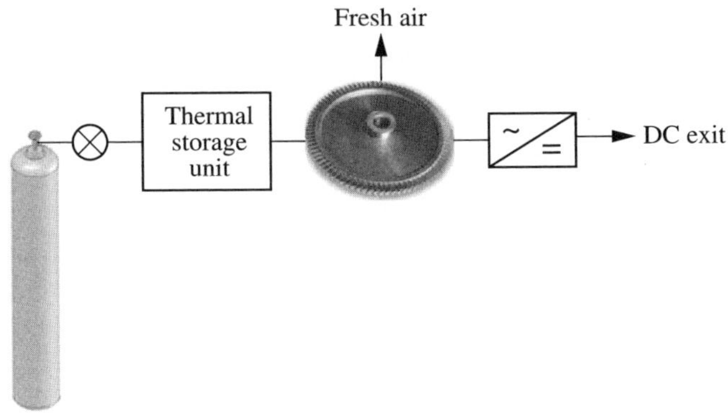

Figure 1.22b. *TACAS[6] system*

During the discharge, electrons move in the external circuit and the positive ions go from the anode to the cathode in the electrolyte.

A battery may be (see Table 1.7):

– Primary: the electrochemical reaction is not reversible (not rechargeable). The chemical transformation slowly destroys one of the electrodes, e.g. Zn-C, Zn-Hg, Zn-Cd, Zn-Air batteries, manganese alkaline batteries or lithium batteries. Lithium batteries are often used in portable equipment.

– Secondary: the electrochemical reaction is reversible. The battery may be recharged and the initial component restored, e.g. Pb-Acid, Ni-Cd, NaS, NiMH, and Li-ion.

Process	Non renewable fuel	Renewable fuel
Non-reversible	Battery (primary battery)	Fuel cell
Reversible	Battery (secondary battery)	Fuel cell

Table 1.7. *Classification of electrochemical components (source: E.Vieil, LEPMI, Grenoble – private communication)*

The density of the stored energy is proportional to the number of electrons per mole (in this respect lithium is a very interesting material) and to the volume mass of its components. Very reductive and very oxidant anode-cathode couples will be considered. The chemical reactions are very temperature-sensitive and each type of battery will have an optimal operating temperature range.

6. CoolAir™ Active Power Co.

The first battery was made by Volta, in 1800, by piling up alternating copper and zinc disks, each pair being separated from the other by a cloth saturated with salt water.

In 1839, the lead or lead-acid battery was invented by Gaston Planté. It was made of lead and lead oxide electrodes (PbO_2). The electrolyte was an aqueous solution of sulfuric acid (Figure 1.23).

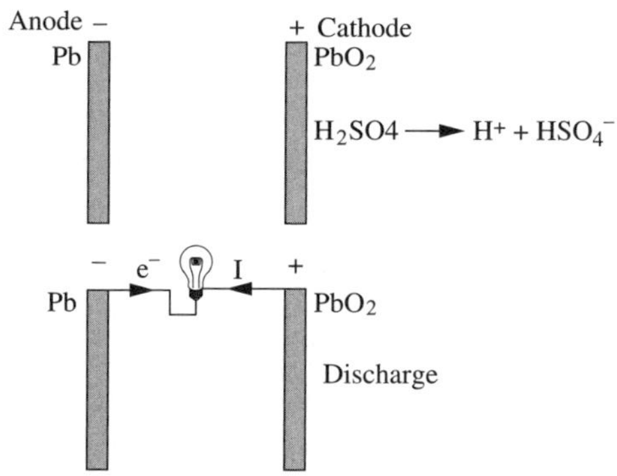

$$Pb + HSO_4^- \longrightarrow PbSO_4 + H^+ + 2e^- \qquad PbO_2 + HSO_4^- + 3H^+ + 2e^- \longrightarrow PbSO_4 + 2H_20$$

$$Pb + PbO_2 + 2HSO_4^- + 2H^+ \longrightarrow 2PbSO_4 + 2H_20 + \mathbf{Q}$$

Figure 1.23. *Lead acid battery during discharge: cathode: PbO_2 reacts with sulfuric acid ==> lead sulfate + water, anode: Pb reacts with sulfate ions ==> lead sulfate, plus electrons, sulfuric acid is consumed. During charge, the lead sulfate is destroyed, making lead and lead oxide; water is consumed and acid is formed*

This type of battery is used extensively, even for equipment of many MWh[7], because of its cost, despite its short life span.

Liquid electrolyte batteries require regular maintenance (level control of the electrolyte).

Batteries requiring no maintenance (valve regulated lead acid (VRLA)) have an electrolyte like a gel which enables the limitation of hydrogen evolution. They are equipped with a regulation valve which allows the gas recombination.

Lead batteries remain in use today and their lifetime is consistently being improved (Figure 1.24).

7. Chino, 40 MWH, 10 MW (California).

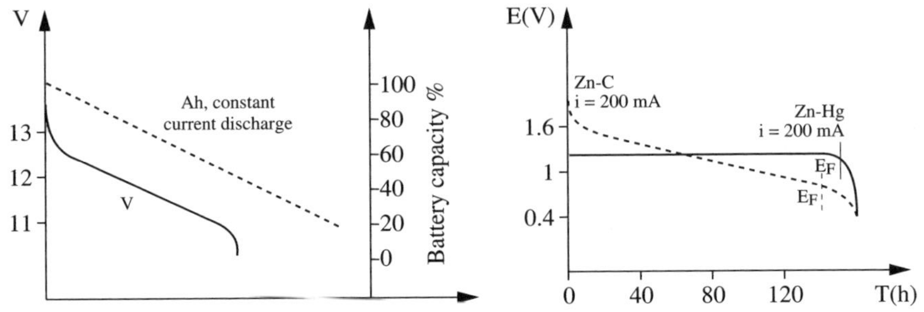

Figure 1.24. *Typical discharge of a lead acid battery and discharge curves of Zn-C and Zn-Hg batteries*

A battery has an internal resistance which varies according to its capacity (Ah), its discharge state and its temperature. At the end of discharge, the limit value E_F of its voltage must not be exceeded, otherwise there is a risk of rapidly draining the battery (in the case of primary batteries) and of reducing its recovering capacity during the charge (in the case of secondary batteries).

The discharge depth is the percentage of the discharge capacity; a battery has self-discharge characteristics due to parasitic electrochemical reactions which represent a loss of capacity and energy.

A number of other pairs may form interesting batteries:

– Ni-Cd: these batteries use a nickel hydroxide cathode $Ni(OH)_2$, a cadmium anode, and a potassium hydroxide electrolyte KOH.[8] This type of battery is expensive and its components are toxic. The discharge reactions are:

- at the anode: $Cd + 2OH^- \rightarrow Cd(OH)_2 + 2e^-$,

- at the cathode: $NiOOH + H_2O + e^- \rightarrow OH^- + Ni(OH)_2$,

- then a global reaction: $Cd + 2NiOOH + 2H_2O \rightarrow Cd(OH)_2 + 2Ni(OH)_2$;

– metallic Ni-Hydrides:[9] metallic hydrides can store hydrogen and may be used as negative electrodes; the positive electrode is a nickel hydroxide. These batteries are of the same type as Ni-Cd batteries, but the hydrogen is used as an active element at the anode. The electrolyte is a potassium hydroxide that must absorb and transport the H^+ (KOH) ions. The batteries have a large stored-energy density in an extended temperature range. They are not toxic. They could be possible

8. 40 MW, Golden Electric Association (Alaska).
9. ElectroEnergyInc 350 V, 500 A 700 W/kg.

replacements for Ni-Cd batteries, but, like them, they do not recover their full capacity if they are not completely discharged (memory effect);

– Li-ion: the electrodes can accept or deliver Li ions and electrons (insertion phenomena) (see Figure 1.25). The electrolyte is a non-aqueous organic component (lithium reacts with water) in which there is no chemical reaction, but it plays the part of a reversible ionic conductor. In this type of battery there are no gaseous elements such as H_2 or O_2 (Figure 1.25). These batteries have very good energy density since lithium is very light. They are thus very compact and can be used in all personal equipment. Li-polymer batteries (ACEP) contain a very thin, solid electrolyte (100 μm) between the lithium electrode (anode) and the vanadium oxide electrode (cathode) which looks like ribbons, leading to combinations with a density > 150 Wh/kg;

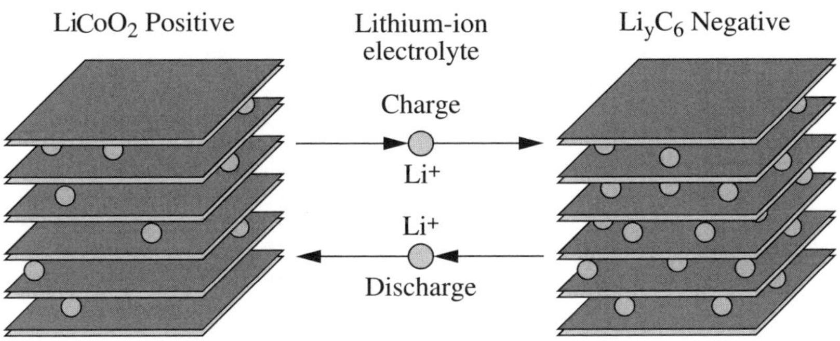

Figure 1.25. *Principle of Li-ion battery operation*

– *NaS*: sodium is interesting because of its high reduction potential (2.7 V), its low cost, and its low toxicity. It can, however, be used in its liquid form only at temperatures higher than 98°C (Na fusion temperature), around 270°C, which leads to thermal and corrosion problems. The batteries are mainly used in the military field or in stationary applications;[10] despite their high-energy densities (> 100 Wh/kg) and their excellent efficiency (90%), they tend to be replaced by Li-ion batteries. The Na-$NiCl_2$ "zebra" system, which also operates at high temperature (300°C) with a ceramic electrolyte which allows the Na^+ ions pass, is more secure than the NaS and can reach mass energies of 85 Wh/kg;

10. 6 MW 1999 (TEPCo Ohito substation).

Anode	Cathode	Reaction	Voltage (V)*	Energy density (Wh/kg)*	Energy density (kWh/m^3)*
Zn	MnO_2	$Zn + 2\ MnO_2 \rightarrow ZnO + Mn_2O_3$	1.5	85	165
Zn	HgO	$Zn + HgO \rightarrow ZnO + Hg$	1.35	100	470
Zn	Air	$Zn + 1/2O_2 \rightarrow ZnO$	1.5	370	1,300
Pb	PbO_2	$Pb + PbO_2 + 2\ H_2SO_4 \rightarrow 2PbSO_4 + 2H_2O$	2	35	70
Cd	Ni	$Cd + 2NiOOH + 2H_2O \rightarrow Cd(OH)_2 + 2Ni(OH)_2$	1.2	35	100
MH	Ni	$MH + NiOOH \rightarrow M + Ni(OH)_2$	1.2	75	240
Zn	Br_2	$Zn + Br_2 \rightarrow ZnBr_2$	1.6	70	60
Li_xC_6	$Li_{(1-x)}CoO_2$	$Li_xC_6 + Li_{(1-x)}CoO_2 \rightarrow LiCoO_2 + C_6$	4.1	150	400
Na	S	$2Na + 3S \rightarrow Na_2S_3$	2	170	345

Table 1.8. *Electrochemical couples (*practical values)*

– redox flow: these batteries use Vanadium ions of different valences. The ions are separated by a membrane which is permeable by the protons. Both reservoirs contain slightly acid electrolytes that are pumped into the cell where they remain separated by an ionic membrane. The electronic exchanges in the aqueous phase near the carbon electrodes correspond to an oxidation state change of the vanadium. (Figure 1.26).

In this type of battery, the functions of energy and power are dissociated; the storage capacity (kWh) is defined by the quantity of electrolytes used; the power (kW) depends on the active surface of the cells (membranes). Many industrial-size prototypes (>400 kWh, >100 kW) have been made[11] along with modular batteries of some kW/kWh for mobile applications. Some models do not have electrolyte circulation.

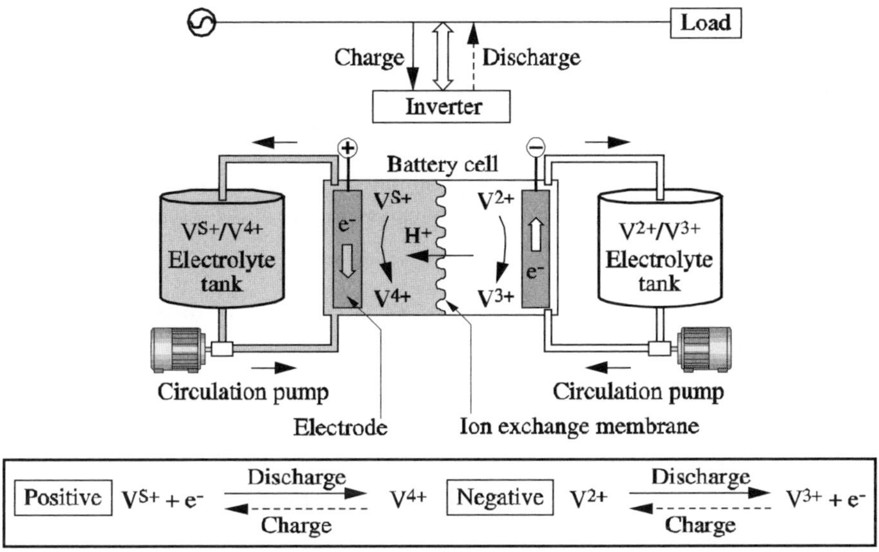

Figure 1.26. *Operation principle of a vanadium Redox flow battery*

Bromine polysulfide (PSB)[12] or zinc-bromine[13] have the same design, with oxide-reduction relations of the type:

$$2\,Br \rightarrow Br_{2} + 2e^{-}$$

$$Zn^{2+} + 2e^{-} \rightarrow Zn$$

Lead batteries are often used and have definite advantages: their cost is low, but their lifetime is limited according to the charge and discharge procedures.

11. Mitsubishi/Osaki Kansai power station; Mitsubischi/Kashima Kita Power Station, Regenesys®: TVA 120 MWh/12 MW; Sumitomo 500 kW/5 Mwh, JPH-NEDO 6 MW.
12. Regenesys 120 MWh, 15 MW Innogy's Little Bradford (UK) station.
13. ZBB 2 MW/2 MWh transportable for PG&E substation (California); VRB Inc King Island.

For other batteries, the main problem is their cost, which is justified for portable applications only.

The main objective is to find new electrochemical pairs, to reduce reliance on active materials, and to use less expensive and less polluting materials.

The redox flow batteries, using Vanadium circulation systems, are attractive for high power output, but the membrane remains a critical point.

Charge and discharge strategies depend on the type of usage. In the case of hybrid vehicles for example, the autonomous operation will be distinguished from the external charge from the grid.

Battery recycling is a necessity because of the toxic level of the material used (Pb, Cd, Br, etc.). Chemical[14] or thermal (cremation) techniques are used.

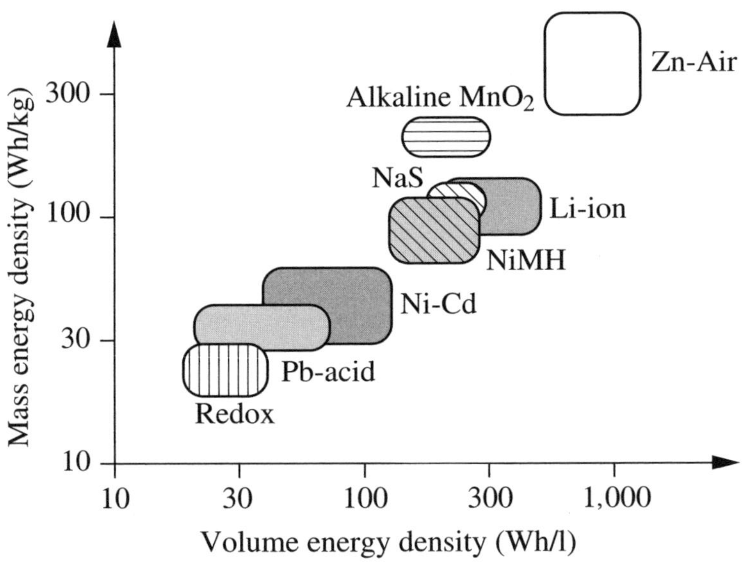

Figure 1.27. *Rangone diagram of batteries*

14. Recupyl.

	Pb-acid	Ni-Cd	NiMH	Li-ion	Li-polymer	V-Redox
Element voltage (V)	2	1.25	1.25	3.7	3.7	1.4
Energy density (Wh/kg)	30–40	45–60	50–90	80–200	150	25–35
Objective 2010	60	100	100	300		50
Power density (W/kg)	200–400	150–1,000	500–100	200–1,000	300	60–166
Objective 2010	600			2,000		
Number of cycles	500	800–2,500	800	1,000	1,000	
Objective 2010	2,000			2 500		
Cost (E/kWh)	50–150	400	400	500	< 250	
Objective 2010	50–120			200		
Lifetime (years)	3–15			3–10		10–15
Objective 2010	20			20		
Energy efficiency (%)	75–85	65–75	65–70	90–95		70–80
Objective 2010				95–97		
Charge time (h)	8–16	1	2–4	2–4	2–4	

Table 1.9a. *Evolution perspectives of battery performance.*
(source: [EU 05])

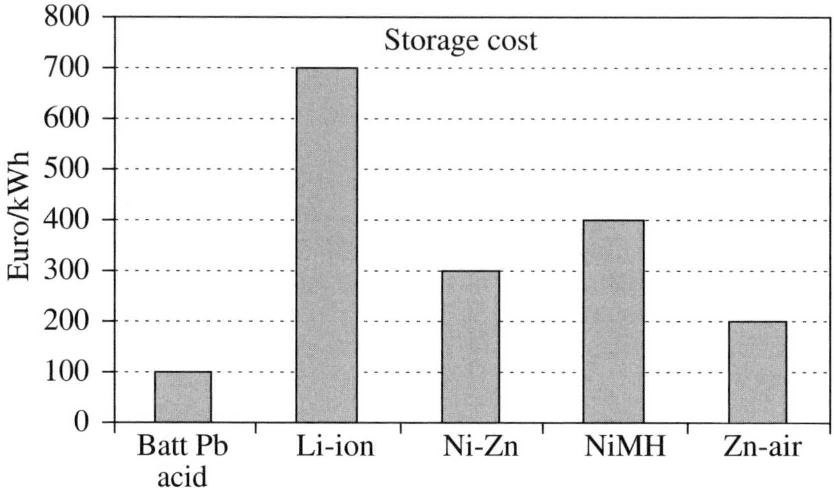

Table 1.9b. *Storage costs*

1.3.4.1. *Hydrogen*

Hydrogen is heavily used in industry (yearly production = 500 GNm^3 = 1.5% of the world's primary energy production). Its energy storage capacity is 30 kWh/kg (three times the gas capacity) which means 3 kW/m^3 (the same order of magnitude as SMES). In its liquid form or under 700 bars, this capacity is about 3 MWh/m^3.

1.3.4.1.1. Production

There are many methods of producing hydrogen:

– fossil hydrogen, carbon or biomass gasification with a risk of secondary production of CO and CO_2;

– water electrolysis from electric sources (nuclear, renewable energies) or at high temperature;

– bio-photosynthesis, fermentation, photo-electrolysis phenomena;

– high temperature thermo-chemistry (800°C to 1,300°C) on metal oxides.

1.3.4.1.2. Storage

For mass storage, natural or artificial[15] caverns may be used. Hydrogen may also be transformed into hydrocarbon or alcohol with a density that enables ease of storage; these components are then recombined efficiently to obtain hydrogen.

In order to store smaller quantities of hydrogen, e.g. for mobile applications, new solutions are under development, but as yet the ideal solution has not been found:

– gas: at 350–700 bars, in composite tanks (metal and glass fiber, carbon) is still expensive. This type of storage still presents problems of reliability, security and, above all, tank durability,[16] and we must take into account the compression energy cost;

– liquid: at 20° K in super-insulated tanks. This technique is less costly than the previous solution but is less attractive due to the financial and energy costs of the liquefaction. It is also less attractive because of evaporation which limits the operation autonomy: even if exchanges can be limited to $1 W/m^2$, taking into account the latent heat ($L = 445$ kJ/kg) and the liquid hydrogen density ($\rho = 70.8$ kg/m^3),
8 g of hydrogen, which represents 110 cm^3 per m^2 of surface per hour, evaporate;

– reversible metal hydrides: absorption and desorption mechanisms and their modeling are not yet fully understood. The quantity of stored hydrogen is still low:

– alloys ($LaNiH_6$, Mg_2NiH_4, $TiFeH_2$, etc.): the cost and mass of which are high (per mass, 6% of H_2 can be stored);

– complex light hydrides: alanates ($NaAlH_4$), hydrides of Mg, hydro-borides alkalis;

– high pressure carbon nano=structures (200 bars);

– porous organo-metallic materials;

– micro-organisms (seaweed), enzymes.

In all cases, the mass densities are put at a disadvantage because of the storage reservoirs (under 350 bars pressure, with a composite reservoir, 1.7 kWh/kg can be obtained).

---

15. Beynes experimentation.
16. Quantum, Dynetek, CEA.

1.3.4.2. *Fuel cells (FC)*

The fuel cell, which was invented by Sir William Grove in 1839, is a renewable cell (see Table 1.6). It generally uses hydrogen and oxygen which are converted into water.

1.3.4.2.1. Operation

A fuel cell is a generator which directly converts the internal energy of a fuel into electrical energy by means of a controlled electrochemical process.

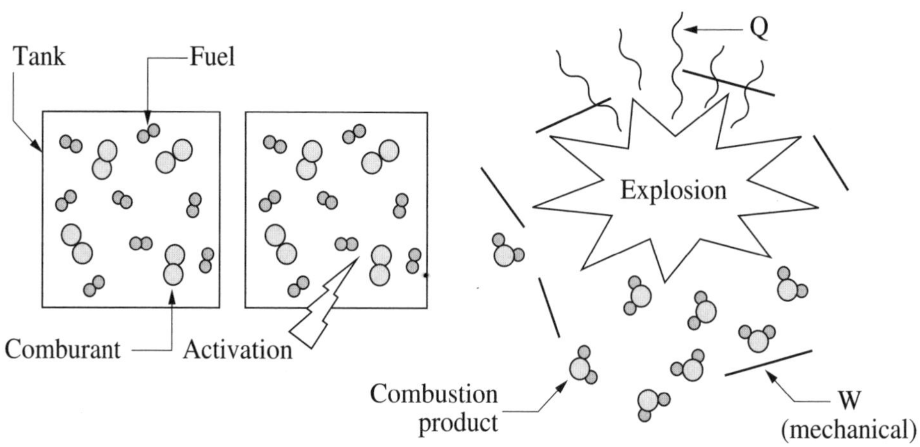

Figure 1.28. *Combustion principle (source: LEPMI, Grenoble)*

Combustion of fuel and oxygen is initiated by an activator. If this combustion is not controlled, it produces an explosion which makes heat and work and gives off combustion waste (Figure 1.29).

This combustion may be monitored, while controlling the exchanges between fuel and oxygen in order to recover heat and electricity. This can be done by separating the electrons from the ions in the electrolyte with a selective membrane. Batteries are made using this method – as are fuel cells, which are renewable – but not reversible batteries (Figure 1.29).

A fuel cell is different from other electrochemical batteries because the agents are permanently renewed and the products permanently evacuated (Figure 1.30).

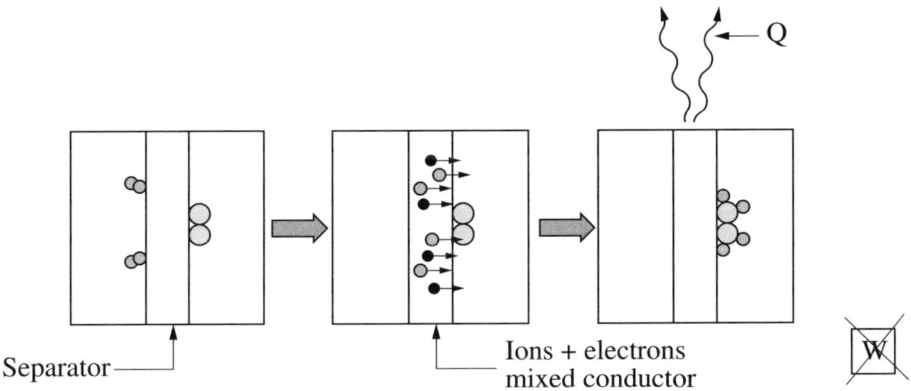

Figure 1.29. *Principle of controlled combustion (source: LEPMI)*

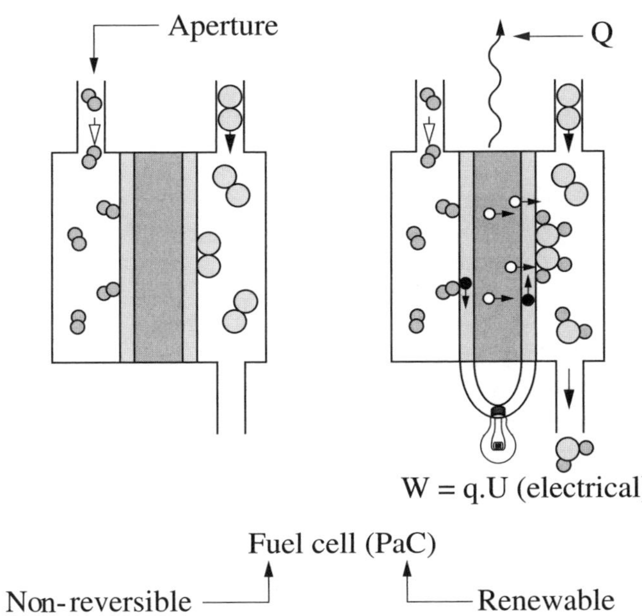

Figure 1.30. *Renewable cell (source: LEPMI)*

The reaction $H_2 + 1/2O_2 \rightarrow H_2O$ may be separated into two half reactions:

$$H_2 \rightarrow 2H^+ + 2e^-$$

$$1/2O_2 + 2H^+ + 2e \rightarrow H_2O$$

The chemical potential of the reaction (-237 kJ/mole) gives a voltage of 1.23 V.

The different types of fuel cells depend on the ions exchanged between anodes and cathodes; they also depend on the fuels and oxidizers, and on the type of membrane and electrolyte.

They can be divided into six categories: AFC (alkaline fuel cell), PEMFC (polymer exchange membrane fuel cell), DMFC (direct methanol fuel cell), PAFC (phosphoric acid fuel cell), MCFC (molten carbonate fuel cell), SOFC (solid oxide fuel cell).

The tables below outline the characteristics of each type of fuel cell and specify the type of electrolyte that gives the operating temperature range.

The high temperature operation of SOFC allows their coupling with gas[17] turbines.

Kind of cell	Anode reaction (catalyst)	Electrolyte membrane	Cathode reaction (catalyst)	T (°C)
PEMFC (proton exchange membrane)	$H_2 \rightarrow 2H^+ + 2e^-$ (To the alloys)	Acid polymer (SO_3H^+) $H^+ \rightarrow$	$\frac{1}{2} O_2 + 2H^+ + 2e^- \rightarrow H_2O$ (To the alloys)	25-80
DMFC (direct methanol)	$CH_3OH + H_3O \rightarrow CO_2 + 6H + 6e^-$ (Pt-Ru, Pt-Mo, etc.)	Acid polymer (SO_3H^+) $H^+ \rightarrow$	$\frac{1}{2} O_2 + 2H^+ + 2e^- \rightarrow H_2O$ (To the alloys)	25-90
AFC (alkaline)	$H_2 + 2OH^+ \rightarrow 6H_2O + 2e^-$ (Ni Ramay, Pt, etc.)	Base polymer (NH_4OH^+) $\leftarrow OH^+$	$\frac{1}{2} O_2 + 2H^- + 2e^- \rightarrow H_2O$ (Ag. to the alloys)	25-80

Table 1.10. *Main electrochemical reactions in fuel cells*

17. Example of Mitsubishi: efficiencies of 65% with natural gas.

Battery type	AFC	PEMFC	DMFC	PAFC	MCFC	SOFC
Name	Alkaline fuel cell	Polymer exchange membrane fuel cell	Direct methanol fuel cell	Phosphoric acid fuel cell	Molten carbonate fuel cell	Solid oxide fuel cell
Electrolyte	Solution KOH	Polymer membrane, proton conductor	Polymer membrane, proton conductor	Phosphoric acid	Li_2CO_3 and KCO_3 melted in a $LiAlO_2$ matrix	ZrO_2 and Y_2O_3
Ions in the electrolyte	OH^-	H^+	H^+	H^+	CO_3^{2-}	O^{2-}
Temperature level	60–80°C	60–100°C	60–100°C	180–220°C	600–660°C	700–1,000°C
Fuel	H_2	H_2 (pure or reformed)	Methanol	H_2 (pure or reformed)	H_2 (pure or reformed)	H_2 (pure or internal reforming)
Oxidants	O_2 (pure)	Air	Air	Air	Air	Air
Efficiency	50–60%	50–60%		55%	60–65%	50–60%
Application areas	Space electric vehicle	Automobiles, portable, co-generation[18]	Portable	Co-generation[19]	Co-generation centralized electricity generation[20]	Co-generation, centralized electricity generation, automobiles[21]
Development stage	Used	Prototypes	Prototypes	Mature technology	Prototypes	Prototypes

Table 1.11. *Properties of fuel cells*

18. Example of Ballard.
19. Example of ONSI Corp.
20. Example of *Fuel Cell Energy*, Ansaldo Fuel Cell SpA, Hitachi.
21. Example of Technology Sulzer Hexis, experimentations: Siemens (100 kW), 220 kW.

1.3.4.2.2. Advantages of fuel cells

Fuel cells have the following advantages:

– fuel cells directly transform chemical energy into electrical energy; their chemical agents are stored outside the cell, thus separating stored energy and power;

– their efficiency does not depend on a Carnot cycle;

– they have a low acoustic emission;

– they allow modular manufacturing;

– they give a co-generation possibility ➔ efficiencies of 50% to 80%;

– they produce water and emit few green-house gasses.

1.3.4.2.3. Disadvantages of fuel cells

Fuel cells have the following disadvantages:

– some technologies require very pure hydrogen;

– their low unitary voltages (≈ 1 V) require series connections (stack) and power electronic interfaces which introduce losses (Figure 1.31) and reduce their efficiency;

– current densities remain in the order of 1 A/cm^2;

– there are losses during transient states;

– current distribution is modified with the generated magnetic field and when the plates surfaces increase;

– thermal and water management ←→ operating point control;

– mass and volume;

– cost: catalysts, membranes, current collectors, waterproof joints (at high temperature);

– lifetime, although theoretically the active material of the electrodes is not consumed.

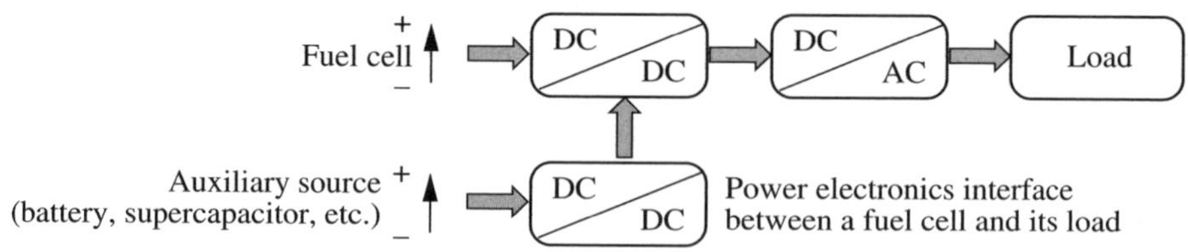

Figure 1.31. *Fuel cell/PE coupling*

	MCFC	SOFC	PEMFC	DMFC	AFC
Investment cost (€/kW)	8,500	15,000	2,000	10,000–100,000	15,000
Objective 2010	1,500	1,500	200	5,000	1,000
Lifetime (h)	21,000	5,000	8,000	1,000	2,000
Objective 2010	40,000	40,000	8,000	5,000	
Efficiency (%)	47	45	45	25	60
Objective 2010	50	50	45	35	60

Table 1.12. *Fuel cell characteristics (source: [EU 05])*

1.3.5. *Storage in the electromagnetic field*

1.3.5.1. *Electric field: capacitors, supercapacitors*

In electronic systems, especially in power electronics, the most common buffer storage elements used are the capacitor and the inductance, either in their usual form or directly integrated into silicon chips (Figure 1.32).

Figure 1.32. *Integration of planar components LCT (inductance, capacitor, transformer)*

Stored energy is weak, in the order of some tens of Wh. The main technologies used are:

– foil capacitors, which in the process of disappearing;

– electrolyte capacitors with a solid or liquid electrolyte. They have good value, but are voltage limited (600 V) and have important leakage currents;

– polymer electrolyte capacitors (polyester, polycarbonate, polystyrene) have low losses and high service voltages. The polypropylene capacitors are used in power electronics because of their good response under a pulse regime;

– ceramic capacitors which have high permittivity ($SrTiO_3$, $BaTiO_3$ $\varepsilon_r = 3{,}000$–18,000) and a storage capacity of some hundreds kJ/m3, which enable operations at high frequency and high voltage.

The electric field stores energy with a volume density equal to:

$$\frac{dW}{dV} = \varepsilon_0 \varepsilon_r \frac{E^2}{2} \quad [1.2]$$

Taking into account the permittivity values ε_0 and ε_r and the acceptable electric fields in the materials, this density remains weak (typically for air, 40 J/m^3) and is limited by the dielectric disruptive voltage E_d. For PVDF dielectric (bi-fluoride polyvinyl), the price remains high, $\varepsilon_r = 10$ and $E_d = 250$ MV/m, and the density reaches 750 Whm^{-3}, less than 1 Wh/kg.

Dielectric	Dielectric rigidity (MV/m)	ε_r
Dry air	3	1
SF_6	30	1
Pyranol	12	4.5
Epoxy	20	3.3
Mineral oil	10	2.2
Mica	40–240	7
Impregnated paper	14	4–7
PET (polyethylene terephthalate)	700	3
Polypropylene	800	2.4
Polyimide	200	3.8
Porcelain	4	6
Glass	25–100	5–7

Table 1.13. *Dielectric properties of materials*

Supercapacitors[22] (double layer capacitors) are capacitors that operate like classic capacitors, but their technology is similar to the operation principle of the electrochemical batteries (Figure 1.33).

22. Becker patent (1957).

They perform better than batteries in terms of power densities and have energy densities higher than their capacities.

The capacitors of the layers between every two electrodes, made of porous carbon, and the electrolyte ions (surface capacitors of 30 μ F/cm^2) are connected in series.

There is no chemical reaction. The energy may be transferred very quickly and as this phenomenon creates charges at the surface of solid faces, the lifetime is very good (>> 10,000 cycles) with an efficiency of 95%. A porous dielectric membrane stops the charge flow between the electrodes. The element voltage remains in the order of 2 to 3 V.

The asymmetric systems, which use one carbon electrode and one metal electrode, have even higher energy densities.

The huge surface (1,000 m^2/g) of the porous electrodes (typically made of carbon powder or fiber polymers) and the very thin separation space (1 nm) enable the storage of very significant energy. Capacities of many kFarads and volume energies of around 10 Wh/kg may be obtained. The supercapacitors that deliver a voltage depending on the charge state are connected to power converters.

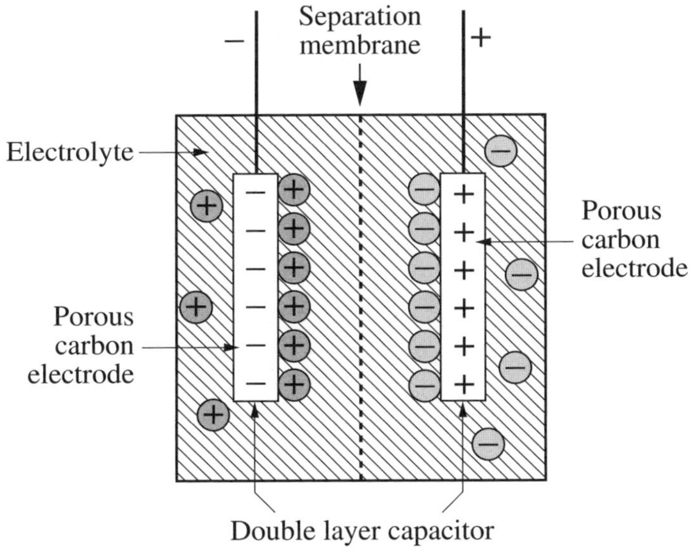

Figure 1.33. *Operating principle of a supercapacitor*

1.3.5.2. *Magnetic field: Superconductor Magnetic Energy Storage (SMES)*

The magnetic field stores energy with a volume density in the air equal to

$$\frac{dW}{dV} = \mu_0 \frac{B^2}{2} \qquad [1.3]$$

which allows, if superconductors are used, densities in the order of 10 kWh/m^3 and in large volumes. Since $\mu_0 >> \varepsilon_0$, then the electric and magnetic fields E and H are in the same order of magnitude, the magnetic storage in the air is then more efficient and enables large systems.

The first SMES used superconductors (NbTi) requiring a liquid helium temperature cooling (4.2 K), which lead to high insulation and cooling costs, as well as high maintenance and operating costs.

The improvements to superconductor materials now enable the cooling of the coils ($Bi_2Sr_2CaCu_2O$) at temperatures of about 20 K in order to optimize the operation of the conductor, thus reducing the necessary cold power.

The absence of resistance from the conductors in their superconductor state allows energy storage in the short-circuited coil, theoretically infinitely.

A converter system charges and discharges the current. The charge and discharge times are limited by the over voltage Ldi/dt which appears at the coil terminals, and are also limited by the component reverse voltage. Valuable power can then be delivered with a very high level of efficiency (> 95%) since the AC losses in the conductors are low (Figure 1.34).

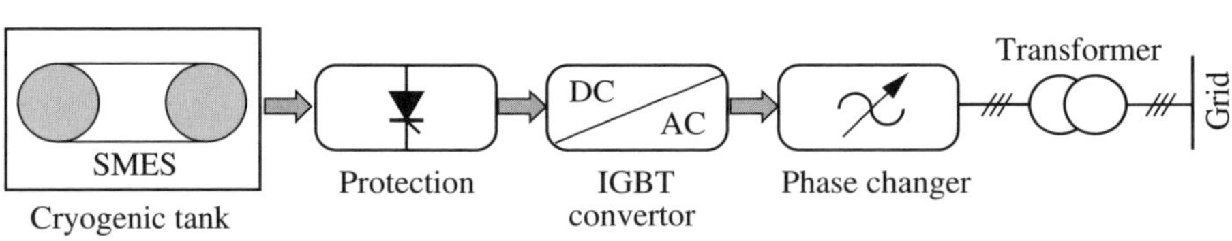

Figure 1.34. *Connection principle of SMES to the grid*

The maximum transferable power varies as $Q_{stored}^{1/2}$.

$$P = UI = L\frac{dI}{dt}I \Rightarrow P_{max} = U_{max}\sqrt{\frac{2Q_{stored}}{L}} \qquad [1.4]$$

To date, this kind of storage has been used for network stability (P and Q control) and to improve the network quality (D-SMES, μSMES)[23]. It has also been used for boost systems (electromagnetic launchers).[24]

Solenoid geometry is used for large volumes, but generates high leakage fields. The toroidal geometry drastically reduces the parasitic fields.

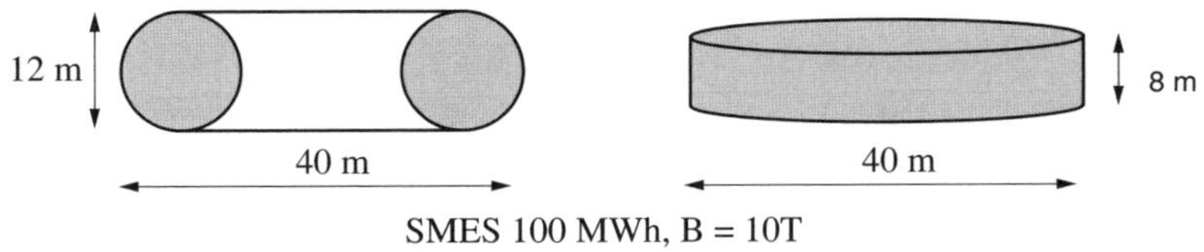

Figure 1.35. *Toroidal and solenoid geometries for a SMES*

The main drawback of SMES cryogenic cooling should disappear if HCT (high critical temperature) super-conductors work at temperatures ≥ 77 K (temperature of liquid nitrogen).

1.3.6. *Heat*

Heat is stored naturally in the ground and is used in geothermal technology (with useful ΔT of 200°C) or in the ocean (with useful ΔT of 20°C) and represents a source used in some accessible sites. Heat storage must be limited to applications directly using this form of energy (buildings). When heat is used, the material characteristics to be taken into account are the thermal capacity (C in $Jkg^{-1}K^{-1}$) and the thermal diffusivity (D in m^2s^{-1}) which will characterize, for a given thermal load, the restoring time of the stored heat. The endothermic latent heat due to the state changes of some materials (L in Wkg^{-1}) may be used; the material absorbs energy when it exceeds the transition temperature and gives it back when it is re-cooled. It is the change of state from solid to liquid that is used. Some of these materials may be integrated into construction materials, e.g. concrete.

Water is often used, e.g. for solar heaters, but its storage capacity remains low (30 kg of water heated with a 30° difference stores little more than 1 kWh). Time constants are rather long and the heat storage may fulfill daily or even season storage needs. Storage temperatures are in the order of 100°C (low temperature

23. Examples of devices achieved by Chubu Electric Co 5 MW/5 MJ and by NEDO 2.9 MJ; Rhineland loop ATC (Wisconsin): 6 units 3 MW, 3 MJ.
24. Example of SMES 20 K, 800 MJ made by LEG.

storage) or higher (high temperature storage) for use with a gas turbine for example. Insulation measures must also be taken to avoid thermal losses. Cold may also be stored for some specific applications.

Materials	Mass heat ($Jkg^{-1}K^{-1}$)	Thermal diffusivity (cm^2s^{-1})	Volume mass (kgm^{-3})	Latent heat ($kJkg^{-1}$)	T of state change (°C)	ΔT ref (°C)	Energy density (kWh/m^3)
Water	4,180	1.4 10^{-3}	1,000	2,250		50	58
Oil	2,700	0.7 10^{-3}	840			50	31.5
Brick	840	5.2 10^{-3}	1,600				
Refractory brick	960	4 10^{-3}	2,600			1,400	970
Granite	820	8–18	2,640				
Phase change material							
Paraffin			820	200	20–60		45
$KF\ 4\ H_2O$				231	18.5		
$Na_2S_2O_4.10\ H_2O$			1,675	254	32.4		118
$CaCl_2.6\ H_2O$				171	29.7		
$Na_2HPO_4.12\ H_2O$				281	35		
$Zn(NO_3)_2\ 6\ H_2O$				147	36.4		
$CH_3(CH_2)_{12}OH$				205	38		

Table 1.14. *Thermal properties of storage materials*

We can envisage a system similar to that of the CAES, which stores energy in its thermal form, in refractory materials electrically heated at 1,400°C during off-peak periods. This stored energy could be recovered in its electric form by means of a temperature turbine with an efficiency of 60%, during peak hours.

The stored energy density is high (360 Wh/kg, 1 MWh/m^3). Storage of 100 MWh of electricity, with an efficiency of 60%, would only occupy a volume of 170 m^3 (Figure 1.36).

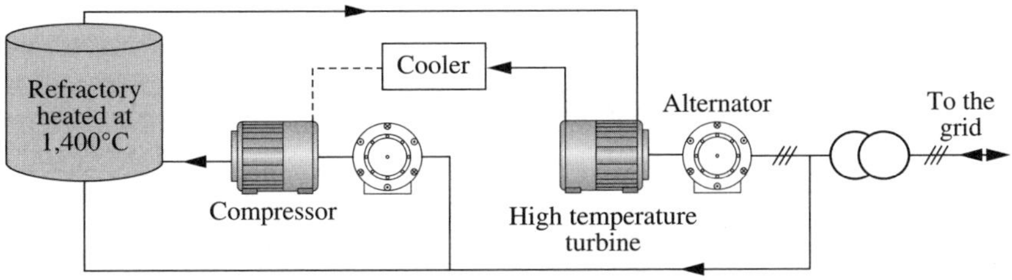

Figure 1.36. *Principle of refractory thermal storage at high temperature. A heat exchanger between the cooler and the compressor would achieve a permanent operation cycle, the storage reservoir being a classical boiler*

1.3.7. *Small-scale storage*

In the field of microsystems, especially for embedded systems, the problems of energy storage are of prime importance for reasons of mass, volume, and autonomy. The smaller the microsystem, the greater the problem.

Lithium mini-batteries only deliver mAh/cm^2 (HEF group) on surfaces of a few cm^2 with thicknesses of a few hundred µm.

Lithium batteries of 200 µAh have been manufactured directly onto a silicon sublayer, and are able to deliver hundreds of $\mu A/cm^2$ per hour (Figure 1.37). These batteries may be integrated in the circuits they supply; they are flexible and may be made with a large range of surfaces and in a large range of shapes.

µfuel cells delivering 0.1 W to 10 W may be associated with these µbatteries to create a complete µsource.

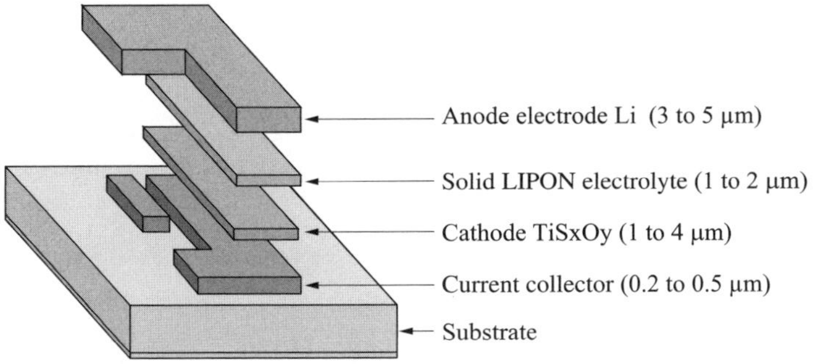

Figure 1.37. *Schematic of a µbattery Li-ion (source CEA)*

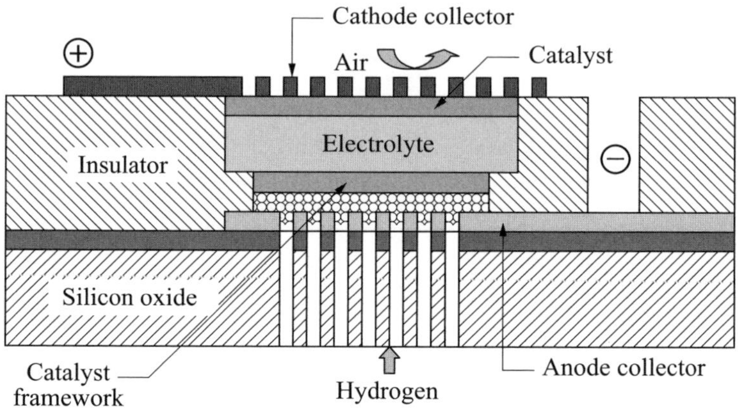

Figure 1.38. *Schematic diagram of a fuel cell on silicon*

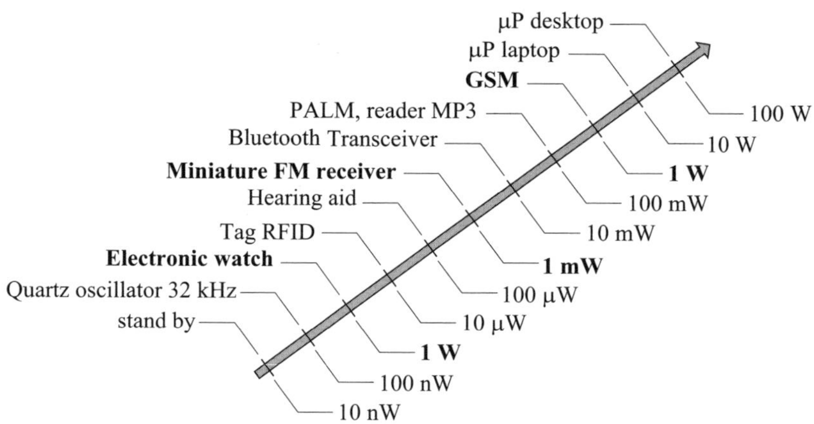

Figure 1.39. *Portable systems consumption*

At these energy scales, the stored energy may be recovered in other, less commonly used forms, macroscopically speaking, particularly the environmental energy:

– thermal energy: the human body used as a heat source for the thermoelectrical effect;[25]

– mechanical energy (body motion, etc.), in the case of piezo electrical materials;

25. For example, Seiko and Citizen watches.

– thermo-mechanical energy in the case of shape memory materials (nitinol, Figure 1.40);

– thermo-magnetic energy in temperature-commutable magnetization materials such as GdCoCu.

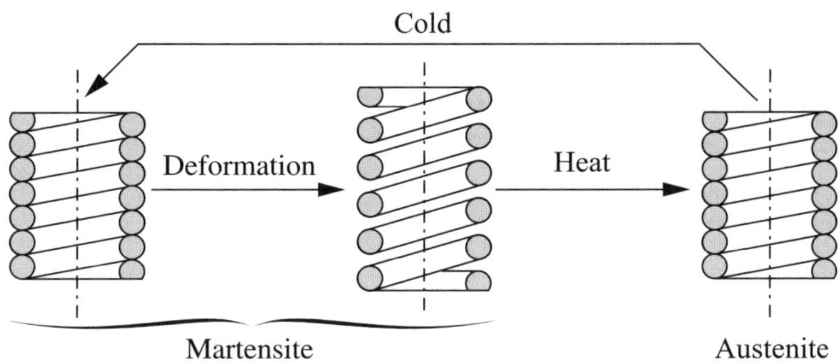

Figure 1.40. *Shape memory alloy*

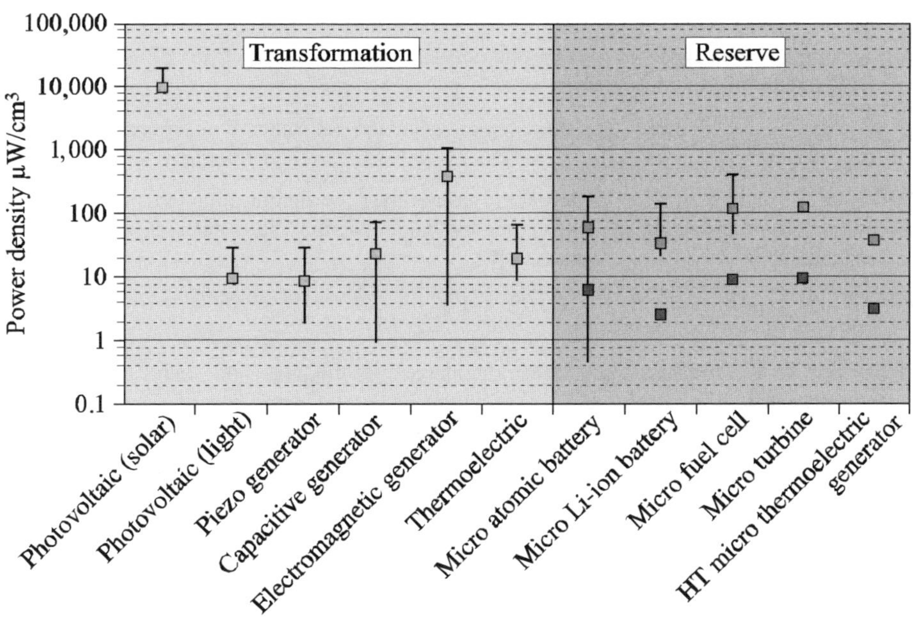

Figure 1.41. *Environmental energy usage*

1.4. Bibliography

[AND 97] M.D. ANDERSON, *et al.*, "Assessment of utility-side cost savings for battery energy storage", *Proceedings of IEEE Trans on Power Systems*, 12–13, 1997.

[BAX 02] J. BAXTER, "Energy storage: enabling a future for renewables?" *Renewable Energy World*, July-August 2002.

[BOB 05] J.L. BOBIN, E. HUFFER and H. NIFENECKER, *L'énergie de demain*, EDP Sciences, 2005.

[CAR 99] F. CARDARELLI, *Materials Handbook: A Concise Desktop Reference*, Springer, 1999.

[CON 99] B.E. CONWAY, *Electrochemical Supercapacitors: Scientific Fundamentals and Technological Applications*, Kluwer, 1999.

[EPR 03] EPRI-DOE, *Handbook of Energy Storage for Transmission and Distribution Applications*, EPRI, Product n°1001834, December 2003.

[EU 01] European Commission, *Storage: a key technology for decentralised power, power quality and clean transport*, Eur 19978 report, European Communities, Belgium, 2001.

[EU 05] European Commission, *Energy Scientific and Technological Indicators and References*, Eur 21611 report, European Communities, Belgium, 2005.

[EVA 98] D. EVANS, "Hybrid capacitor applications", *Proceedings VIII International Seminar*, *Double Layer Capacitor and Similar Energy Storage Device*, Florida, US, 7-9 December 1998.

[FAU 98] J.F. FAUVARQUE, "Présentation générale des super condensateurs à double couche électrochimique", *Journées d'études sur les super condensateurs*, Paris, 1998.

[GER 02] O. GERGAUD, Modélisation énergétique et optimisation économique d'un système de production éolien et photovoltaïque couplé au réseau et associé à un accumulateur, PhD Thesis, Cachan, 2002.

[HER 05] W.A. HERMANN, *Quantifying Global Energy Resources*, Science Direct, Elsevier, 2005.

[KIN 99] A. KINSHI *et al.*, "Micro thermoelectric modules and their application to wirstwatches as energy source", *XVIII International Conference on Thermoelectrics*, Baltimore, 1999.

[LAS 01] J.C. LASSÈGUES, *Super condensateurs*, Techniques de l'Ingénieur D3334, 2001.

[LEV 02] A. LEVASSEUR *et al.*, *Les micro accumulateurs*, Journées du club EEA, Cachan, 2002.

[LIN 01] D. LINDEN and T.B. REDDY, *Handbook of Batteries*, 3rd edition, McGraw-Hill, 2001.

[LUC 02] P. LUCCHESE, *Pile à combustible, possibilité de miniaturisation*, Journées du club EEA, Cachan, 2002.

[MAR 98] A. MARQUET *et al.*, *Stockage d'électricité dans les systèmes électriques*, Techniques de l'Ingénieur D4025, 1998.

[MUL 03] B. MULTON and J. RUER, "Stocker l'électricité: oui, c'est indispensable et c'est possible! Pourquoi, où, comment". *Club Ecrin, Conference énergies alternatives*, Paris 2003.

[MUL 96] B. MULTON and J.M. PETER, "Le stockage de l'énergie: moyens et applications", *Revue 3EI*, vol. 6, June 1996.

[OHS 04] H. OHSAKI *et al.*, "Development of SMES for power system control: present status and perspective", *Physica C*, 412, 2004.

[RIB 01] P.F. RIBEIRO *et al.*, "Energy Storage Systems for Advanced Power Applications", *Proc of IEEE 89*, December 2001.

[SAR 01] C. SARRAZIN, *Piles électriques*, Techniques de l'Ingénieur D3320, 2001.

[SUR 05] C. SURDU *et al.*, "La centrale virtuelle, un nouveau concept pour favoriser l'insertion de la production décentralisée d'énergie dans les réseaux de distribution", *Revue 3EI*, vol. 40, March 2005.

[VAN 03] S. VAN DER LINDEN, "The commercial world of energy storage: a review of operating facilities", *Annual Conference of the Energy Storage Council*, TX, US, March 2003.

[ZIN 97] J.C. ZINK, "Who says you can't store energy", *Power Engineering*, vol. 101, no. 3, March 1997.

Chapter 2

Nuclear Fission, Today and Tomorrow: from "Renaissance" to Technological Breakthroughs

Summary

In this chapter, the following topics will be considered:

– What are the big challenges that nuclear fission will face with in the framework of the sustainable energy policies in the short-, medium- (2010), and long-term (2030)?

– What European Community research is directed at these challenges, particularly those concerning future nuclear systems and fuel cycles?

There are two types of challenges in the present "renaissance" of nuclear power in Europe:

– Scientific and technological (S/T) challenges, in which the European Community research plays a direct role.

– Economic and political (E/P) challenges, in which the European Community research plays an indirect role as a support to different decisions.

One of the main challenges the future generations will be faced with is the strong growth of energy consumption – not only of electricity, but also of high-temperature

---

Chapter written by Georges VAN GOETHEM.

industrial heat – a challenge that is already being faced, but which will become greater when the "low-carbon economy" is built.

In Europe these challenges in the energy field require a European Community approach. The European Union (EU) is in charge of the creation of the framework that will enable us to face these energy challenges. This will be done, among other things, by means of research and innovation within the European Community, as mentioned in the Green Book on Energy, "A European strategy for a sustainable, competitive and secure energy" (8 March 2006) [EUR 06].

This document analyzes, among others things, the science and technology challenges linked to nuclear fission and to solutions suggested by the international research (in particular that by the Euratom Research and the EU's Training Framework Programs (FP)). This research concerns both the public and private sectors and is naturally oriented towards industrial applications. Its sphere is very wide:

– in space: the community concerned by the nuclear fission is larger than the EU and covers all the scientific disciplines;

– in time: if we look towards 2010 and 2030, we will be able to distinguish three generations of nuclear technologies (called II, III, and IV).

To each of these three generations (GEN) a couple of challenges are linked:

– GEN II (1970–2000): supply security and environmental protection;

– GEN III (around 2010): safety and competitiveness improvements;

– GEN IV (around 2030): co-generation and integral actinide recycling.

The applications of the Euratom research are illustrated by practical examples related to nuclear systems and to the fuel cycles of GEN III and IV, including partitioning and transmutation technologies of minor actinides. The aspects of sustainability, economy, safety, and proliferation resistance receive particular attention.

2.1. Introduction: all energy options kept open in 2006 Green Book

Many discussions are taking place about the European Project, in particular about the European response to the global threats to energy, economy, and citizen safety. Energy supply is one of the pillars of the European economy. In fact, energy has played a key role in the development of the EU. Two of the early treaties dealt with energy, namely the Paris Treaty, which in 1951 established the European

Community of Coal and Steel (CECA), and the Treaty of Rome, which on March 25, 1957, established the European Community of the Atomic Energy (EURATOM).

In the 27 Member States of the EU there are approximately 500 million consumers and 25 million firms. The EU is the second largest energy market in the world. However, the EU must import 50% of its energy needs (which represents an annual bill of €240 billion) and its needs keep growing every year. As a matter of fact, the increase in energy demand is about 2% per year. If nothing is done, the imported energy will thus reach 70% of the EU's total energy needs in 2030.

Generally speaking, at the world level, the situation of energy supplies changes according to the following elements: unpredictable costs of fossil fuels; climate changes (let us recall, for example, the consequences of Hurricane Katrina in New Orleans in August 2005); world demographic growth; increasing populations in emerging countries like India, China, Brazil; use of high technology products in developed countries. In addition, there is greater uncertainty in the energy market because of large industrial acquisitions. Let us also remember the serious disturbances of energy supplies that have occurred in recent years, particularly those in the United States of America and Canada (north-eastern USA and Ontario and Quebec in Canada during the summer of 2003) and in the EU (Italy, Sweden, and Denmark, in the autumn of 2003); we should also keep in mind the Russian–Ukrainian gas crisis during recent winters.

These facts prove that energy must not be considered as "a given." On March 8, 2006, the EU published a Green Book [EUR 06] entitled "A European strategy for a sustainable, competitive and secure energy." This Green Book suggests joint actions in the following areas (actions which will also affect the control of energy supply and demand):

– *internal market*: working towards an entirely competitive internal energy market; for example, a new energy regulator to help with the completion of the internal market;

– *internal policy of energy supply and solidarity between Member States*: for example, new legislation on fuel and gas stocks;

– *energy mix*: diverse, efficient, and sustainable; for example, an examination of the energy strategies in the EU, to be regularly presented to Member States' governments;

– *environment*: an integrated approach to climate change; for example, a low-carbon economy and an action plan to help the EU save 20% of energy from now to 2020;

– *energy technologies and innovation*: to make Europe a world leader in low-carbon technologies and to reinforce the EU infrastructures;

– *foreign relations*: a coherent external energy policy; for example, a new long-term pact with Russia on the energy supply.

The energy strategy proposed for Europe is based on the balance between supply security, sustainable development (especially, a low-carbon economy), and competitiveness.

It is, in fact, a mix of technologies which covers the three primary energy sources, namely:

– *fossil fuels*: clean coal (which means combined with carbon capture and storage), oil, and natural gas;

– *renewable energies*: hydraulic electricity, wind and sea energy, solar heating, photovoltaic, and biomass;

– *fissile fuels*: for nuclear fission.

All types of energy are also taken into consideration, particularly: electricity, different hydrocarbon fuels (of bio or synthetic origin), and hydrogen; all of them are produced from the three above-mentioned energy sources.

In relation to the nuclear energy (fission and fusion), we must remember the research mission given to the European Community by the Euratom Organization mentioned above: "in order to accomplish its mission, the European Community will promote notable research and will ensure the diffusion of information.

2.2. Nuclear energy: 50 years of industrial experience

2.2.1. *Four generations of fission reactors: a continuous technological evolution*

In the EU, nuclear fission is a fact: at the end of 2007 there were a total of 146 reactors in 15 of the 27 Member States, representing an installed capacity of 135 GWe.

In relation to the *electrical production* per primary energy type (production in tons of equivalent petroleum (TEP) i.e. 42 10^9 J), nuclear is first (32%), followed by coal (21%), natural gas (20%), oil (15%), along with hydraulic energy (9%) and other renewable energies (3%), which means that in the EU 56% of the electrical production increases CO_2 emissions.

In the EU, in relation to primary energy consumption (which means fossil, renewables, and nuclear), nuclear represents 15% of the total consumption (in TEP) and is thus an important factor in the EU policy for the security of energy supply. It is important to remember that the consumption of the other types of primary energy in the EU is as follows: oil (37%), natural gas (24%), coal (18%), hydraulic energy (4%), and other renewable energies (2%) – which means that in the EU about 80% of the primary energy consumption increases CO_2 emissions. Approximately two-thirds of these primary energies (including all the nuclear energy) are used to generate electricity: the remaining one-third is directly used in its fuel form.

Universally, in relation to the production of electricity per primary energy type (production in TEP), nuclear power plants produced 16% of all electricity, compared to coal (40%), hydraulic and other renewable energies (19%), gas (15%), and fuel (10%) – which means that 65% of the electricity production increases CO_2 emissions.

In relation to nuclear fission, the reactor designs are classified according to their technological generation. Each generation generally brings evolutionary types of improvements, or even revolutionary types of improvements when a technological breakthrough occurs.

2.2.1.1. *Generation I (1950–1970, USA, Russia, France, and the UK)*

These reactors are prototype commercial reactors built during the 1950 to 1960 period (after "Atoms for Peace" speech delivered to the UN in 1953 by the US President Dwight D. Eisenhower). They mainly use natural uranium as fuel (thus avoiding the enrichment operation), graphite (or heavy water) as moderator, and CO_2 as coolant. Few of them are still operating in the EU (the Magnox is now operating only in the UK and it's use will end by 2010). Generation I reactors are not discussed further in this book

It must be mentioned that during the period 1950–1970, the development of the sodium-cooled fast neutron breeder reactors first occurred, namely: Enrico Fermi in 1963 (USA), Rapsodie in 1967 (France), BOR-60 in 1968 (Russia), and Joyo in 1978 (Japan). The aim was that these fast neutron reactors made better use of natural resources (transformation of the fertile U-238 into fissile Pu-239), and also to contribute to more efficient management of nuclear waste.

2.2.1.2. *Generation II (1970–2000, 31 countries worldwide)*

These are commercial reactors developed over the years from 1970 (following the OPEC oil crisis in 1974) until 1990, and still operating today. They use enriched uranium as a fuel and they are water-cooled and moderated. The lifetime of the fuel in the plants was three years in the 1960s and in the 2000s it is approximately 5 years. They are the light-water reactors (LWR), which include the boiling-water reactors (BWR) and the pressurized-water reactors (PWR). The fuel (uranium dioxide, of the ceramic type) is generally introduced in long zirconium alloy tubes. The uranium-235 enrichment factor goes from 0.7 (original fraction in the natural uranium) to 3.5–5%.

At the end of 2007, there were 439 nuclear power stations in the world. Their average age was 23 years (approximately their half lifetime). However, 50 reactors are more than 30 years old and eight are more than 40 years old. All of them are operating in 31 countries (and about 90% are of generation II type and 10% of generation I type) and have a total operational time of more than 10,000 reactors-years, 4,000 of which were in Europe.

Their classification is as follows: 198 pressurized-water reactors, 66 VVERs (PWRs of Russian design), 94 boiling-water reactors (BWR), 43 pressurized heavy-water reactors (PHWRs, for example CANDU), 18 gas-cooled reactors, 12 LWGRs (which means that they are graphite-moderated, with pressurized tubes for the boiling water coolant), and four fast breeder reactors. In addition, a total of approximately 280 reactors are operating in 56 countries for scientific research purposes, and approximately 220 ships and submarines are propelled by nuclear reactors.

By the end of 2007, a total of 35 new power stations were under construction in the world (some of them of Generation III type such as the first EPR (European Pressurized Reactor) in Olkiluoto, Finland).

Finally, a total of 35 firm orders have been placed for reactors namely: 11 PWRs (including one EPR in France), 10 VVERs, nine BWRs, two PHWRs, two FBRs (Fast Breeder Reactors) in Russia and one other reactor. A total of 26 reactors (the site and technology type of which are known) were under negotiation.

The technologies of second generation nuclear fission have been proved to be safe and efficient. Let us remember that the great majority of these technologies come from designs originally used for naval propulsion. However, as is the case in every technology, improvements are always possible. This chapter demonstrates this, especially in the following fields: reactor safety and reliability, economy (efficiency of the power stations), sustainability (environment respect, optimal use of fuel, and waste processing), and proliferation resistance.

2.2.1.3. *Generation III (2010, "renaissance" of the nuclear reactor with continuing improvements)*

These are reactors designed in the 1990s. Some are already operating (the first one in Japan) and others are under construction (like the above-mentioned EPR). These reactors are Generation II developments with improved safety and efficiency (which are qualified as evolutionary). In particular, reactors must anticipate events that could damage the core of the reactor. They must also protect against consequences of the core fusion (though it is highly unlikely to happen). The human factor (man–machine interface) is also better addressed in this generation. Some of these reactors have advanced technological characteristics (like a more simple design for its construction and its operation, and underlying safety systems), which are the precursors to technology of the Generation IV reactors.

2.2.1.4. *Generation IV (2030 forecast technological breakthrough for the reactor and the fuel cycle)*

These systems will aim to produce not only electricity, but also heat, for example for the following applications: industrial usage where high temperatures are needed (particularly in the petrochemical industry), and synthetic fuel production and hydrogen production using minimum greenhouse gases. Medium temperature heat will be used for, among other things, the desalination of seawater.

These systems produce both heat and electricity (co-generation). These systems will probably be used commercially by about 2030. They should include improved safety systems and be highly economical. They also should produce a minimum of waste (which is feasible because of very innovative fuel cycles) and be strongly resistant to proliferation. Many of the reactors will be fast breeder reactors. The fuel cycle will be completely closed; not only the uranium and plutonium will be recycled, but also all the minor actinides. In the case of the fast breeder reactors, the combination of operation and recycling on-site seems to be especially promising (hence the description "revolutionary").

The International Atomic Energy Agency's [IAEA 3w] definition, distinguishes between two groups of advanced nuclear reactors

The first group is called "evolutionary". Its aim is to improve a well-established technology and experience – that of the Generation III – so a prototype is not necessary for its industrial deployment. From the safety point of view, the two objectives of the evolutionary reactors are a reduction of the frequency of severe core damage (to be achieved by, for example, way of passive safety systems) and a limitation of the consequences were a severe accident to happen (to be achieved by, for instance, reinforcing the containment function).

The second group is called "revolutionary" or "innovative". These reactors use improved or innovative systems relating, in particular, to the thermodynamic efficiency, the safety, and the processing of wastes. These are the Generation IV reactors, and prototypes are necessary before they can be deployed in industry. From the safety point of view, the main objective of these reactors is to eliminate all severe core damage as far as is practicable, for example by further optimizing the margins and relying on passive safety systems.

In the EU, during the next 15 years, the current reactors will continue to operate (at least, in 12 of the 15 Member States with nuclear power stations – let us remember that in 2007 three countries were still under nuclear moratoria: Germany, Belgium and Sweden). Since most of these reactors have reached their half lifetime (that is, about 20 years), the industrial and political authorities in many countries have started investigating the advantages and disadvantages of replacing them on their nuclear fleet. As far as new constructions are concerned, it seems that at least until about 2020, the new nuclear power stations will be of the evolutionary type (Generation III). These will usually be developments using light water as both coolant and moderator.

2.2.2. *Other applications of nuclear energy: space exploration and fusion (International Thermonuclear Experimental Reactor (ITER))*

2.2.2.1. *Nuclear energy sources for interplanetary space missions*

Since the beginning of activities in space, the search for energy sources has been focused on the only two known sustainable energy options:

– solar energy (of the photovoltaic type); and

– nuclear energy (spontaneous decay of radio-isotopes and nuclear fission after neutron absorption).

On Mars the solar energy flux is half that on Earth. A flux of only 600 watts per square meter reaches Mars (which means about 150 electric watts generated with the most efficient solar panel presently available). Consequently, distant space exploration (which means beyond Mars) cannot take place except with nuclear energy sources. A nuclear power source (NPS) is a device that uses the heat produced during nuclear reactions.

In current applications, this heat is delivered by one of the two following nuclear reactions:

– *The spontaneous decay of unstable radioisotopes* (which release alpha, beta or gamma radiations to reach a more stable state). The radioisotope heating units

(RHU) are used as heat sources (of a few watts) for thermal regulation. In the range of 1 W to 1 kW, the radioisotope thermoelectric generators (RTG) convert the decay heat of plutonium-238 (0.56 W/g, half lifetime = 88 years) into electricity. The RHUs and the RTGs have been used by the Russians and by the Americans for their missions to the Moon and to Mars and for their interplanetary space missions.

– *The fission of heavy nuclei after absorption of neutrons* (same principle as for terrestrial nuclear reactors). In the range of 1 kW to 10 MW, the nuclear fission reactors that produce electricity in space have a distinct cost advantage compared to the RTGs. The heat produced by fission is converted into electricity for on-board electric current applications (for example, by means of thermoelectric systems similar to the RTGs, or Topaz-type thermo-ionic systems) or for nuclear electric propulsion (NEP). Many low-power fission reactors (that produced electricity in the range of 1–6 kWe) were launched during the 1970s to 1980s by the Russians as multiple purpose sources for space missions orbiting the Earth. In the USA, a NEP reactor in the 100 kWe range is under development (Prometheus program) for the exploration of Jupiter's moons and as a lunar surface reactor. In the range of 100 MW to 10 GW, nuclear thermal propulsion reactors (NTP) to be used in space could generate power to directly heat a hydrogen propellant which is stored as liquid in cooled tanks. The gas, which is very hot (about 2,500°C), is then expelled at a supersonic speed through a nozzle in order to produce the necessary thrust for the launching from or the landing on a planet.

In Europe, the most important developments took place in Germany at the beginning of the 1960s, and in the UK and in France during the 1980s, with emphasis on radioisotopic heating and space fission reactors. The French Erato (Exploratory Research for Advanced Technology Office) program has investigated turboelectric systems (20 kWe) based on sodium-cooled fast breeder reactors (SFR) (UO_2 fuel, 670°C) and high-temperature reactors (HTR) (ultrafine coated particle fuel, 840°C). Some European groups still have contracts with the European Space Agency (ESA) for nuclear applications in space exploration.

2.2.2.2. *Perspectives for the fusion energy (Generation V): ITER, IFMIF and Demo*

In the longer term, fusion energy (sometimes called Generation V) should make a great contribution to our energy needs. Fusion energy is particularly attractive from the environmental protection point of view. It does not generate any long-life radioactive wastes; only the reactor materials near the plasma, which will be activated by the released neutrons, will produce radioactive wastes. Its fuel is easily available: deuterium is extracted from ordinary water (natural concentration 16.2 mg/liter in oceans) and tritium is produced from lithium (natural concentration of 20 mg/kg in the Earth's crust).

Fusion energy may be produced in sufficient quantities to meet the demands of cities and industries. The traditional technique is based on the magnetic confinement of the plasma in a *tokamak* (Russian acronym for vacuum chamber and toroïdal magnetic coil) in contrast to inertial confinement using high-power lasers or other beam devices for the implosion of material particles. The first large-scale experiment of this type was the Joint European Torus (JET), in Culham, UK. In 1997, a fusion power of 16 MW was produced using a heating power of 25 MW, and it lasted one second. The next step will be the *tokamak* ITER. In addition to ITER (initially working in pulse mode), an alternative design is being studied: the Wendelstein 7-x stellarator (normally operating under continuous mode). This project, which is the responsibility of the Plasma Physics Max Planck Institute, is under construction in Greifswald, Germany.

The ITER program (in Latin *iter* means journey or direction) is a worldwide collaboration which aims to demonstrate the technological and scientific feasibility of fusion: a self-maintained reaction called hot plasma, made of inductive-type deuterium-tritium discharges, must be produced. The ITER experimental reactor will be able to work for long periods of time and under sufficient power levels so as to be close to industrial conditions.

The main characteristics of the ITER reactor, as planned in 2006, are: heating power = 70 MW and fusion power produced = 500 MW (power amplification factor Q near 10), including a 100 MW alpha produced power; discharge duration during the first operational phase = around 400 seconds; plasma exposed surface = about 680 m^2 (maximum thermal flux = 20 MW/m^2); temperatures from approximately 200 million °C on one side to close to absolute zero (–271°C) on the other side. The ITER reactor is being built in Cadarache, France (as was decided by the Parties, June 2005, Moscow) [ITER 3w].

The ITER Agreement which established the ITER Organization (which currently has seven members) was signed in May 2006 in Brussels. The ITER program should last 30 years – 10 years for construction of the reactor and 20 years for its operation. Its cost should be about €10 billion over 30 years. The EU and France will finance 50% of the project. In fact, it is the second most expensive scientific project after the International Space Station (ISS). The first plasma production is scheduled around 2017.

Demo should be the next step after ITER, and aimed at producing electricity with roughly the same physical dimension (and cost) as ITER. However, its fusion power should be increased to obtain an electric power of at least 500 MWe, as the norm in today's standard electric power stations. Practically, the fusion power produced by Demo should be increased up to 3,000 MW depending on technological developments. The power amplification factor Q could be between 30 and 50. In the most optimistic scenario, the Demo reactor could even be considered as a prototype (Proto) that might generate electricity around 35 years after ITER started. This would bring the beginning of electricity production by fusion to about 50 years from now.

The ITER results concerning the plasma and material behaviors will be essential but not sufficient to start Demo. A high-energy, high-intensity neutron source will be necessary to test and check the performance of materials that are subject to intense radiations of the fusion neutrons (typical kinetic energy of 14 MeV). The International Fusion Material Irradiation Facility (IFMIF) might be built in Japan. The basic material for the design of the European Demo blanket is a Eurofe- type material (reduced activation ferritic-martensitic RAFM), possibly reinforced by oxide dispersion strengthened (ODS) steels which further increases the resistance to neutron radiation.

It is expected that after about 100 years, the activation level of internal component materials will be low enough to make their storage and even their recycling easier. The coolant considered in the European reference designs of the cover is helium (He) and the breeding materials are Li containing ceramics or liquid eutectic Li-Pb. The materials Eurofer, He and Li-Pb are also interesting for GEN IV fission.[1]

2.3. Main actors: common needs, international vision and strategic instruments

Generally speaking, five main categories of actors are found in the field of research and development (R&D) in nuclear fission:

- research organizations (public and private);
- systems suppliers (e.g. nuclear vendors, engineering companies, etc.);
- energy providers (e.g. electric utilities, heat and/or hydrogen vendors);
- regulatory bodies and associated technical safety organizations (TSO);
- educational and training institutions, in particular, the universities.

1. The fusion energy is not described in this document, see the following sites:

– www.ITER.org/, ITER and its partners, i-e the European Union, the People's Republic of China, Federation of Russia, and Japan, South Korea Republic, The United States of America + India since December 2005.

– www.efda.org/, *European Fission Development Agreement* (EFDA) and its associations i-e cooperation in the European Union and in Switzerland + Romania.

One of the major customers of nuclear research is obviously industry, in particular, systems suppliers and energy providers (electricity and/or heat).

In general, the systems suppliers sell nuclear installations such as electric power stations and fuel cycle factories, as well as, in the future, heat and electricity co-generation installations, factories of synthetic fuels and hydrogen applications, desalination factories, etc. These companies also work for the nuclear-waste treatment agencies.

Included in energy providers are the utility (electricity) and heat supply companies related to urban or industrial heating, as well as, much later, to synthetic fuels or hydrogen, etc.

In order to update the energy infrastructures – an urgent need because of their age – the industry would want the governments to play a triple role:

– To initiate ambitious research and development programs: public financing should be devoted to visionary research programs in strategic fields, as well as to large infrastructures, and to appropriate training.

– To finance investment: to make industrial investment easy, especially during transition periods before technological breakthroughs (for example, during the transition from the carbon-intensive economy to the (low-carbon) combined economy of electricity and hydrogen).

– To regulate in order to ensure equal treatment: to harmonize the rules for the lifetime extension of nuclear installations and to establish best practice guidelines (for example, codes and standards for GEN IV).

The nuclear industry shares a common vision as far as a political and legal framework is concerned. Due to the financial and technological risks involved and to the sociopolitical challenges related to each big innovation, the ideal work environment should be as follows:

– objective, which means a clear distribution of responsibilities between industry and regulators, as well as between innovators, researchers, and customers;

– coherent, which means harmonious treatment of every nuclear partner – private and public – and the convergence of good practices for nuclear safety;

– predictable, which means no unexpected requirements from authorities or the market, a stable international social and political framework, stability of public opinion, etc.

As far as the national regulatory authorities are concerned, it must be remembered that one of their missions is to propose legal and technical rules related to nuclear safety and security (in particular, in relation to proliferation resistance) and to ensure the implementation of these rules in relation to the nuclear installations.

We bear in mind that nuclear safety is defined by the IAEA as "the achievement of proper operating conditions, prevention of accidents or mitigation of accident consequences, resulting in the protection of workers, the public and the environment from undue radiation hazards" establishing the foundations of the "defense in depth" [IAE 93].

There are two types of proliferation resistant measures: intrinsic and extrinsic.

The intrinsic measures are technological measures. They concern the chemical form of nuclear materials, the radiation field, the heat generation, the generation rate of spontaneous neutrons, the complexity of the civilian military conversion, the mass and volume of the nuclear materials, and the characteristics of the nuclear systems which limit the access to nuclear materials.

The extrinsic proliferation resistance measures are legal measures. They represent the IAEA's international non-proliferation regime. The anti-proliferation policies consist of a combination of technical designs, operational modes, institutional measures, and safeguard measures related to nuclear installations.

The objective of these policies is to meet the requirements generated by the two following threats:

– The development of nuclear weapons programs by states other than the five states that are signatories to the UN non-proliferation treaty and that have nuclear weapons in accordance with the requirements of the IAEA.

– The acquisition of fissile materials by terrorist organizations.

Before examining how the Euratom research program meets some of nuclear fission's big challenges, we need to examine the general criteria of the EU policies in relation to nuclear fission research and development.

As is shown by the history of the construction of the European Community, there are three necessary, but not sufficient, conditions that ensure the success of any long-term Community action (including those related to research), namely:

– *Significant consensus on the common needs* (*users* level): based on the identification of the common interest domains of R&D between private and public sectors in different Member States (for example, after a major consultation with users).

– *Shared vision as to European-level solutions* (*decision-maker* level): based on a convergence of the main decision-makers in order to find an agreement on reasonable long-term objectives (quantified if possible) and to propose European-level objectives.

– *Instruments to implement the proposed solutions* (EU level): based on legal and financial incentives (for example the Framework Program) to satisfy the above-mentioned users and decision-makers.

In the nuclear fission research field, a general consensus on the common research needs exists and the EU proposes instruments to implement research programs.

As far as a shared vision is concerned, however, it must be recognized that divergences concerning advantages and drawbacks of nuclear fission power exist among the Member States of the EU.

2.4. On the eve of a technological breakthrough: six challenges for research and development

In relation to the safety of reactors, a weak point in both past and current technologies (Generations I and II) may be the exclusive presence of active safety systems. These systems are based on the active electrical and mechanical control of equipment like probes, valves, pumps, accumulators, heat exchangers, and auxiliary energy systems. It is also recognized that the many redundancies that characterize these reactors increase the complexity of the system, and the construction and maintenance costs. The next generations of nuclear reactors will have more simple systems, while continuing to improve the safety of the systems. The main difference with respect to second-generation reactors is the systematic use of passive safety systems, which do not need any active control or manual intervention to manage accidents. These systems are based on gravity, natural convection, spring or compressed gas systems, or even on their resistance to high temperature.

In terms of economy (particularly, thermodynamic efficiency), the relatively low outlet temperature of the core – about 350°C – is the weak point in current technologies in light water reactors. This means that the thermodynamic efficiency is quite limited (about 35%). Generation IV technology should achieve significantly higher thermodynamic efficiencies, and consequently should lead to the reduction of production costs. If other coolants are used, temperatures such as 400–600°C for CO_2, 500–700°C for liquid metals, and 700–900°C for gas (helium) could be reached. For example, a temperature of 900°C at the core outlet leads to a thermodynamic efficiency of up to 44%, i.e. around one-third higher than the standard LWR.

A weak point in current technology (based on thermal neutron spectrum and non-MOX standard fuel) in relation to use of fuel (sustainable development) is its limited capacity to extract the potential energy from the uranium fuel.

The natural uranium extracted from the Earth's crust is mainly composed of two isotopes: 99.3% is non-fissile U-238 and 0.7% is fissile U-235. Consequently, in Generation I and II (without MOX) thermal neutron reactors, less than 1% of the potential energy of uranium fuel is used. Let us remember that the used fuel still contains, in general, 1% of U-235 and 1% of fissile Pu and 95% of fertile U-238, i.e. 97% of reusable materials.

With the introduction of Generation IV, especially with fast reactors, this unused energy will be exploited, in practice increasing by a factor 60 the quantity of energy extracted from the same quantity of uranium. The fast reactors can convert the fertile U-238 into fissile Pu-239 faster than they can consume it. Very low U-concentration minerals will then become profitable, including the U contained in the oceans (natural concentration of 0.0032 mg/liter).

As for waste treatment (sustainable development), one of the weak points in the present technologies for the used fuel is how highly radioactive waste is processed.

Most of the radionuclides (mainly the fission products) decay quickly in such a way that after 50 years their collective radioactivity is reduced to less than 0.1% of their initial level. The main long-life actinides are the heavy transuranium radionuclides (also called TRUs, and with an atomic number greater than U*92) that result from neutron capture; these are Np*93, Pu*94, Am*95 and Cm*96. The minor actinides are Np, Am and Cm, whereas the major actinides are U and Pu. These transuranium elements have isotopes, some of which remain highly radioactive for tens of thousands years and/or are very radio-toxic, such as: neptunium-237 (half-lifetime of 2.1 million years), plutonium-239 (24,000 years), americium-241 (430 years, gamma emitter), and curium-244 (18 years, alpha and gamma emitters).

In the future, when using Generation IV systems, in particular fast neutron reactors (kinetic energy near 2 MeV), all actinides (major and minor) of used fuel will be separated from each other and introduced again in the fission reactors. Partitioning and transmutation technologies (P&T) are part of the integral actinide recycling strategy of Generation IV. The final objective of P&T is to close the fuel cycle by burning all the minor actinides, including those stored in the past. The destruction of long-life fission products such as technetium-99 (210,000 years) and iodium-129 (16 million years) needs a solution, which is also investigated.

In principle, these GEN IV and P&T revolutionary technologies will produce clean waste (which means only small quantities of fission products with a relatively short lifetime) and will recycle the dirty fuel (which contains all the transuranium isotopes) in the nuclear plants. In some fast neutron Generation IV systems, all the actinides will be reprocessed on site. By doing this, the maximum of energy will be extracted from the mineral resources while producing clean waste.

As far as the proliferation resistance is concerned, a weakness in the present technology is the risk connected to the plutonium (Pu) and MOX (mixture of U and Pu oxides) transportation from the reprocessing installations to the power-generating stations. A solution must also be found for the disposal of the reserves of military nuclear materials. In 1997, the USA and the Federation of Russia admitted that their set of nuclear installations greatly exceeded military objectives in the new international context after the Cold War. Consequently, they have dismantled thousands of nuclear warheads, declared hundreds of tons of their fissile material reserves (especially plutonium) as being far beyond their military needs, and promised that these materials will never be used for weapon production.

To burn plutonium, transmutation in LWRs or in fast reactors (in particular for non-fissile isotopes, as Pu-240) may be used. Accelerator-driven systems (ADS) may also be used. They consist of dedicated hybrid systems coupling a particle accelerator (for example, high-energy protons) and a sub-critical reactor (for example, fast neutron gas or lead-cooled reactor). The Generation IV fast reactors will not, however, eliminate the need for an international non-proliferation regime (including safeguards), but they will facilitate it by burning the continuously generated plutonium. Moreover, the use of on-site reprocessing will make any illegitimate extraction of fissile materials highly impracticable.

Research and innovation in nuclear fission are thus in the forefront of significant world-level challenges over three timescales: 1970–2000, around 2010, and around 2030. These timescales correspond to three generations of nuclear energy technologies – II, III, and IV – with which the following challenges are associated:

– GEN II (1970–2000): (1) supply security and (2) environmental protection;

– GEN III (around 2010): (3) safety and (4) competitiveness improvement;

– GEN IV (around 2030): (5) co-generation and (6) integral actinide recycling.

For the moment, we are faced with a nuclear "renaissance", mainly in countries where questions of security and safety and waste processing (linked to the operation of the reactors and the fuel cycle) are in the process of being addressed. This tendency is clear, particularly in Finland and France where decisions have been taken to build new nuclear plants (Generation III – evolutionary type).

Following a public debate and government approval in December 2003, the Finnish company TVO ordered the construction of a Generation III power station, namely a 1,600 MWe EPR at Olkiluoto (contracts with Areva NP and Siemens AG). The construction certificate was issued in February 2005 and the grid connection is scheduled for 2011.

Following similar public debates in France, in May 2006 the company EDF decided to start the construction of the first unit of an EPR series – 1,600 MWe – at Flamanville (Basse-Normandie) with grid connection scheduled for 2012.

Some countries are already preparing for the construction of Generation IV reactors based on nuclear technological breakthroughs. Relating to this subject, let us remember the announcement made by French President Jacques Chirac in his 2006 New Year message to industry and trade union managers. Mr Chirac announced "the immediate starting of a CEA program of a prototype reactor of fourth generation operating in 2020". With the construction of an innovative nuclear reactor, France plans to remain a world leader in nuclear energy. However, Mr Chirac did not mention the type of reactor: gas (VHTR/GFR) or metal-cooled (SFR) which are the two GEN IV systems presently favored by the CEA and French industry.

The above announcement must be situated in the context of the public debate in France on the management of high- or medium-activity long-life radioactive waste. Any new nuclear construction is possible politically only if the problem of waste is solved. In June 2006, the French National Assembly and Senate adopted the Program Law on sustainable management of radioactive materials and wastes. As an illustration of how a law can guide research toward results that will serve as the basis for important political decisions (for example, the management of the back-end of the fuel cycle):

> The research and the implementation of the means necessary for the final safety of the radioactive waste are undertaken in order to prevent or to limit the burden that will be supported by future generations. ... Research and investigation of this waste are pursued according to three complementary directions as follows:
>
> (1) The partitioning and the transmutation of long-life radioactive elements. The corresponding investigations and research are carried out together with those undertaken on the new generation of nuclear

> reactors … along with those undertaken on accelerator-driven systems dedicated to the transmutation of waste. These studies will be made available in 2012 through an assessment of the industrial prospective in view of a prototype to be built before the 31st of December 2020.
>
> (2) The reversible storage in deep geological layers (depth of about 500 m). The corresponding research and investigations are undertaken in order to choose a site and to design a storage center. According to the investigation results, the request to authorize the storage center … will be made in 2015 and the operation of the center will start in 2025.
>
> (3) The temporary storage. The corresponding research and investigations is to be undertaken, no later that 2015, to develop new storage installations or to modify existing installations to meet the needs, especially in terms of capacity and duration …

In the same way, an American initiative addressed to the world must also be mentioned. In the United States, in January 2006, President George W. Bush launched a global nuclear energy partnership (GNEP) after discussions with the UK, France, Russia, Japan, and China. It limits sensitive technologies (i.e. manufacturing, enrichment and reprocessing of fuel) and reinforces the international safeguards regime in the framework of increased use of nuclear energy, while ensuring the supply to partners. GNEP also aims to solve the problem of waste in the United States, and thus allow once again the reprocessing of used fuel and fast neutron reactors (especially the advanced burner reactors). The Carter doctrine of the 1980s, which has forbidden the reprocessing of the plutonium, has been abandoned. The GNEP strategy contains the following seven elements:

– the construction of a new generation of nuclear-power-generating stations in the United States;

– the development and the implementation of new technologies for fuel recycling;

– an efficient processing of used nuclear fuel and, in the near future, its storage in the United Sates:

– the design of advanced reactors for the transmutation of waste (advanced burner reactors) that produce energy from reprocessed nuclear fuel;

– a program to make nuclear fuel available to emerging countries in order for them to acquire and use the nuclear energy economically while reducing the risk of nuclear proliferation;

– the development and manufacture of small reactors dedicated to the needs of emerging countries;

– the improvement of security measures in order to better resist proliferation and to achieve an optimal safety level within the framework of increased use of nuclear energy.

2.5. Generation II: supply security and environmental protection

2.5.1. *Challenges facing the current nuclear reactors*

Two challenges facing the current nuclear reactors (Generation II) are particularly important for Euratom research, namely:

– (1) security of energy supply (availability, diversification, independence);

– (2) environmental protection (low carbon economy).

In the following sections, we will show how nuclear fission is a very reliable source of energy with a low-carbon content that has a security of supply.

2.5.1.1. *Security of energy supply (availability, diversification, independence) – challenge 1*

As previously seen, the nuclear fuel cycle and electrical production industry is, from a commercial point of view, technically mature and well known

Particularly in the EU, the nuclear industry currently provides about one-third of the EU's electricity needs. In other words, the electricity used by one family or one factory out of three is of nuclear origin. With a capacity factor often close to 90% (which means about 8,000 hours per year), nuclear energy is a remarkably reliable energy source for the production of base-load electricity (which means the amount of electricity necessary at any time, all year long). The capacity factor is the ratio of delivered energy to the energy that would be generated if the production system were exploited at its maximum capacity over one year. A ratio of 90% far exceeds the efficiency all other means of electricity production.

Moreover, uranium may be stored in a relatively easy manner and at low cost and the reserves are sufficiently high. As far as the future is concerned (ratio reserve/production or R/P), the International Energy Agency and the Organization of Economic and Cooperation Development [IAE 05a], statistics from 2005 give the following figures for the exploitable uranium resources at a cost less than US$130/kg:

– 85 and 675 years of consumption for the conventional resources, those currently known and the total (including phosphates);

– the largest known conventional resources are: 23% in Australia, 18% in Kazakhstan, and 12% in Canada, which are politically stable countries where the European companies in the uranium field hold significant shares in the resources.

The above data for the fissile fuel (uranium) reserves must be compared to fossil energy sources, namely:

– oil: 40 years with an expected peak around 2030;

– natural gas: 70 years with an expected peak around 2070;

– coal: 160 years with an expected peak around 2150.

This does not mean that the primary energy source will disappear after the peak, but rather that, after the peak, production will start to decrease in quantity and in energy content, whatever the production means used.

It is also important to emphasize that combustible fossil fuels (especially oil and natural gas) have many industrial applications beyond their use for electricity generation. Fossil energy sources are used not only for goods and passenger transportation, but also for agriculture which uses a lot of hydrocarbons for fertilizers and pesticides. The use of uranium in nuclear reactors enables the preservation of precious and limited reserves of fossil fuels.

Nowadays, the core of a typical LWR power station of a 1 GWe power contains 100 tons of U-235 fuel (enriched by approximately 4%, with burn-up rate of 40 GWd/t of heavy metal, and with a load cycle of 4 years). The future Generation IV will enable a drastic improvement of this performance; with the reintroduction of fast neutron reactors, in particular, it is possible to multiply by a factor up to 60 the generated energy from uranium, while reducing the quantity of high-activity wastes.

2.5.1.2. *Environmental protection – challenge 2*

The Kyoto Protocol (negotiated in March 1998) is obviously very important for the EU energy policy. This agreement under the auspices of the United Nations finally came into force on 16 February 2005, seven years after its negotiation.

As a result, the governments that signed the agreement have a low-carbon economy policy; this means that energy efficiency and production are generated from clean, intelligent and competitive sources for the benefit of industry, transport and people.

The EU has adopted a realistic approach in the battle against greenhouse gases. The clean technologies alone will not win this battle. Political institutions need to create a market and to launch research and development programs. Let us remember that the main Kyoto objective is to reduce, by 2012, greenhouse gas emissions by 5.2% compared to their level in 1990.

Nuclear fission must be discussed in the context of the battle against greenhouse gases. Some CO_2 is released during both manufacturing of nuclear fuel and power plant operation; but on a global scale, carbon releases per generated energy unit are extremely low (in fact 20 times lower) compared to fossils fuels. Without nuclear energy, CO_2 releases in the EU would increase by 30%. Moreover, we must remember that nuclear energy does not generate any acid rain or gas (or NOx-type fog) or aerosols. Nuclear energy is therefore part of the solution in the context of a sustainable energy mix, which indeed contributes to supply security and environmental protection.

2.6. Generation III: continuous improvements in safety and competitiveness

2.6.1. *Challenges for the next generation of nuclear reactors (current renaissance)*

On the basis of already known technologies in the reactors and nuclear fuel field, some countries have started to design and build Generation III reactors, which are of the evolutionary type.

Those reactors rely on the present industrial infrastructure (notably the fuel cycle) and have a list of improvements to achieve by 2010, such as:

– standard designs in order to speed up the licensing procedures in order to reduce the investments costs and the building time;

– more robust designs, thus enabling better management of the power plants;

– a higher capacity factor and lifetime – in general 60 years (competitiveness);

– safety improvements (in particular, a reduction of severe core-damage probabilities);

– minimal effect on the environment.

It is obvious that any decision to build a new power plant is the responsibility of the commercial sector and therefore must meet the requirements of the private investors (in addition to the above-mentioned requirements) – in particular a good return on investments – and must also meet the requirements of the industrial operators (such as easier power-plant operation). However, such a decision requires interaction with governmental organizations and recognition of public opinion.

Two challenges to the present nuclear "renaissance" (Generation III) are particularly important for Euratom research, namely:

– continuous improvement of the safety and reliability of installations (challenge 3);

– greater competitiveness and better management of power plants (challenge 4).

2.6.1.1. *Continuous improvement of the safety and reliability of installations – challenge 3*

As far as nuclear safety is concerned, we note a general concern in the world, particularly after two severe accidents:

– Three Mile Island (Pennsylvania, United States, 1979), where there was a serious loss of coolant with an important fusion of the core and hydrogen flames, but where there was no spread of radioactive substances outside the concrete containment building;

– Chernobyl (Ukraine, 1986), where there was a very severe accident caused by a positive neutronic feedback coefficient for core voiding in a reactor for which the quality of design and the management norms were very low.

After these two severe accidents, the building of new power plants was stopped – completely in the United States and in Europe, but not in Asia.

The vulnerability of the power plants to severe accidents and terrorist attacks is still considered by some people to be an unsolved problem. Confidence in safety remains weak; therefore, a severe accident – even without victims – would have a very negative impact on the future of nuclear energy. Although strict national and international rules tend, at least, to reduce any risk, the potential consequences of severe accidents (the probability of which is low) could be very high. In order to regain the confidence of public opinion and of investors, technical safety must be continuously improved in every phase of the power-plant cycle (from construction to dismantling). The same comments can be made in relation to nuclear security. It must be improved in the new nuclear power plants and the international non-proliferation regime must be improved at the same time.

For all new reactor designs, in-depth defense (with high safety margins) is still the fundamental basis for safety. This approach is deterministic and is based on the concept of design basis accident (DBA). In fact, in order to better solve the inherent uncertainties related to innovative designs of Generation IV, a deterministic and probabilistic study will be necessary; the probability accidents should be quantified and defense measures should be introduced at the design level in order to ensure that the possibility of accidents is reduced to a minimum.

According to the European Utility Requirements (EUR), when the designing of a reactor begins, any situation that could lead to unacceptable levels of core damage frequency (CDF) or to frequent early and heavy rejection of fission products into the environment should be eliminated. The design of Generation IV reactors will therefore require methodologies based on risk evaluations. Those methodologies will also be used for safety analysis and for various interventions (systems replacement, etc.), as well as for inspection.

2.6.1.2. *Greater competitiveness and better management of power plants – challenge 4*

As we remember, the fixed costs of nuclear energy are quite high because of the high initial investment, but the variable costs are low because of the low fuel cost, leading to a global cost per delivered MWh that is generally far below those from other electricity production sources. The fuel costs in nuclear plants are, indeed, rather low; they are nearly negligible in hydraulic installations, and are very high in coal, oil, and gas installations. A Finnish study on EPR has analyzed the impact of fuel cost on the electricity production cost: to double the fuel cost would increase the MWh cost by 9% for electronuclear energy, 31% for coal, and 66% for gas.

Other international studies on EPR have shown that with a series of 10 units sharing the costs of the first one, the investment cost is estimated at €2.6 billion per unit, which corresponds to €1,630/kWe. The cost of EPR electricity production, which is estimated at €28/MWh, is more competitive than that of more modern installations which have a conventional cycle combined with natural gas.

To be fair, for all sources of energy we should compare the costs once the dismantling and waste treatment costs have been included, as is the case in the EC study NEEDS. These external costs are what the society must pay for the impact that technology has on health and the environment. In the present market conditions, nuclear energy seems to be very competitive, in particular if it is compared to fossil fuels.

As far as Generation III (2010 forecast) are concerned, the calculation is more complicated because the actual capital costs (based on the life expectancy of the power plant life) and the investment income required by the Stock Exchange may be very different. This issue is considered in the annex of the 2006 Green Book 6 energy "What are the Stakes? – Information Document on the Green Book – A European Strategy for Sustainable, Competitive and Secure Energy", which says on the topic of cost calculations:

> The capital costs for nuclear electricity are about €50/MWh if we base calculations on 6,000 operating hours per year and an interest rate of

15%, which is the reference rate in the energy field. It is difficult to be more competitive. However, if the cost is based on 8,000 hours per year with a profit of 6.5% (reflecting the present interest levels, but not necessarily the uncertainty of electricity sales in a very competitive market), and on a lifetime expectancy of 50 years, the cost will be €16/MWh, which is highly competitive. Another important cost factor is the price of CO^2 emissions. If we use €20/ton of CO^2, we must add about €20/MWh to electricity based on solid fuels and on lignite. There arc good reasons to think that the cost of electricity production for the future generation of technologies will tend toward about €60/MWh, a cost level that is sufficient to justify significant investments in normal market conditions.

Good management of power generating stations also requires an analysis of the aging mechanism (which is the topic of many fundamental and applied research projects) and an investigation as to their potential life expectancy. On the technical side, there are two limitations on life expectancy:

– aging of the two major, and not replaceable, components of the reactors: the tank and the surrounding wall;

– obsolescence of the implemented technologies and the acceptable safety level.

In the EU in general (unlike the United States), it is particularly hard to imagine a scenario where the lifetime of the current power generation stations (Generation II) would extend beyond 40 years.

2.6.2. *Selection of 12 Generation III-type reactors*

A dozen designs for Generation III reactors are ready for the market (they are described below). There are many evolutionary type developments in relation to PWR, BWR and CANDU (Canada Deuterium Uranium). For the moment, the main reactors of this type are operating in Japan. There are four ABWR (advanced BWR, General Electric-Hitachi-Toshiba) of 1,300 MWe; the first two have started operation in Kashiwazaki (with simplified systems, including the containment building, and constructed in 48 months); two ABWR are in the final phase of construction in Taiwan.

In the EU, four advanced designs of PWR and BWR have been developed to meet the European utility requirements (EUR) already mentioned:

– (1) EPR (European pressurized reactor), the US commercial version of which is considered to be an evolutionary type reactor: Areva NP (formerly Framatome

ANP, France and Germany) developed this big reactor (1,600 MWe) in the 1990s as the new French reference and received the final approval in 2004. The EPR is an evolutionary type development, based on the designs of N4 (such as Civaux, 1,450 MWe, 1999) in France and of Konvoi (such as Isar 2, 1,400 MWe, 1988) in Germany. This evolutionary approach, as well as the high outlet power, ensures the competitiveness of this design. The EPR will operate in load-following mode; it will have a fuel burn-up rate of 65 GWd/t, a core accepting up to 100% MOX, and a thermal efficiency higher (36%) than the other light water reactors. Its capacity factor should be about 92% for a 60-year lifetime, thus reducing the costs of the generated electricity.

– (2) SWR-1000 (SiedeWasserReaktor or BWR): Areva NP has also developed an evolutionary design BWR (1,200 MWe). The SWR-1000 is a combination of improved and innovative passive safety systems. These passive safety systems are emergency condensers (which have worked successfully for many years in the reactors of Gundremmingen A in Germany and in Dodewaard in the Netherlands), containment wall condensers, passive pressure pulse injectors, and other gravity-driven systems. The main design was developed in 1999. The reactor has a simpler design than other generations and will use high burn-up rate fuels; the reloading cycles of the core will be approximately 24 months.

– (3) ESBWR (European or economical simplified boiling water reactor): General Electric has developed this 1,550 MWe BWR with passive safety systems from its advanced design ABWR, previously mentioned. In this type of reactor, in the highly unlikely event of a severe accident, the residual power would be removed without danger from the containment building, thanks to long duration passive safety systems.

– (4) BWR 90 +: a Westinghouse (Sweden) design for a 1,500 MWe evolutionary type reactor. It contains improved safety systems, compared to BWR-90, with advanced digital systems, including reactor protection systems. This will relieve the pressure on the operator. Savings will also be made as a result of space and wiring reduction.

In the Russian federation, many advanced designs of PWR with passive safety systems exist, or are under development, such as V-392 (VVER-1,000,under construction of two units in India) and V-448 (VVER-1,500) developed by Gidropress.

In the United States, many PWR and BWR advanced designs of the evolutionary type are proposed for the international market, in addition to the three designs mentioned above: EPR, SWR-1000 and ESBWR. These are AP-1,000/AP-600 (PWR, Westinghouse, originally designed in 1997, NRC-certified) and ABWR (advanced boiling water reactor already mentioned, four units operating in Japan,

NRC certified). A innovative step towards Generation IV is being made by the development of high-temperature gas-cooled reactors (HTR) with direct cycle gas turbines. It is PBMR (pebble bed modular reactor of 165 MWe, Eskom in South Africa and Westinghouse, and construction is planned to start in 2014) and GT-MHR (gas turbine–modular helium reactor, 285 MWe, General Atomics and OKBM, planned for Seversk/Tomsk-7). We are aware that Areva is developing an advanced HTR: Antares. Finally, we mention an integrated PWR design, which is quite close to Generation IV: IRIS (international reactor innovative and safe, units of 5 MWe, Westinghouse) will be marketed 2015.

In the Republic of South Korea, there is the APR-1 400 reactor which is an advanced PWR, developed by KEPCO and KHNP, and NRC certified (with two 1,450 MWe units under construction, to be finished by 2013); this reactor is based on KSNP-1,000 (an optimized power reactor in Korea) which is an evolution of the CE 80+/Westinghouse design, built in 1999.

In Japan, in addition to the above-mentioned ABWR, APWR (an advanced PWR with a simplified operation), derived from Sizewell B in the UK, Westinghouse and Mitsubishi Heavy Industries (1,500 MWe) is under construction in Tsuruga.

In the People's Republic of China, there is CNP-1 000 (PWR with a burn-up rate of 65 GWd/t and a loading cycle of 24 months), designed by the Chinese national nuclear company (CNNC, with some AREVA technology transfer).

In Canada, AECL (Atomic Energy of Canada Limited) proposed two CANDU reactor designs of the evolutionary type with passive safety systems: CANDU 6/9 (750/925 MWe) and ACR (700 to 1,000 MWe); the first is certified; the second is under certification in Canada.

2.7. Generation IV (2030 forecast): technological breakthroughs in competitiveness and sustainability

2.7.1. *International collaboration (GIF with US-DOE, INPRO with IAEA, Euratom in the EU)*

Many international climate and energy policy specialists are investigating what may happen to planet Earth during the next decades. What are the scenarios for a "low carbon economy" within 40 years? These experts predict the advent of a new hydrogen-based economy as a possible solution to the question of finding a replacement for carbon intensive economies.

Since the thermodynamic efficiency of the present and future nuclear power generating stations (Generations II and III, respectively) is relatively low for electricity generation (around 35%), the announcement of a new nuclear system Generation IV raises fresh hopes. The objective is not only to generate electricity, but also to develop applications, such as high temperature heat generation, for industry, particularly the petrochemical industry, for manufacturing synthetic fuels for transportation, and for hydrogen production from water. High temperature heat is generated from very efficient thermodynamic cycles in innovative systems which also reduce the toxicity and the volume of high activity wastes in a safe and non-proliferating manner. Another application of combined generation (nuclear co-generation) is sea water desalination, which is of interest to many countries, such as Argentina, Canada, China, Egypt, France, India, Indonesia, Republic of South Korea, Morocco, Pakistan, Russia and Tunisia

The above-mentioned challenges are of worldwide relevance. In 1999 a group of nine countries (led by DOE/ONEST in the United States and comprising Argentina, Brazil, Canada, France, Japan, South Africa, Republic of South Korea, the UK, and the United States) started an international initiative the objective of which was to select a series of mature nuclear systems for industrial deployment by 2030. In 2001 this group became the Generation IV International Forum (GIF). In February 2002, Switzerland became a member of the forum, and the European Atomic Energy Community (Euratom) joined in July 2003. China and the federation of Russia became partners of the GIF in 2007. Six innovative nuclear systems were selected in 2002 after the evaluation of more than 100 different concepts of nuclear systems by more than 100 experts from a dozen countries. The technical secretariat of GIF is located at the Nuclear Energy Agency (OECD (Organization for Economic Cooperation and Development)/NEA (Nuclear Energy Agency)).

The GIF is concentrated on the advanced R&D aspects and mainly represents the interests of the manufacturers of innovative systems. In February 2005, five members of GIF signed the intergovernmental framework agreement for the international collaboration in research for and development of nuclear energy systems of the Generation IV type. This framework agreement was signed by Euratom at the beginning of 2006 (a decision made by the Council on December 20, 2005). Consequently, a large number of research centers in public and private sectors and industry in EU countries with nuclear power-generating stations participate in this GIF program, either directly or through Euratom.

Another international initiative, complementary to GIF, that we must mention is INPRO (International Project on Innovative Nuclear Reactors and Fuel Cycles).

This was launched by the IAEA in 2000 with results expected for 2050 This initiative was proposed by the president of the Millennium summit and confirmed by the General Assembly of the United Nations in 2001. In June 2008, INPRO had 28 members from developed as well as emergent countries: Argentina, Armenia, Belarus, Belgium, Brazil, Bulgaria, Canada, Chile, China, the Czech Republic, France, Germany, India, Indonesia, Japan, Morocco, the Netherlands, Pakistan, the Republic of South Korea, the Russian Federation, Slovakia, South Africa, Spain, Switzerland, Turkey, Ukraine, United States of America, as well as the European Commission (Euratom). Taking into account its unique mission in the field of nuclear technology, of security and safeguards regime, the IAEA invited all the interested member states to consider joint action in relation to nuclear innovation. INPRO concentrates on the needs of the users of innovative nuclear systems that generate energy and which should be extremely competitive and safe. INPRO has not chosen any specific technology. It produces a catalog of principles for the evaluation of these systems in six areas (economy, feasibility, environment, safety, waste treatment, proliferation resistance) and deals with some lateral questions, such as access to research infrastructures.

2.7.2. *GIF criteria: sustainability, economy, safety, reliability and proliferation resistance*

A series of technological criteria for industry and society have been identified by the GIF in four fields where the Generation IV nuclear energy systems must demonstrate their performance:

– Sustainable development: these systems will supply energy in a sustainable manner, which means in accordance with the objectives of non-atmospheric pollution, of availability of these long-term systems, and of efficient fuel use for energy generation at world level. Moreover, they will reduce and manage their nuclear wastes in an optimal manner (reduction of volume, radioactivity, residual power, etc.).

– Economy: these systems will have an obvious life-cycle cost advantage compared with other energy sources (production cost per kWh). They will also have a financial risk level comparable to other energy generation systems (investment cost per installed kWe). The burden of maintenance will be reduced in the long term, while improving protection for public health and environment.

– Safety and reliability: these systems will optimally work from a safety and reliability point of view. They will also have a very low probability of reactor core damage, which in addition would be limited . Moreover, they will not need any off-site evacuation plan in the highly unlikely case of a severe accident.

– Proliferation resistance and physical protection: these systems will be so efficient that the theft of material usable for weapons will be extremely difficult and there will be maximum physical protection against terrorist attacks.

A certain number of methodologies have been developed by the GIF in order to assess the performance of the different nuclear systems according to the above performance criteria. For each of these criteria, the GIF-associated research programs (in particular Euratom) propose different solutions that are discussed hereafter.

2.7.2.1. *Sustainable development, e.g. use of fissile resources and waste reduction*

– Develop fast breeders that convert fertile U-238 or Th-232 into fissile Pu-239 or U-233, respectively. In the case of thorium, there is less production of actinides, such as Pu.

– Keep together in the fuel, Pu and/or U, two important components extracted from used fuel during the recycling operation.

– Develop an entirely closed fuel cycle by destroying all transuranics: transmutation of long-lived radioactive elements, either in heterogeneous or in homogeneous processes, with secular disposal of high-level and short-lived wastes as the ultimate goal. This visionary strategy relies on research in Generation IV (see section 2.7.4) and in accelerator-driven systems (ADS) devoted to waste transmutation, following the work initiated in 1993 by Nobel Prize winner, Carlo Rubbia.

2.7.2.2. *Economy, e.g. thermodynamic efficiency and systems simplification*

– Penetration of new energy markets, e.g. by generating both electricity and high temperature heat (necessary, especially for synthetic fuel and hydrogen generation), as a result of very high core outlet temperatures.

– Use of advanced approaches, such as probabilistic, deterministic mixed investigations for the design of the reactor, and the development of Brayton-cycle gas turbines to improve electricity generation, with or without co-generation;

– Reduction of constraints on containment structures, e.g. by using coolants with very high boiling points or by developing the containment function in a non-conventional manner.

2.7.2.3. *Safety and reliability, e.g. thermal, neutronic, mechanical and chemical balance*

– Practical elimination, at design level, of any severe accident, or more precisely of any situation that could lead to severe core damage and/or to large, early releases of fission particles.

– Design reactors with negative neutronic feedback coefficients for temperature and voiding, leading to the loss of reactor power during an accidental temperature increase, or the loss coolant in the core.

– Passive removal of residual heat as a step beyond the last in-depth stage defense, as well as the use of materials of high thermal capacity in the core, and of coolants which do not react chemically with air or water.

2.7.2.4. *Proliferation resistance and physical protection, e.g. against external attacks*

– To develop a completely closed fuel cycle (see Section 2.7.2.1 above);

– To operate at a reactor rate close to one (no need of uranium blanket).

– To use systems that make it difficult to extract fissile material from used fuel, although this may complicate the fissile material accounting.

Of Generation IV's challenges two are particularly important for Euratom research and may lead to technological breakthroughs:

– co-generation (electricity–heat) for new industrial requirements, particularly for synfuels and hydrogen, challenge number 5;

– integral recycling (sustainability of the fuel cycle and better exploitation of unused energy), challenge number 6.

2.7.3. *Co-generation (electricity–heat) and integral recycling (sustainable development)*

2.7.3.1. *Co-generation (electricity–heat) for new industrial needs (in particular, for synthetic fuels and hydrogen) – challenge number 5*

"The stone age finished, not because of lack of stones, and the oil age will finish, but not because of lack of oil" is a famous prediction made in September 2000 by Cheyk Zaki Yaman who was the Saudi Arabian oil minister from 1962 to 1986, i.e. during the oil embargo during OPEC crisis in 1974.

From another point of view, Jules Verne, the creator of the futuristic scientific novel, in his book *The Mysterious Island* (1874) writes: "'yes my friend', says Cyrus Smith, 'I believe that one day, water will serve as fuel, that hydrogen and oxygen contained in water, used alone or together, will provide an inexhaustible energy and light source, of which intensity coal is not capable, and I believe that when the coal resources are exhausted, we will heat ourselves thanks to water. Water will be the coal of the future'".

For the "low carbon economy," hydrogen seems very promising. In other words, the water engine of Jules Verne is not exactly a writer's dream. Thanks to hydrogen, which can be produced from water, and which, when it burns in air, produces water, fiction is on the point of becoming reality.

Hydrogen (H_2) is not a primary energy source, like fossil, renewable or nuclear energy. Hydrogen, or more precisely, the family of energy vectors based on hydrogen, could however become of prime importance as an alternative which may provide an answer to the questions of energy security and environmental protection. The competition between electricity, hydrocarbon, and hydrogen as energy vectors will probably be very fierce, and generate many debates. Furthermore, these three vectors could continue to coexist far into the future, but in very different proportions compared with today.

Let us remember that hydrogen and oxygen may combine in a fuel cell to produce electrical energy. A fuel cell (eg. a battery) uses a chemical reaction in order to deliver an external electrical voltage, but is different from a battery in the sense that the fuel is continuously supplied as oxygen and hydrogen gas. The fuel cell may generate electrical energy with a higher efficiency than just burning hydrogen to produce the necessary heat needed to drive an electrical generator. The fuel cell is not subjected to the thermal-type constraints of the second thermodynamic law (Carnot efficiency). Its only by-products are breathable oxygen, drinkable water, and heat – thus it does not pollute. The inverse process of a fuel cell is water electrolysis. When delivering energy, water may be broken into diatomic molecules of hydrogen (H_2) and oxygen (O_2).

Moreover, hydrogen (see Chapter 4) may be stored in large quantities, unlike electricity, and may eventually be converted into electricity in fuel cells (useful, for example for innovative transport technologies). Until now, hydrogen has been used for very specific industrial energy needs, e.g. for fertilizers or oil refineries. The fixed and mobile uses are:

– fertilizers and petrochemical industry uses (fixed uses): industrial processes consuming not only hydrogen, but also heat – presently around 50% of the total hydrogen is used in ammonia production (NH_3 for fertilizers) and 37% in oil refineries;

– passenger and goods transportation (mobile uses): presently, 8% of all hydrogen is used in methanol production and 1% as fuel in space programs. The 4% remaining is used for other purposes. Some electric motor vehicles already use fuel cells, which are based on hydrogen conversion into electricity, e.g. using methanol.

Molecular hydrogen does not naturally exist in large quantities so it cannot be extracted or pumped. Hydrogen is everywhere, but not available anywhere. It is a synthetic energy vector which must be produced:

– either from fossil fuels (C_xH_y) or from biomass ($C_xH_yO_z$). Conventional technologies are: natural gas reforming (methane) with water steam (steam forming), working at 500–950°C (85% of the total H_2 production today); oil transformation (7%); coal gasification in the presence of steam (4%);

– or from water (H_2O). This uses the conventional technology of water electrolysis: for example, using electricity during the off-peak periods (4%) – remember, the temperature of direct water splitting is higher than 2,500°C.

In the future, many new fixed uses of hydrogen will probably appear – such as power production for electricity and/or heat, ranging from kW (domestic production) to MW (industrial production) – whereas the mobile uses (particularly portable equipment in W and transport in kW), using fuel cells, will increase considerably.

The hydrogen market should develop at a rate of about 4% per year. The limited market of mobile hydrogen (13% today) grows at a rate of about 10% per year. In 2004, hydrogen represented 1% of the total world energy consumption (it is planned to reach 20% in 2050) and world hydrogen production reached 50 million tons, split between fixed and mobile uses.

There are, however, many problems to solve before an economy based on hydrogen becomes a reality. One of the major problems is its production in large quantities by means of low carbon technologies. Another problem is public acceptance, in particular, in relation to safety. The nuclear energy systems of Generation IV (some of them with the temperature at the core outlet > 900°C) could be the only clean technologies with an energy density sufficiently high enough to produce significant quantities of hydrogen from water, while generating electricity (co-generation).

In fact, there are only two clean technologies, i.e. they do not use hydrocarbons from fossil fuels, for producing large quantities of hydrogen. Of course, production of hydrogen from biomass is also a sustainable approach, but it is still at a laboratory stage. The two clean industrial technologies are based on the splitting of water (H_2O) using an external heat source:

– thermochemistry or thermochemical water split cycles (TCWSC): the cycle of iodine-sulfur (IS): water steam + SO_2 + I_2, also called a Bunsen reaction, at 180°C in a cycle of four reactions, the last of which produces SO_2 at 830–900°C) – the efficiency of this cycle goes from 30% at 750°C to 60% at 1,000°C,which is in the temperature range of many GEN IV systems;

– high temperature electrolysis steam (HTES): water steam + solid electrolyte of ceramic type (inverse process of solid oxide fuel cells (SOFC)), the ionic conductivity of which is satisfactory only beyond 700°C, using electricity generated by a peak gas turbine, e.g. Brayton cycle, in a GEN IV system (three advantages: energy efficiency, compactness, economy).

During the transition period until the low carbon economy is fully established, the production of advanced energy vectors seems to be a promising approach, notably, synthetic hydrocarbon fuels which are useful for transportation, – from a synergy of primary energy sources (fissile and fossils). These synergetic processes will combine chemical and nuclear systems. The synthetic fuel – as opposed to the biofuel – is derived from a fuel other than oil. There are three categories of liquid synthetic hydrocarbon fuels: the CTL (coal to liquid), GTL (natural gas to liquid) and BTL (biomass to liquid); all of them may also be hydrogen enriched.

The nuclear energy fission, especially in Generation IV with its co-generation systems (electricity and high temperature industrial heat), therefore offers a response to the above challenges. In order to address these challenges, nuclear energy will penetrate new energy markets beyond electricity, exploiting, in particular, the promising possibilities of co-generation. This technological breakthrough will also have an important favorable impact in terms of compliance with the Kyoto requirements. Nuclear energy, which today plays a role only in the electricity market, will contribute, in a sensitive way, to the reduction of dependency on fossil fuels, which is one of the major causes of climate change. The ratio of nuclear energy in the energy mix, which is today 15% in the EU (or 6% worldwide), will then increase significantly.

2.7.3.2. *Integral recycling (sustainability of fuel cycle and better exploitation of unused energy), challenge 6*

In 2007, there were about 200,000 tons of used nuclear fuel in the world, including a separated civil plutonium reserve estimated at 200 tons. The quantity of used fuel in the world increases at a rate of about 10,000 tons per year.

In the context of sustainable development, recycling seems to be an attractive option for improving the efficiency of the management of natural resources and reducing the accumulation of radioactive waste. Traditionally, the objective of recycling is to extract the fissile materials in order to recycle them and consequently to reduce the volume and heat of high activity waste. The main objective is to properly solve, and in a non-invasive manner, the congestion problem which has left thousands of tons of nuclear waste stored near reactors, often around big cities, and to reduce its radio-toxicity and volume. At present, very innovative technologies for recycling are under development. There is emphasis on ending the fuel cycle so that only small quantities of relatively short-lived, high activity waste remain.

It is worth noting that in the EU in 2007, the waste generated each year by a typical 1,000 MWe reactor consisted of 30 tons of spent nuclear fuel (SNF) which was further transformed into approximately 4 m^3 of High Level Waste (HLW) and up to 400 m^3 of Low and Medium Level Waste (LMLW). As a consequence, in the EU some 50,000 m^3 of radioactive waste (i.e. 100 cm^3/person) and 500 m^3 (i.e. 1 cm^3/person) of HLW are generated each year. Those figures are to be compared to the 1,000 million m^3 of industrial waste and 10 million m^3 of toxic industrial waste in the EU, 2.02 million cm^3/person) [EUR 98]. A total of 2 million m^3 of radioactive waste has been disposed of in the EU so far, most of it in surface or near-surface facilities. The major fuel cycle strategy for LMLW is storage in surface or deep sites for 30–100 years, following the local policy for plant decommissioning. As far as HLW is concerned (most of it already vitrified), the total volume of waste in the 27 EU Member States, from the reprocessing of used fuel, fission products, and minor actinides stored since 1950 has been about 6,000 m^3. The disposal option for long-lived and high-level waste (including spent fuel) is deep geological disposal.

As far as actinide treatment is concerned, it is as well to remember some figures: every year, in the European Union, a total of 25 tons of plutonium (civil), 3.5 tons of minor actinides (conventional burn-up rate of nuclear fuel = 50 GWd/t) and 3 tons of long-life fission products (out of a total of 100 tons of fission products) are included in the 2,500 tons of used fuel.

In many countries, studies are being carried out on types of technology and the options for the future of the fuel cycle. Up to now, at the global level, no solution has been found by the political decision-makers regarding the very long-term management of the high-activity radioactivity waste. Today there are no deep-layer geological storage sites for high-activity waste. The only noticeable exceptions are Finland, where geological storage is planned (in Olkilutoto by 2020), and Sweden.

As of today, there are three main options for the back-end of the fuel cycle:

– 1: Once-through or open cycle (final disposal of spent fuel): strategy of direct geological disposal of spent fuel, often with the possibility of retrievability, as is the intention in most EU countries and in the USA; notice that direct disposal requires up to five times more space than vitrified waste disposal and that "plutonium mines" could be formed with relatively easy access after 100 years.

– 2: Recycling or closed cycle: reprocessing as in France (25 years of experience, La Hague), UK, Russia (RT-1 facility for radiochemical reprocessing of SNF), and Japan, as well as in China and India, and which is also being reconsidered in the USA. "Reprocessing" or "partitioning" operations deal with the separation of various physical and chemical elements. According to the above balance, more than 96% of the spent nuclear fuel is actually re-usable if fertile material is converted into fissile: fertile U-238 can indeed be reprocessed for storage or re-usage, and the less than 1% of Pu can be used in MOX fuel for recycling in existing or future reactors. The most advanced technique for waste recycling today is the separation of U and Pu from spent fuel with no recovery of minor actinides and fission products. The Purex reprocessing technology, used in France and Japan, separates spent fuel constituents into three main streams: approximately 95% of U; 1% of Pu; 4% of fission products and minor actinides. The last 4% is vitrified as HLW in view of its storage in the final disposal. Through reprocessing, seven spent UOX fuel assemblies can be recycled in one spent MOX fuel assembly (15% of natural resources are saved).

– 3: Long-term/interim storage, ideally 60 to 100 years: this consists of temporarily placing these substances in a specially laid-out installation at the surface or at a low depth, e.g. on-site in spent fuel pools and in dry storage systems until the fuel is recycled or disposal is possible. Interim storage allows heat and radioactivity levels to decrease. This is actually an interim solution, usually referred to as "wait-and-see".

An increasing number of countries worldwide are concerned about heavy actinides and fission products that will not decay to safe levels for hundreds of thousands of years. From a sustainable development perspective, recycling seems to be an attractive option for improving the efficiency of natural resource management and reducing radioactive waste accumulation. The research community and policy-makers are discussing sustainable solutions for the long-term disposal of long-lived high-level radioactive waste along the following lines:

– the destruction of the TRUs, i.e. transmutation of long-lived radioactive elements (TRU and long-lived fission products) either in single or in double strata, with secular disposal (high-level and short-lived waste) as the ultimate goal. This visionary strategy relies on research in Generation IV and in accelerator-driven systems devoted to waste transmutation following the work initiated in 1993 by Nobel Prize winner Carlo Rubbia. ADS are systems composed of a linear accelerator and a spallation target, providing fast neutrons to a sub-critical reactor. Special efforts are necessary, in particular to set up the technical specifications and safety requirements, in view of industrial exploitation;

– the search for a geological environment that will remain stable for the specified period, i.e. reversible or irreversible disposal in deep geological formations, that goes together with the development of waste forms that can contain these elements for that period. The most advanced technique for high-level waste today is vitrification in waste packages for geological disposal. This reduces the amount of waste by a factor of up to five, and the radio-toxicity by a factor of 10. A number of issues, however, are still open for research in geological disposal, such as climate change, the slow deformation and desaturation of rocks, the slow degradation of waste containers (hydrogen production by corrosion), the slow release of radionuclides, groundwater movement, and the transport of radionuclides in groundwater and their impact on public health.

In conclusion, the integral recycling of nuclear fuel is a very ambitious, but feasible, objective. Then the only high-activity waste resulting from recycling or manufacturing will have a relatively short life and quantities will be small.

Future recycling technologies will be able to produce clean waste, i.e. containing only short-life fission products, while recycling dirty fuel in Gen IV nuclear systems, i.e. containing all the transuranics and possibly some short-life fission products.

By this process, maximum energy will be generated from mineral resources, which is the equivalent of better exploitation of the unused energy, and only fission products will be included in the waste, all of which contributes to the sustainable development.

2.7.4. *Choice of six Generation IV-type nuclear systems and Euratom research*

In 2002, the GIF chose six nuclear systems (Table 2.1) which were then proposed to the laboratories of the GIF member states, in particular Euratom. A roadmap has been established for each of the following three phases:

– technological and scientific feasibility, which aims to solve the major S/T challenges, with a phase of international research, of total duration = 5 to 15 years;

– performance validation, which aims to optimally satisfy the GIF criteria – sustainability, economy, safety, proliferation resistance. Duration: 5 to 10 years;

– demonstration, which aims to convince the electronuclear industry, in particular the systems suppliers and the energy providers. Duration: 3 to 6 years.

The R&D GIF budget for the six systems is estimated for the first 15 years to be roughly 6 billion US dollars.

	Neutronic spectrum	Coolant	Temperature (°C)	Fuel	Fuel Cycle	Power (MWe)
Sodium Cooled Fast Reactor (SFR)	fast	sodium	550	U-238 and MOX	closed	50 2,000
Very High-Temperature Reactors (VHTR)	thermal	helium	1,000	prisms or pebbles of UO_2	open	275
Gas-Cooled Fast Reactors (GFR)	fast	helium	850	U-238	closed	300 1,500
Supercritical Water-Cooled Reactors (SCWR)	thermal or fast	water	625	UO_2 or MOX	open (thermal) or closed (fast)	1,700
Lead-Cooled Fast Reactors (LFR)	fast	Pb-Bi or Pb	480–800	U-238	closed	10 – 100, 600
Molten Salt Reactors (MSR)	epithermal	salt fluoride	700–800	UF in salt	closed	1,000

Table 2.1. *Selection of the six nuclear systems of Generation IV type*

Below, the six GIF nuclear systems are briefly described, with emphasis on electricity and/or co-generation, reference power, type of neutronic spectrum, coolants, fuels, S/T challenges, etc.:

– sodium-cooled fast reactors (SFR), demonstration scheduled in 2020;

– very high-temperature reactors (VHTR), demonstration in 2020;

– gas-cooled fast reactors (GFR), demonstration scheduled in 2025;

– supercritical water-cooled reactors (SCWR), demonstration in 2025;

– lead (alloy) cooled fast reactors (LFR), demonstration scheduled in 2025;

– molten salt reactors (MSR), demonstration scheduled in 2030.

2.7.4.1. *Sodium-cooled fast reactors (SFR)*

– Electricity production and full actinide management (enhanced fuel utilization, efficiency close to 40%, core outlet temperature of 550°C)/reference power = modules of 50–150 MWe or plants of 600–1500 MWe. Earliest demonstration in 2020.

The SFR systems have more than 300 reactors-years' experience thanks to five decades of fast neutron reactors in eight different countries, e.g. in France, Phoenix and Superphénix; in Europe, the European Fast Reactor project (EFR, 1988–1995); in the Federation of Russia, BN-600; in Japan, Joyo and Monju.

The cycle of SFR fuel is closed with fertile uranium conversion and integral recycling of all the actinides. Monju should become the pilot installation to test these characteristics. There are many fuel options: metal (U – TRU – 10% of Zr alloy) for 50 MWe units (burn-up rate of 80 GWd/t) and MOX (containing TRU) for 2,000 MWe reactor (burn-up rate of 50 GWd/t). In every option, the temperature of the coolant at the core outlet is 550°C and the system pressure is 0.1 MPa. Electric generation is obtained by means of a secondary sodium circuit. The following safety points must be noted: use of well-known ferritic and martinsitic steels for the structures, large thermal inertia, large boiling margin, primary circuit operating at about the atmospheric pressure, and intermediate sodium circuit which separates the radioactive sodium from the primary circuit.

2.7.4.2. *Very High Temperature Reactors (VHTR)*

– Co-generation of high-temperature process heat and electricity (efficiency 45–50%, core outlet temperature up to 1,000°C); no actinide management. Once through cycle/reference, power = 600 MWth/300 MWe. Earliest delivery of demonstrator in 2020.

VHTR systems are graphite-moderated (thermal neutron spectrum), with helium cooled core and open fuel cycle. There are two types of core combinations which result from long industrial experience:

– prismatic blocks such as the operating Japanese HTTR of JAERI, or new designs under development, such as General Atomics's GT-MHR, NGNP from US-DOE, and ANTARES from AREVA NP;

– pebble bed, such as the former European AVR (Arbeitsgemeinschaft VersuchsReaktor, Juelich, 1967), the Chinese HTR-10 operating at the Institute of Nuclear Energy Technology (INET) and the future South African PBMR under development by Westinghouse (165 MWe).

The fuel of VHTR is made of microspheres or TRISO particles with an external diameter less than 1 mm, containing UO_2 (uranium oxycarbide), or even thorium. The TRISO particle fuel, which means tri-isotope, is made by coating layers made of high and low density parolytic carbide and of silicium carbide (SiC), or made of an advanced ZrC alloy. For the future, Pu or a mixture of minor actinides is being studied as oxides, with or without dilution in another matrix (burn-up rate objective equal to a maximum of 200 GWd/t).

There are many possible options for fuel, but no integral recycling of actinides as done in Generation IV. The objective is an outlet temperature of the core of 900–1,000°C with a pressure of 5–9 MPa. This gives a thermodynamic efficiency close to 50% and allows the co-generation, with hydrogen production (up to 200 tons per day, either by thermochemistry TCWSC, or by electrolysis HTES). The cycle will be either direct (helium), or indirect (nitrogen). From the passive safety point of view, the refractory characteristics of the fuel (1,600°C) and the large thermal inertia of the core which allows the passive removal of the heat in case of severe accident must be noted.

2.7.4.3. *Gas-cooled fast reactors (GFR)*

– Co-generation of electricity and process heat (efficiency close to 45%, core outlet temperature of 850°C); full actinide management (enhanced fuel utilization)/reference power = 1,000 MWe. Earliest delivery of demonstrator in 2025

Like other helium-cooled reactors, such as VHTR, GFR systems will be high temperature units (850°C at the core outlet for the direct cycle, system pressure of 5–7 MPa). This will give a very high thermodynamic efficiency and will allow the co-generation of electricity and heat with synthetic fuels or hydrogen production. For high-temperature co-generation, helium gas will directly drive a gas turbine (Brayton cycle turbine with high efficiency and well compacted). The indirect cycle (helium-supercritical CO_2) is also under investigation.

The fuel cycle will be closed. The used fuel will be reprocessed, probably on site, and all actinides will be recycled in order to minimize the production of long-life radioactive waste. Many types of fuel will be considered for their high-temperature qualities and capacity to retain fission products. The reference fuel for actinides (U, Pu and MA) will be of type C, N or O_2. The fuels will be of dispersion, solid solution, or particle types. The assembly will be of plates or blocks, needles or particle bed types.

Economy and sustainability require higher power densities whereas lower values would be more desirable for safety. The value of about 100 MWth/m^3, chosen for GFR, seems to be a good compromise between the values of VHTR and liquid metal fast generators, respectively around 10 to 200 MWth/m^3. The technological challenges include structural materials inside and outside the core. These should be able to resist fast-neutron damage and very high temperatures (up to 1,600°C),and low thermal inertia of the core in case of evacuation of residual power due to low helium pressure.

GFRs are in some ways an extension of VHTRs and will therefore take advantage of the experience acquired by VHTRs). They will share many technological challenges, such as extremely resistant materials and high efficiency turbo-machines for the conventional part of the reactor.

2.7.4.4. *Supercritical water-cooled reactors (SCWR)*

– Electricity production at high temperatures (improved economics, next step in LWR development, efficiency close to 45%, core outlet temperature of 510°C); full actinide management in the fast version. Reference power = 1,700 MWe. Earliest delivery in 2025.

SCWR systems are very high-pressure water-cooled reactors with thermal neutron spectrum working above the thermodynamic critical point of water with a thermal efficiency of 44%, which is a third more than today's LWRs. There is no steam production as the reactor works above the critical point of water, i.e. 22 MPa and 374°C, at which steam is generated (monophasic coolant whatever the thermal regime). The supercritical water in these systems (25 MPa and 625°C at the core outlet) directly drives the turbine without any secondary circuit of steam. No extra turbine development is necessary as the technology already exists in power stations using advanced fossil fuels. Systems providers, such as Siemens PG, are already using this technology in advanced coal-fired power stations (28.5 MPa and 620°C, thermodynamic efficiency 46%). The main objective of these SCWR is basic electricity generation at low cost, which could also be used for synthetic fuels or for H_2 (by electrolysis).

The SCWR aims to optimize two well-known technologies: LWRs which are the most commonly used power reactors in the world (80% of the world's nuclear industry), and fossil fuel power plants using supercritical water, a certain number of which are already in operation. One of the major challenges is the big decrease of coolant density in the core and the resulting instabilities. Corrosion due high-pressure steam is another interesting R&D challenge. There are two designs: pressurized tubes, such as CANDU, or pressurized vessels like those in Europe. The passive safety characteristics are similar to those of simplified boiling water reactors. The fuel is uranium oxide, enriched in the case of open fuel cycle. The SCWR system may however be built as a fast reactor (MOX fuel for Pu recycling) with integral recycling of actinides (closed cycle).

2.7.4.5. *Lead-cooled fast reactors (LFR)*

– Co-generation of process heat and electricity (full actinide management, a possible small turnkey plant, efficiency close to 45%, core outlet temperature up to 800°C). Reference power: batteries of 10–100 MWe and plants of 300–600 MWe. Earliest delivery 2025.

One of the characteristics of the LFR system is the cooling of the core (by natural convection) by means of lead or an alloy lead bismuth eutectic (LBE). The fuel is impoverished metallic uranium or nitride with integral recycling of actinides, from reprocessing installations.

The two power options being considered are batteries of 10–100 MWe with very long cycles of fuel reloading (15 to 20 years, which is an interesting characteristic for proliferation resistance) or a single 600 MWe plant. Battery refers to long duration units, manufactured in factories, and does not refer to electrochemical-energy batteries.

The characteristics of these reactors are designed for the co-generation of electricity and heat for small-size markets and for emerging countries that are interested in electro-nuclear energy, but who do not wish to develop their own fuel cycle industries. The fuel cycle is closed: the fuel is MOX, or one derived from nitride, containing fertile uranium and minor actinides. The outlet temperature of the core is 480°C at a pressure of 0.1 MPa (thermal efficiency 40%), with some advantages concerning corrosion, creep, and thermal shocks under transients.

This technology corresponds to the Russian technology of the fast reactor BREST-300 (planned in Beloyarsk, Research and Development Institute of Power Engineering (RDIPE)) which is based on 40 years' experience with eight submarines (LBE coolant, 155 MW, being dismantled). With advanced materials, temperatures close to 800°C are even considered for LFR; this could enable production of hydrogen.

2.7.4.6. *Molten salt reactors (MSR)*

– Co-generation of process heat and electricity (full actinide management, efficiency 45%, core outlet temperature 700°C, breeding in thermal spectrum using Th and in fast using U-Pu). Liquid or solid fuel. Reference power: 1,000 MWe. Earliest delivery 2025.

In MSR systems, the most innovative of GIF, uranium fuel is dissolved in a liquid salt mixture, e.g. Li_7F-BeF_2 or sodium fluorides, flowing in the channels of a graphite core as a moderator with an epithermal neutral spectrum. The primary circuit crosses a heat exchanger, which exchanges with a secondary salt circuit. This in turn exchanges with a tertiary circuit for the co-generation of electricity and heat. The production of synthetic fuels and hydrogen by the thermal chemical way is possible if a temperature of 800°C is reached.

MSR technology, tightly linked to partitioning and transmutation, allows a quasi-continuous recycling of fuel in a closed cycle, either as a dedicated incinerator or coupled with a fast reactor. Fission products are continuously extracted and the actinides are completely recycled, whereas plutonium and other actinides may be added to U-238. The outlet temperature of the coolant is 700°C at a pressure of 0.5 MPa. The boiling point of most of the molten salts is greater than 1,400°C. Some research programs investigate the fuel cycle Th-232/U-233 which minimizes the production of long-life nuclear waste, compared to the cycle U-238/Pu-239.

Research on molten salt reactors was intense in the 1970s with the molten salt reactor MSRE (Oak Ridge National Laboratory; fuel: liquid solution of Li_7, beryllium, thorium, and uranium fluorides). The interesting characteristics of the MSR fuel cycle are the production of clean waste (only short-life fission products in small quantities) and passive safety, thanks to natural convection at every power level.

Of the six GIF selected systems, three have a fast neutron spectrum: SFR, GFR, and LFR. The first version of SCWR will be a thermal reactor, followed by a fast reactor version. The two gas-cooled systems are tightly linked: GFR is a natural extension of VHTR; the MSR system is a non conventional one.

Apart from VHTR, all these systems will operate with a closed fuel cycle. As far as the future hydrogen economy is concerned, any of the six GEN IV systems will be able to produce synthetic fuels and hydrogen. Thermochemistry, in particular, can be used in four systems, namely, VHTR, GFR, MSR, and LFR, with coolant temperature at the core outlet of around 1,000, 850, 700, and 480–800°C, respectively.

Finally, we must mention the possible use of the thorium cycle which does not require fast neutrons. The two nuclear systems VHTR and MSR may work with thorium fuel. To recap, although not fissile itself, Th-232 can absorb slow neutrons to produce uranium-233, which is fissile.

2.7.5. *List of Generation IV-related Euratom R&D projects*

The countries who signed the GIF agreement (including Euratom) committed themselves to carry out research on the six innovative systems described above. They share their results, in particular those regarding co-generation.

It is, therefore, natural that part of the Euratom research activity deals with Generation IV and with nuclear co-generation systems for future industrial needs, as a support to the international GIF program. The main objective is ambitious, with sustainable development objectives such as the fast reactor core, refractory fuels, high outlet core temperatures, direct conversion with advanced gas turbine, and integral actinides recycling. These objectives are linked to a fuel reprocessing plant, integrated if possible on-site. In addition, some fusion energy material technologies may be interesting, such as ferritic-martinsitic steel and ODS.

The Euratom R&D projects that deal directly with GEN IV systems are:

– Raphael for the VHTR system (reactor for industrial heat, hydrogen and electricity generation). The project was financed for four years with a total budget of €20 million, including €9 million from the EU. It started in April 2005 and was coordinated by Areva NP Paris-Erlangen;

– GCFR for the GFR system (gas-cooled fast reactor). The project was financed for four years with a total of €3.6 million, including €2 million from the EU. It started in March 2005 and was coordinated by AMEC-NNC Manchester;

– HPLWR for the SCWR system (high performance light water reactor). The project was financed for 3.5 years with a total budget of €4.6 million, including €2.5 million from the EU. It started in September 2006 and was coordinated by FZK Karlsruhe;

– Elsy for the LFR system (European lead-cooled system). The project was financed for 3 years with a total budget of € 6.5 million, including € 2.95 million from the EU. It started in September 2006 and was coordinated by Ansaldo Nucleare Genoa;

– Eisofar for the SFR system (roadmap for a European sodium-cooled fast reactor). Strategic study financed for one year with a total budget of €0.5 million, including €0.25 million from the EU. It started in 2007 and was coordinated by CEA Cadarache;

– Alisia for the MSR system (molten salt evaluation for innovative applications). Strategic study financed for one year with a total budget of €0.5 million, including €0.25 million from the EU. It started in January 2007 and was coordinated by CEA Saclay.

As far as the integral recycling of nuclear fuel is concerned, in line with the GEN IV approach, the Euratom R&D program develops techniques that ensure cycle sustainability, including proliferation resistance, and better use of unused energy. It deals with partitioning (P) and transmutation (T). The Euratom R&D projects that are directly dealing with P&T:

– Europart (European research program on minor actinides partitioning of high activity waste from spent fuel), Project financed for four years with a total budget of €10.3 million, including €6 million from the EU. This project started in January 2004 and was coordinated by the CEA;

– Eurotrans (European research program on the transmutation of high activity nuclear waste in hybrid systems with accelerators/ADS). Project financed for four years with a total budget of €43 million, including € 23 million from the EU. This project started in April 2005 and was coordinated by FZK;

– Red-Impact (impact of partitioning, transmutation, and waste-reduction technologies on the final storage of nuclear waste). Project financed for three years with a total budget of €3.5 million, including €2 million from the EU. This project started in March 2004 and was coordinated by KTH Stockholm.

2.8. Education and training: main objectives (modules, mutual recognition, and mobility)

Since 2000, nuclear education and training have been particularly important in the Euratom R&D framework program. Obviously, all the sectors of nuclear fission are involved: nuclear engineering (covering Generations II, III, and IV), radioactive waste management (including P&T), and radiation protection (including applications of ionizing radiation).

Education and training are defined as follows:

– education is a basic and permanent activity. Education covers a wider spectrum than training; it guarantees that future generations will remain skilled and able to innovate. It consists mainly of transmitting a fundamental type of knowledge, which is generally the function of university institutions;

– training consists of offering the necessary qualifications to achieve a particular result. Training offers other activities than those of university institutions. It generally consists of transmitting know-how and of teaching good practice in order to train professional experts in a specific area.

The Euratom strategy of education and training has identified three goals:

– a modular approach and common qualification criteria in order to identify the best training modules and to deliver quality labels;

– a mutual recognition system throughout the EU, especially by the Bologna/Erasmus ECTS in order to stimulate excellence;

– to facilitate mobility for students and teaching staff, especially by financing private-public partnerships at European level.

In order to achieve the above objectives, a non-profit-making association (in compliance with the French law of 1901) was created in September 2003. It is the European Nuclear Education Network [ENEN 3w], which today includes approximately 50 institutions. This association may be considered as a step towards the development of a virtual nuclear university which would connect all training programs in the nuclear fission sector throughout Europe. Cooperation with the World Nuclear University (WNU), an initiative of the World Nuclear Association (WNA) under the auspices of the IAEA, which was started in September 2003, must also be mentioned.

2.9. Conclusion: nuclear energy – a part of the solution in a sustainable energy mix

At a global level, nuclear fission is currently enjoying a certain renaissance (Generation III, 2010) and could soon experience a technological breakthrough (Generation IV, forecast 2030). This has been demonstrated in this chapter which describes the scientific and technological challenges linked to this evolution:

– today, with Generation II: supply security and environmental protection (challenges 1 and 2), supply of clean electricity in large quantities;

– by 2010, with Generation III: safety, reliability and competitiveness improvements (challenges 3 and 4), evolutionary improvements;

– by 2030, with Generation IV: co-generation (electricity-heat) and integral recycling (challenges 5 and 6); revolutionary innovation for reactors and fuels.

The results of the research carried out in the Euratom Framework Programs (FP) are emphasized. The total value of all these projects under FP-6 (2003–2006) is €158 million including the EU contribution of €78 million. The main results were discussed at the FISA 2006[2] conference. A detailed description of all the other Euratom R&D projects in nuclear fission is available on the Cordis website [CORDIS 3w]. More recent results of Euratom research were discussed at the FISA 2009 conference.

This type of research and technological innovation is included in the framework of the EU energy policy which aims to deliver to future generations a balanced mix of energy. It must meet three objectives at the European level:

– ensure security of supply;

– protect the environment in a sustainable manner;

– ensure the competitiveness of Europe.

2. FISA conference: Conference on EU Research and Training in Reactor Systems Luxembourg, March 13, 2006.

It must be noted that the four Generation IV criteria (sustainability, economy, safety and reliability, proliferation resistance) are fully compatible with the above criteria.

Future generations will certainly need new technologies, not only for electricity generation, but also for very high temperature industrial heat generation. Even if we do not believe in a low-carbon economy, future applications will cover many areas, such as production of synthetic fuel for transportation, better performances by petrochemical factories, production of clean hydrogen (if possible from water) in large quantities, and protection of the environment. Medium temperature heat will also be useful, for example, for sea-water desalination.

In Europe these challenges in the energy field need a European Community approach. The EU is in charge of the creation of the framework to face these energy challenges using European research and innovation as recommended by the Green Book on 2006 energy "A European strategy for a sustainable, competitive and secure energy" [EUR 06]:

> The seventh program framework (2007–2013) admits that there is no unique solution to our energy problems, but it deals with a broad portfolio of technologies: sustainable energy, clean coal capture and sequestration, economically acceptable bio-fuels for transportation; new energy vectors, such as hydrogen, and that are respectful towards the environment (for example, fuel cells), energy efficiency, advanced nuclear fission and fusion energy (through ITER agreement).

Nuclear fission is thus one part of the solution: it is an important element in a mix of energy technologies to be transmitted to the future generation.

As far as the large-scale building of Generation III reactors (2010 forecast) is concerned, nobody knows when the nuclear industry and investors will make firm decisions. This type of decision depends not only on the scientific/technological innovation, but also on the economic and politic context.

We have briefly analyzed the need to have an international environment which is at the same time objective, coherent, and foreseeable. As far as the construction of Generation IV (forecast 2030) is concerned, the R&D activities could take many decades before producing results that are of interest to industry. The important decisions will depend, principally, on the future of the low carbon economy. If the national industrial policies support the nuclear renaissance, public opinion must be assured that the highest levels of safety, security, and environmental protection will always apply.

The European Community research and innovation program for GEN III and GEN IV contributes to the general interest objectives. Moreover, nuclear education and training at EU level guarantees a solid base of competence for future decades.

In the Euratom programs, there is a unique synergy between research, education and innovation and this has been illustrated by concrete implementations. The market of nuclear knowledge is thus a reality: nuclear knowledge and expertise have been created, transmitted, and used in a sustainable manner in the EU.

In conclusion, the renaissance that we are facing today (Generation III) and the technological breakthrough that is being prepared in the laboratories for the future (Generation IV) do require significant space and time perspectives:

– in space: the R&D in the nuclear fission field is increasingly a worldwide activity which requires a wide multidisciplinary knowledge;

– in time: industries and politicians must make very long-term commitments covering the total lifetime of the installations, in reality at least 60 years.

Finally, we must remember that the success of technological breakthroughs (such as Generation IV) depends not only on international research on a worldwide scale, mainly funded by the public sector (such as the FP-6 in the EU or the DOE projects in the United States), but also on the scientific cooperation and financial participation of the private sector. The players involved form a large community:

– the research institutes (public and private);

– the system manufacturers and the engineer-architects;

– the energy suppliers (or electric utilities) and associated organizations;

– the regulation authorities and the associated technical safety organizations;

– the education and training institutions (mainly the universities).

2.10. Bibliography

[EUR 98] *Radioactive Waste Management in the EU*, European Commission, 1998.

[EUR 06] *A European Strategy for a Sustainable, Competitive and Secure Energy: Green Book on Energy*, European Commission: COM, 105 of 108, March 2006.

[IAE 93] *Fundamental Principles of Safety: the Safety of Nuclear Installations*, *Safety Series*, volume 110, IAEA Vienna, 1993.

[IAE 05a] *Red Book Uranium 2005*, IAEA and OECD, Paris, 2005.

[IAE 05b] *Developments in Uranium Resources, Production, Demand and the Environment*, IAEA TECDOC Series No. 1425.

Web Sites

[ITER 3w] www.ITER.org

[IAEA 3w] www.iaea.org

[ENEN 3w] www.enen-assoc.org

[CORDIS 3w] www.cordis.europa.eu/fp6-euratom/library.html

20.11. List of acronyms

ABWR = advanced boiling water reactor

AGR = advanced gas-cooled reactors

AHTR = advanced high temperature reactor

BWR = boiling water reactor

CANDU = CANadian Deuterium Uranium heavy-water moderated and natural uranium fuelled (original design in Canada)

CDA = core disruptive accident

DHR = decay heat removal system

EFR = European fast reactor

EMWG = economics modeling working group (GIF crosscutting methodology)

EPR = European (or Evolutionary) pressurized power reactor

FBR = fast breeder reactor

FBTR = fast breeder test reactor (Kalpakkam, India)

GCR = gas-cooled reactor

GFR = gas-cooled fast reactors

GIF = Generation IV International Forum – http://www.gen-4.org/GIF/Governance /index.htm

HFR = high flux reactor (Petten, Netherlands)

HLW = high-level waste

HTR = high temperature gas reactors

IFR = integral fast reactor (USA)

ITER = international thermonuclear experimental reactor in Cadarache (FR)

LFR = lead-cooled fast reactors

LMFBR = liquid metal fast breeder reactor

LMLW = low- and medium-level waste

LMR = liquid metal reactors

LWGR = boiling Light-Water-cooled, Graphite-moderated, pressure-tube Reactors (RBMK)

LWR = light water reactors

MOX = mix of U and Pu oxides

MSR = molten salt reactors

NOx = nitrogen oxides

PBFR = prototype breeder fast reactor under construction in India (1,250 MWth)

PBMR = pebble bed modular reactor (Republic of South Africa)

PHWR = pressurized heavy water reactors (e.g. CANDU)

PWR = pressurized water reactor

RBMK = Reactor Bolshoï Moshchnosti Kalani (also called LWGR)

SCWR = supercritical water-cooled reactors

SFR = sodium-cooled fast reactor

UOX = uranium oxide

VHTR = very high temperature gas reactors

Chapter 3

Co-generation

3.1. Co-generation

3.1.1. *Co-generation principle: heat, force, and Carnot efficiency*

Co-generation, or combined heat and power (CHP) is the technique of energy conversion that comes after the simultaneous energy conversion into two forms of useful energy, i.e. heat and work.

The work most often described in this chapter is the mechanical energy of rotation and electrical energy. As work is a combination of force and displacement, the term "heat-force" is used in Dutch and German; the word "co-generation"[1] is used in French and in English. The generation of heat[2] does not generally cause a problem; usually, boilers are used for this purpose. On the other hand, work is more difficult to generate. Starting again with fossil fuels, there are two possible ways to generate work. The first is the electrochemical method used in fuel cells (which will be developed later in the chapter), but it has not yet made a commercial breakthrough. Fuel cell technology seems to be promising, but is far from being considered as a mature technology. The second technology is at the thermal level and is based on the classic laws and concepts of thermodynamics. There are

---

Chapter written by William D'HAESELEER and Patrick LUICKX.

1. In German, we speak about *Kraft-Wärme Kopplung*; in English, the term Combined Heat and Power (CHP) is used.

2. The word heat must not be strictly interpreted. In this chapter the distinction is not always made between heat in the form of thermodynamic energy transfer and (although erroneously used) the heat content of a body. More properly, the latter should be described by a *body* with internal energy at a higher temperature.

technologies of the combustion engine and of electrical power generating stations. These are currently the basic technologies for work generation.

Although the first law of thermodynamics claims the conservation of energy, in which we have assumed since Einstein that mass is a condensed form of energy, the second law reveals that not all forms of energy have the same value. It means that work can be converted into heat without limitations, whereas the inverse process will never happen.

Starting with a calorific amount Q_1 from a tank at high temperature T_1, a desired work W and a lost heat Q_2 at low temperature T_2 are obtained. The higher the tank temperature is, the greater the work will be. As for motors and electric generating stations, the generated work is emphasized; the efficiency of a thermal engine is defined as.[3]

$$\eta \equiv \frac{\text{useful work}}{\text{heat intially available}} = \frac{W}{Q_1} \qquad [3.1]$$

The maximum efficiency a thermal motoring machine can theoretically achieve is given by the Carnot cycle and is called the "Carnot efficiency".

$$\eta = 1 - \frac{T_0}{T_1} \qquad [3.2]$$

in which T_1 is the temperature of the hot source, and the temperature of the cold reservoir is chosen as being the ambient temperature T_0. Both temperatures are expressed in Kelvin (0°C = 273 K). With T_1 = 500°C, the Carnot efficiency is equal to 63%. With T_1 = 1,000°C, this value rises to 77% (in both cases, T_0= 15°C has been chosen).

The feasible efficiencies are noticeably lower in practice. In summary, the following efficiencies may be assumed:[4]

---

3. "Thermal engine" is an expression that describes the full range of combustion motors and electric plants. Sometimes, the expression "electric motor' is used.

4. The reader must pay attention to the term "thermal efficiency." In the context of a thermal engine, it is usual to speak of thermal efficiency when speaking of the described output in [3.1]. This means the efficiency of the thermal cycle to generate work. However, in the context of co-generation, this expression has a different meaning. For the described output in [3.1], the term "mechanical efficiency" will be used in this chapter dealing with co-generation. After the conversion of mechanical power into electrical power by means of an alternator, we will refer to the "electrical efficiency of the thermal engine-alternator" module.

Gas and gasoline engines		$\eta \sim 30$–40%
Diesel engines		$\eta \sim 40$–45%
Gas turbines:	power > 10 MW	$\eta \sim 30$–40%
	power < 10 MW	$\eta \sim 15$–30%
Modern coal plant		$\eta \sim 40$–47%
Nuclear plant		$\eta \sim 33\%$
Combined-cycle power station		$\eta \sim 55$–60%

For each of these technologies, the more efficient machines perform best. In order to deliver as much work as possible, the maximum has been done to reach these efficiencies. However, in practice, about 40 to 60% of the initial energy is dispersed in the environment.

For thermal engines, the environment is usually taken as the low-temperature tank. When the ambient temperature decreases, the Carnot efficiency increases (equation [3.2]). As the efficiency of the real cycles is dependent in the same way on the low-temperature tank, so efficiency will also increase when the temperature T_0 decreases. This explains why the efficiency of a thermal power generating station is higher in winter than in summer.

If in an optimized electric power generating station, the working fluid is not cooled down to the ambient temperature, but rather to a higher temperature T_2, e.g. for processing purposes, this will have a price: the working efficiency will decrease. This loss of efficiency will happen in every steam-turbine plant in which process heat is extracted from the thermodynamic cycle in the form of a purge or of a backpressure turbine.

This is the case in standard thermal power generating stations, and in the combined-cycle power plants. The heat losses in these types of plants are already at their minimum because considerable efforts have been made to use all the high temperature heat. Although such a power station pours out 40 to 60% of its heat into the environment as described earlier, this happens at a low temperature, just above the ambient temperature. The only way to obtain useful process heat at higher temperatures is to interfere with the thermodynamic cycle, thus reducing the work efficiency.

The reciprocating engines and the simple gas turbines, on the other hand, enable heat recovery without lowering the power efficiency: the average temperature heat lost in the other devices can, in this case, be recovered without any intervention in the thermodynamic cycle. The energy of the exhaust gases coming from gas turbines and piston engines is an example of this form of heat recovery, as well as the heat from oil cooling and from the casing of piston engines.

3.1.2. *The quality of energy and the concept of exergy*

Heat and work clearly have different qualities. Work is the most advanced form of energy since it may be entirely transformed into heat, whereas heat can be converted into work up to a certain point. As previously described, the maximum part of heat that can be converted into work is defined by the Carnot efficiency. This means that with a tank at a given temperature T, the efficiency increases with this temperature: a Carnot cycle may then be fed by this heat to give power efficiency equal to $(1\text{-}T_0/T)$. So, the ability to deliver work increases with the tank temperature. A high-temperature tank will have energy of higher quality than a low-temperature one (taking into account equal calorific capacity); the quality of the energy content of a state is called exergy.

More scientifically speaking, exergy is defined as being the maximum quantity of work that a system can deliver (while evolving towards equilibrium with the environment). Fuels before combustion, electricity, and work have a high quality and therefore a relative exergy content of 1; a body at temperature T has a relative exergy content equal to $(1\text{-}T/T_0)$.

The quality or the exergy content of the heat poured into the environment for an optimized power station is very limited. Therefore, the drained heat is of practically no value.

3.1.3. *Co-generation, an intelligent way to generate heat*

Present-day boilers dedicated to heating are at first sight very efficient devices. Energy efficiencies of 90–95% are obtained. However, in their classic use, as space heating, the energy quality is badly used. The starting point is a high-quality fossil fuel; the exergy content is 1, whereas hot water at 80–90°C is made available to be finally used to heat a space to 22°C.

Thus, the fuel exergy is reduced to only 5% of its initial energy content. In other words, the exergetic efficiency of a central-heating system is barely 5%. This means that by heating spaces in such a way, the intrinsic quality of a fuel is largely wasted. Space heating by means of central-heating boilers is a inefficient process from an exergy point of view. There must be a more intelligent means to generate heat at low temperature.[5]

The exergetic efficiency of a common power station is roughly equal to its work efficiency. Assuming that the heat poured into the environment has no more exergy content, the exergy efficiency of a combined-cycle power plant, for example, is 55–60%.[6] This demonstrates the high performance of the combined-cycle power plant technology.

As described above, co-generation has nothing in common with heat recovery in electric power stations. Co-generation should instead be considered an intelligent way to generate heat. Instead of leaving a high-quality fuel to deteriorate in order to generate heat at low temperature, it would be better to connect conversion equipment, such as a turbine or a motor, during deterioration in order to extract a little more work. From this point of view, the co-generation is thus a method by which electricity is obtained as a by-product, whereas heat (otherwise generated in a regular boiler) is generated anyway.

As far as reciprocating engines and simple gas turbines are concerned, co-generation may be considered as a heat recovery technique, which completes the generation of mechanical power. However, it must be added that electricity generation with engines is inferior to that of combined-cycle technology and only makes sense when heat is recovered. Gas turbines[7] are only used to generate electricity during peak hours of electricity consumption where a rapid start-up is important. In principle, heat recovery at this level is possible, although it limits the flexibility of the peak turbine.

The above discussion clearly shows that co-generation makes sense from the energy point of view only if all the recovered heat is utilized. If the heat generated by the co-generation installation is wasted, the result will be a small electric generation unit with an efficiency lower than one of today's combined-cycle plants. Co-generation cannot then be justified from an energy point of view when heat is not used for a specific purpose.

---

5. The heat pump is thermodynamically very efficient. It pumps the ambient heat up to the temperature required to heat a space.

6. Strictly speaking however, if we consider the minimum exergy content of the elapsed heat, the exergetic output will be a slightly higher than the work output.

7. As isolated units of electricity generation, not as a combined-cycle power station.

3.2. Overview of existing technologies

3.2.1. *Currently used technologies*

The standard prime movers for co-generation applications that are currently used most often are of the following types:

- steam turbine;
- gas turbine;
- reciprocating engine, especially gas and diesel engines.

All these pieces of equipment are characterized by their fuel type, the kind of combustion process, their power efficiency, and the quantity and the temperature of the recoverable heat. One important characteristic of a co-generation installation is known as the heat/power ratio $\dot{Q}/\dot{E}$ for a given operating point; it represents the ratio between the power included in the heat $\dot{Q}$ and the electrical-work power $\dot{E}$.[8]

3.2.1.1. *Steam turbines*

Steam turbines produce power by means of mechanical rotation in a thermodynamic Rankine cycle. They are used in present-day power-generating stations.

The principle schemes for heat extraction in steam turbines are presented in Figures 3.1 and 3.2.

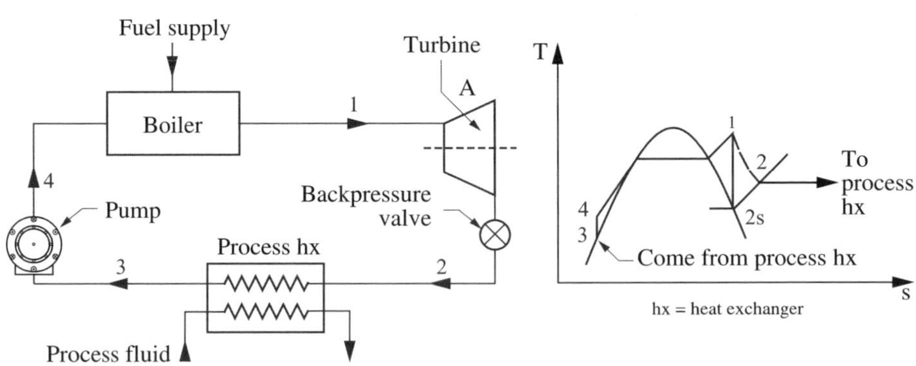

Figure 3.1. *Sketch of heat delivery by steam turbines, by mean of a back-pressure turbine*

8. A power (energy by time unit) is expressed by a point above the symbol. E indicates the electrical energy; $\dot{E}$ is the power.

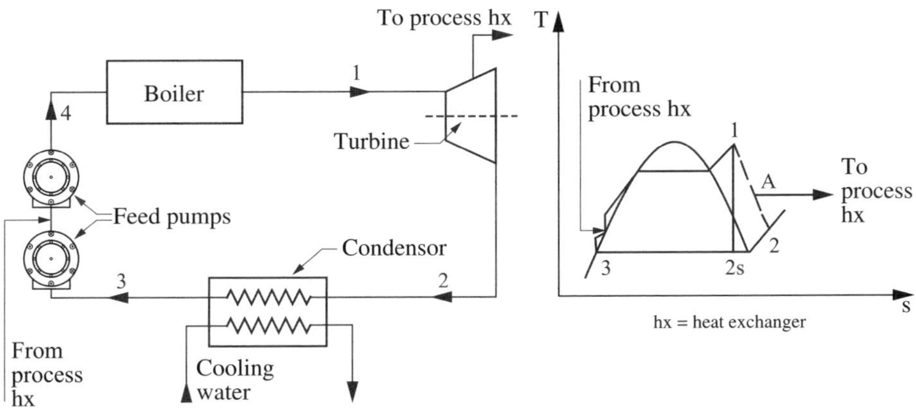

Figure 3.2. *Sketch of heat delivery by steam turbines, by means of steam bleeding*

The more the quality of the available steam increases, the more the mechanical power generated by the turbine will decrease. Depending on the chosen pressures and temperatures, the heat/power ratio can adopt values going from 3:1 to 10:1 (for back-pressure turbines, heat normally prevails with $\dot{Q}/\dot{E}$ ratios of 4–5:1 up to 10:1; for steam bleeding, there is less usable heat, leading to a $\dot{Q}/\dot{E}$ ratio of the order of 3–4:1).

Steam boilers that supply the steam turbines are systems based on external combustion. The combustion chamber and the thermodynamic process fluid are physically separated. The major advantage of this type of system is that any kind of fuel may be used. The drawback is an efficiency limitation due to the heat transmission processes.

Historically, this type of co-generation was widely used in industry. Most suppliers used, and are still using, this co-generation technology. The installations of back-pressure turbines are found with electric powers ranging from 50 kW_e to over 100 MW_e. Typical power efficiencies reach 10–20% (greater for bigger turbines). Regarding steam purging, the usual electric power is that of a classic steam power station (some 100 MW_e).

3.2.1.2. *Gas turbines*

The principle scheme for heat recovery in gas turbines is illustrated in Figure 3.3.

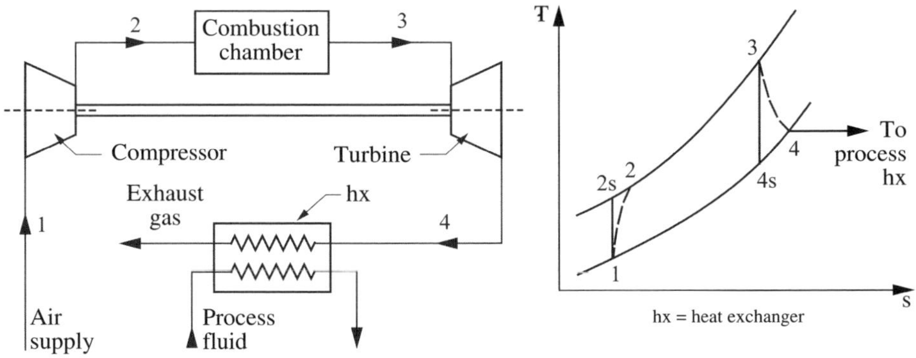

Figure 3.3. *Diagram of co-generation with a gas turbine*

The aim of this heat recovery scheme is the recovery of the high temperature heat contained in the exhaust gases in a recuperation boiler, which will make steam available for process purposes.

As regards gas turbines, it is sufficient to say that it is an internal combustion mechanism in which the thermodynamic fluid is the reactive solution air-fuel. The drawback is that the fuel alternatives are limited: either a high-quality liquid fuel, such as kerosene, or a gas, e.g. natural gas or propane.

The gas turbines used in co-generation applications tend to be of the simple cycle type, without any energy recovery to maximize the working power. The disadvantage of this is that the gas turbine is less sophisticated, although it is more reliable.

Another advantage of co-generation applications is that the usable heat of the exhaust gases is of higher quality, but lower mechanical work efficiency is the price to be paid.

Turbines for co-generation are now available and competitive in the power range from about 100 kW_e up to more than 100 MW_e.[9] According to the specific design of the turbine-recuperation boiler combination, the typical $\dot{Q}/\dot{E}$ ration ranges from about 1:1 to 3:1. A simple 2:1 ratio is typical.

Gas turbines are supposed to work at full load, since their efficiency at partial load is drastically reduced. Empirically speaking, a gas turbine should never work at a power level below at least 2/3 of its full load operating point. As the exhaust gases

9. The current use of gas turbines for co-generation applications is between 1 to 100 MW_e.

contain excess air, it is possible to have an additional combustion in the recuperation boiler without excessive effort: simple fuel injection is enough.

3.2.1.3. *Reciprocating engines*

As typical piston engines for co-generation applications, diesel engines are used to generate high power (100 kW_e–60 MW_e), whereas gas engines, which in fact are slightly modified versions of gasoline engines, are used for lower power levels of 5kW_e to a maximum of 4 MW_e.[10]

A typical scheme for co-generation using engines is presented in Figure 3.4.

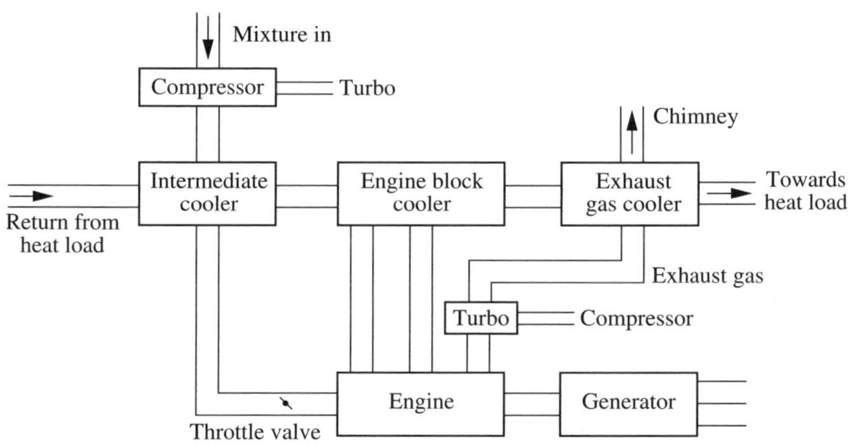

Figure 3.4. *Diagram of co-generation unit based on a reciprocating engine*

Typical temperatures of the heat to be recovered are around ± 50°C, ± 80°C, and ± 100°C respectively for the intermediate cooler, the cooling lubricating oil, and the cooling water. The temperature of the exhaust gases rises to about 400–500°C. These temperature levels explain the engine choice as being dependant on the temperature required for the heat application and of the return heat which follows. If, because of the process, a temperature of more than 150°C is necessary, only the exhaust gases can deliver heat. In such a case, the recoverable thermal heat-power of the engine is rather limited.

Taking into account a return temperature of 70°C, the $\dot{Q}/\dot{E}$ ration of small gas engines (< 1 MW_e) is about 1.5:1. Large gas engines have a ratio of 1:1 to 3:1. With

10. Most of the diesel and gas engines for co-generation applications are between 200 kW_e and 2 MW_e.

return temperatures of 90°C, the $\dot{Q}/\dot{E}$ ratio for gas engines is typically between 0.5: 1 and 1: 1.

The exhaust gases of diesel engines must not be cooled too much; overcooling causes the fumes to condense, which results in corrosion. The typical heat-power ratio is from 1: 1 to 1.2: 1 for medium-power diesel engines and 0.5: 1 for higher-power diesel engines, which allows for a return temperature of 70°C.

The partial load characteristics of reciprocating engines are more advantageous than those of the gas turbines. Both gas and diesel engines still work properly at up to half their nominal power. Under such circumstances, the efficiency of electric generation in a gas engine never falls below 80% of its full load value; a diesel engine keeps more than 90% of its full load value.

The $\dot{Q}/\dot{E}$ ratio of gas engines increases slightly at partial load, whereas the $\dot{Q}/\dot{E}$ ratios of diesel engines will decrease slightly compared to their nominal operating point.

3.2.1.4. *Combined-cycle power stations*

A combined-cycle power station is a system in which a gas turbine is combined with a recuperation boiler. The steam from this boiler is then used to generate electrical and mechanical power. When optimized for electricity generation, this will be a high-efficiency power-generating station for work generation, ranging up to 55–60%.

A combined-cycle power station completely recovers the heat from the exhaust gases of the gas turbine, thus maximizing electricity production. By means of steam purge or using a back-pressure steam turbine, it is also possible to obtain some high temperature heat from the recuperation boiler. The drawback is less efficient electricity generation. A co-generation installation with a gas turbine and a steam purge is consequently a hybrid form; combined-cycle principle on one hand, and a co-generation installation principle with a gas or steam turbine as generator, on the other.

3.2.2. *Prospective technologies*

3.2.2.1. *The Stirling engine*

The Stirling engine is based on a thermodynamic cycle which theoretically reaches an efficiency equivalent to the Carnot efficiency. The Stirling cycle takes place between two isotherms and two isochors.

During the isothermal expansion, external heat is added in order to preserve the isothermal state change. During the isothermal compression, heat must be carried away outside. The fact that the exchanges which happen during the isochor state changes are internally compensated is typical of this cycle. This compensation is made by storage in a regenerator; the extracted heat from one isochoric state change is absorbed in the other. In this way, it is possible to exchange only the external heat with the constant temperature tanks during the isothermal state change, with an efficiency level which will, theoretically, be equal to that of the Carnot cycle.

In theory, the Stirling cycle is more efficient than the Carnot cycle because the work ratio, defined as the ratio between the delivered net work and the gross work, is considerably higher than that of the Carnot cycle. More specifically, this work ratio has the same value as its efficiency, being the Carnot efficiency. The Stirling engine also has the same characteristics as another cycle: the Ericsson cycle, which is located between two isotherms and two isobars.

One advantage of the Stirling engine compared to the internal combustion engine (as the Otto and diesel cycles) is the free choice of the heat source. There is no dependency on gasoline, fuel oil, or gas. This enables, for instance, the use of a Stirling engine in which the heat comes from biomass. There are also versions of Stirling engines in which solar energy is used as the heat source.

A typical Stirling engine is incorporated in a boiler, where the motor itself will generate the electric power and where the boiler's residual heat will be used for calorific applications. The combination of a Stirling engine and a boiler is considered to be a co-generation installation. In this case, it is clear that heat is the primary product of the installation and that the electrical and mechanical power comes from the incorporation of the Stirling motor in the boiler.

The practical, realizable efficiencies of the Stirling engine are quite low. The question then is whether Stirling motors will have a significant impact in the next 20 years.

3.2.2.2. *Fuel cells*

Fuel cells are electrochemical devices which directly transform chemical energy into electricity. However, just as an electrochemical reaction, such as a hydrogen isotherm oxidation, is exothermal, heat is also generated by the fuel cell. A fuel cell must then be appropriately cooled in order to obtain an optimal stationary operating temperature for the electrochemical reactions. The heat extracted by means of the cooling circuit may be used in heating applications; if the heat available is used well, a co-generation system is then in place.

Research is being carried out on five types of fuel cells, all based on H_2 as the cell fuel. A sixth type, which would directly use methanol as the cell fuel, is still in the very early stages of development. Depending on the type of fuel cell, the temperature of the generated heat varies from 70°C to about 100°C.

The alkaline fuel cell (AFC) and the proton exchange membrane fuel cell (PEMFC) are typical fuel cells with low temperatures between 70°C and 80°C. The phosphoric acid fuel cell (PAFC) is a medium-temperature cell with an operating point of about 200°C. The molten carbonate fuel cell (MCFC) and the solid oxide fuel cell (SOFC) operate at temperatures of about 650°C and 1,000°C, respectively. The idea is to combine these high-temperature fuel cells with a recovery boiler and a steam turbine in order to generate electricity from a combined cycle.

Fuel cells are potentially promising for future co-generation applications. However, before really making a breakthrough, an important technological optimization is required. Moreover, from an economic point of view, fuel cells have a long way to go before being competitive.

3.3. Co-generation installation dimensioning

Although the co-generation installation can, at any time, deliver heat as well as electricity, with a characteristic $\dot{Q}/\dot{E}$ ratio depending on a full load or partial load regime, heat and electricity demands are seldom equal; sometimes a consumer might need electricity, but no heat.

A co-generation installation is most efficient when it can deliver its maximum constant power without wasting heat, without working at partial load too often, and without frequently being stopped.

To find this optimally designed installation, we must consider the user's chronological heat and electricity requirements. Figure 3.5 presents the energy needs of a hypothetical hospital in the month of February, and shows that for this consumer, the heat demand is always higher than the electricity demand. The inverse situation may also occur.

Starting with the chronological development of the heat and electricity demand, as illustrated in Figure 3.5, it is possible to create a load duration diagram, sometimes also called a monotone diagram. Such a diagram represents on its horizontal axis the annual number of hours where the energy power demand exceeds *A* (see Figure 3.6).

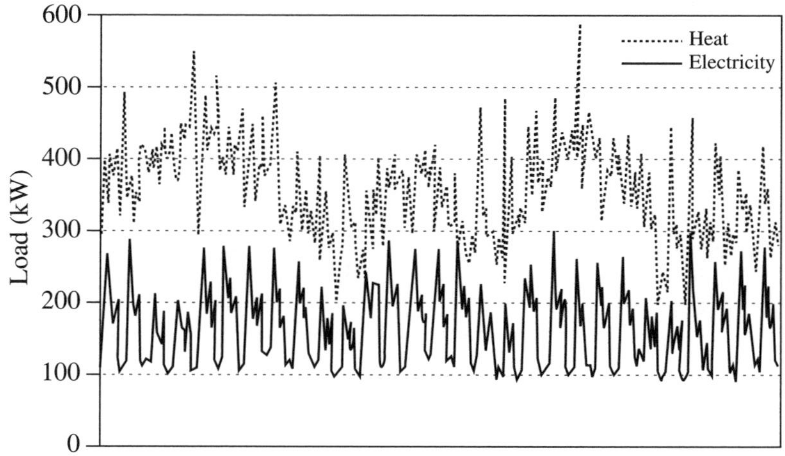

Figure 3.5. *Example of heat and electricity demand during February in a hypothetical hospital*

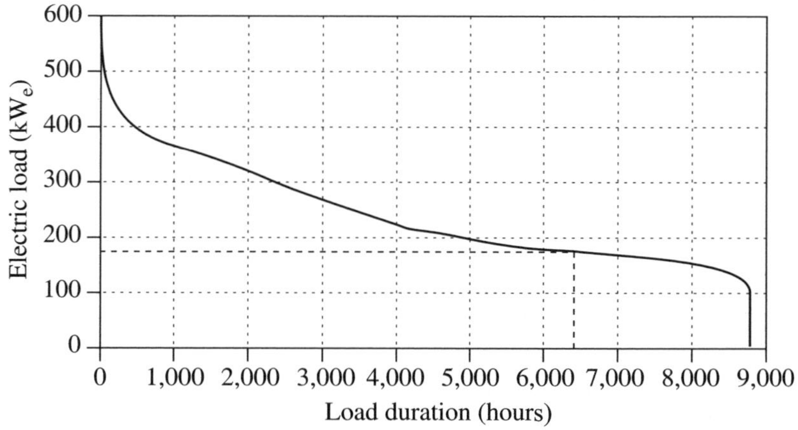

Figure 3.6. *Hypothetical example of a load duration diagram for electricity. The time axis in this figure is only for illustration and has no relation to the time axis of Figure 3.5. In principle, the electrical power peak of Figure 3.5 must correspond to the highest value in Figure 3.6*

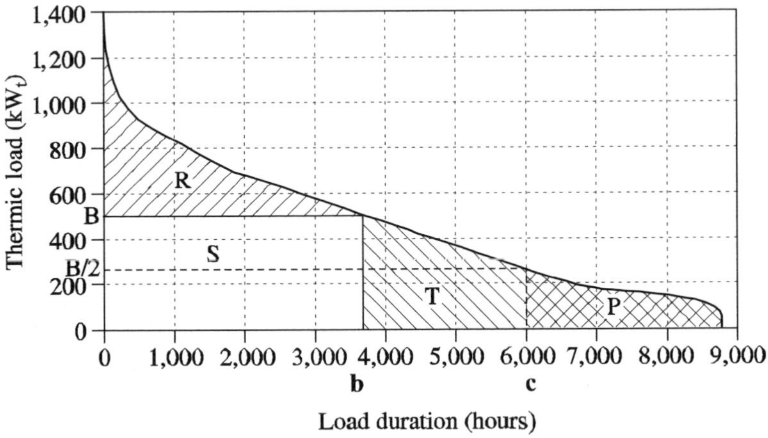

Figure 3.7. *Hypothetical example of a load duration diagram for heat. The time axis in this figure is for illustration purposes only, and has no relation to the time axes in Figure 3.5. In principle, the heat peak of Figure 3.5 must correspond to the highest power value of Figure 3.7*

The best method for determining the dimensions of a co-generation installation is to find the rectangle with the maximum surface in the monotone heat diagram. We will assume that area S in Figure 3.7 represents this rectangle. On the ordinate axis, the thermal power of the co-generation plant to be installed is deduced (point B). The chosen installation will generate, at full load, a power of B kW_{th} over b hours and will generate a thermal energy quantity Bb kWh_{th}. When the appropriate generation technology is chosen for the application, the electric power to be installed can be deducted from the $\dot{Q}/\dot{E}$ ratio at full load.

By determining the dimensions in this way, the operation of a co-generation installation is dictated by the heat demand. When too little electricity is generated, electricity is bought in from the public grid; when the generation of the co-generation plant is too high, the excess is sold to the grid. Either way, there is no heat waste, which is the aim with such a system.

Surface S in Figure 3.7 represents the thermal energy generated by a full load co-generation plant. During most of the *b* hours, the heat demand is greater than the power delivered by the installation. Surface R represents the additional energy that must be delivered by boilers. Surfaces T and P also represent the additional heating per boiler to meet the heat demand when the co-generation plant is inactive.

When operating a motor-driven co-generation plant, a partial load operation may be justified and the motor may be active during a time period c, which corresponds to the half-load B/2 point. The co-generation plant will then generate a thermal energy quantity equal to surface S+T. Thus, the machine will be off for (8760-c) hours; the additional heat is necessary to cover the surfaces R and P.

As far as gas turbines are concerned, they normally have to operate exclusively at full load, and consequently are active during *b* hours. For some applications, a partial load operation could be justified, as long as the electric efficiency is high enough in order to save energy by using a co-generation plant instead of a separated generation

We must however note that the abscissa axes in Figures 3.6 and 3.7 represent only the required power as being higher than a given value and without any chronology. Figure 3.7 does not mean that the motor co-generation plant works throughout the period 0→ c. The machine is regularly shut down for a total period of (8760-c) hours.

As Figures 3.6 and 3.7 do not mention anything about the shut-off frequency, it is possible that the machine is frequently disconnected and that it does not operate for the total *c* hours/year, in order to avoid additional wear. It is also possible that the machine is active for generating electricity, without any heat demand.

3.4. Assessment of the energy advantage of co-generation

3.4.1. *Fuel use rate*

It is usual to define the following efficiencies when dealing with co-generation.

Electric efficiency, α_E, is the ratio between the useful delivered electric power $\dot{E}$ and the fuel power, $\dot{F}$.[11,12] The ratio between the generated useful thermal power $\dot{Q}$ and $\dot{F}$ is called the thermal efficiency, α_Q, in the context of co-generation. The efficencies α_E and α_Q represent, respectively, the reduced values of the quantities of electric generation and thermal generation per input fuel. In order to compare the combined energy conversion in a co-generation plant with that of

11. F comes from "fuel", and as an input energetic value for F, it is common to choose the lower heating value (LHV) (in kJ/kg). Assuming a fuel low equal to $\dot{m}_f$, the result is that

$$\dot{F} = \left(\dot{m}_f\right)\ PCI.$$

12. This is the output which has, until now been called (electrical) work efficiency.

separated generation in an electric plant on one hand, and a boiler on the other hand, we consider the scheme in Figure 3.8.

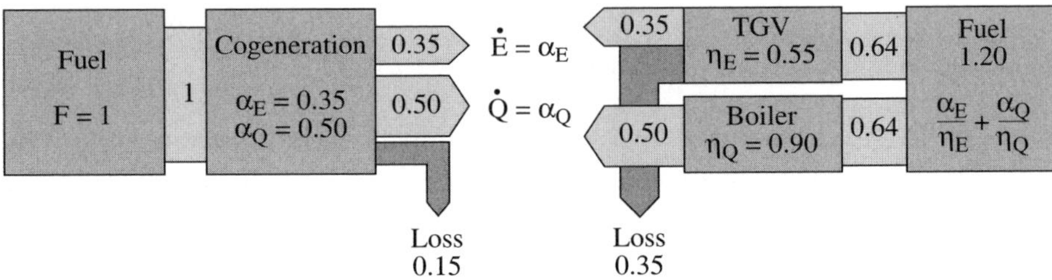

Figure 3.8. *Simplified representation of the energetic balance between a CHP unit (left) and separated production (right)*

The scheme starts from an electric power demand $\dot{E}$ and from a thermal power $\dot{Q}$. On one hand, $\dot{E}$ and $\dot{Q}$ are supplied by a co-generation plant (left in Figure 3.8); on the other hand, $\dot{E}$ is supplied by an electric-power generating station, whereas $\dot{Q}$ is generated by a boiler (right in Figure 3.8). The figure compares two plants: co-generation and separated production, which deliver exactly and simultaneously the required power $\dot{E}$ and $\dot{Q}$.[13]

The electric efficiency of an electric plant is represented by η_E, the thermal efficiency of the boiler by η_Q. In the figure, some typical values are used as an example:

Co-generation: α_E=0.35 (or 35%)

α_Q=0.50 (or 50%)

Separated: η_E=0.55 (combined-cycle)

η_Q=0.90 (high-efficiency boiler)

As shown in Figure 3.8, the fuel supply $\dot{F}$ represents one unit; the power ratios of the co-generation plant are identical to their respective efficiencies:

13. The situation where the energy demand does not correspond to the generation with a co-generation installation, and where the whole electric generation system must be considered instead of a reference plant will be discussed later on.

$$\dot{E} = \alpha_E$$
$$\dot{Q} = \alpha_Q \qquad [3.3]$$

To produce the same quantity, $\dot{E}$, the separated electrical station needs a fuel provision of $\dot{E}/\eta_E$. The boiler, in addition, requires $\dot{Q}/\eta_Q$ fuel units.

In elementary co-generation books, the rather misleading expression "total efficiency of a co-generation installation" is sometimes used. This would refer to the ratio between the total usable power and the necessary fuel for the plant:

$$\alpha_{TOT} = \frac{\dot{E} + \dot{Q}}{\dot{F}_{WKK}} = \alpha_E + \alpha_Q \qquad [3.4]$$

This value is then compared to the efficiency of the electrical generation in a power plant, i.e. η_E. As an example, in Figure 3.8, the total efficiency would reach 85%, whereas a combined-cycle power plant reaches an efficiency of only 55%. Thanks to the co-generation, the performance would be improved by more than half ((85–55)/55).

What is not often mentioned is that the efficiency of a co-generation installation is, by the same reasoning, lower than that of the high-efficiency boiler, with $\eta_Q = 90\%$.

The definition of equation [3.4] is not the total efficiency in the thermodynamic sense. This equation adds two energy quantities which have very different qualities. As mentioned above, heat and work are certainly not equivalent. We recommend the use of the term "fuel-use rate α_{TOT}" of a co-generation plant, which means that it deals with energy use of the fuel and nothing more.

A little more judicious comparison between co-generation, on one hand, and the combined cycle power plant and the boiler, on the other hand, may be found as the value-weighted fuel utilization ratio α^{TOT*}. With this performance indicator, it is possible to estimate the value of the heat generated by the co-generation unit, by calculating how much electricity would have been generated with the same fuel using the separated method. To generate α_Q heat units, the high-efficiency boiler needs α_Q/η_Q fuel units. With this fuel quantity, the combined-cycle power plant would have generated η_E (α_Q/η_Q) electricity units; therefore:

$$\alpha^{TOT*} = \alpha_E + \alpha_Q\left(\frac{\eta_E}{\eta_Q}\right) \qquad [3.5]$$

In the example of Figure 3.8, η_E/η_Q reaches 0.6, and α^{TOT*}=0.35+0.3=0.65. This value must be compared to α^{TOT}=85%.

3.4.2. *Exergetic efficiency of a co-generation installation*

The most accurate method of characterizing the performance of a co-generation installation is to use the exergetic efficiency. If the useful heat has been generated at a temperature T, the exergetic efficiency is equal to[14]

$$\alpha^{ex}_{WKK} = \alpha_E + \alpha_Q\left(1 - \frac{T_0}{T}\right) \qquad [3.6]$$

where T_0 indicates the ambient temperature. For an ambient temperature of 15°C and for the above values of α and η, we have the following accurate exergy comparison of the technologies given in Table 3.1.

α^{ex}(%) / T (°C)	Co-generation	CC	HE boiler
15	35	55	0
50	40	55	10
100	46	55	21
200	55	55	35
300	60	55	45

Table 3.1. *Exergy efficiency of the co-generation installation, combined-cycle (CC) and high efficiency boil r (HE boiler), as shown in Figure 3.8, for different temperatures of the heat process with an ambient temperature of 15°C. For co-generation, the formula in equation [3.6] is used; for a CC plant,* $\alpha ex \cong \eta E$ *; for a boiler,* $\alpha ex = \eta Q(1-T0/T)$

14. In fact, the Carnot factor to be used here is slightly more complicated because not only the temperature, but also the pressure of the available heat and return heat, are important. Instead of T, $T_{mi} = \Delta h / \Delta s$ should be used, where Δh and Δs represent the enthalpy and the entropy difference between input and output of the heat consumer, respectively.

Table 3.1 shows that the exergetic efficiency of a co-generation installation gradually increases with the process temperature T. For the temperatures of the example, the exergetic efficiency of a co-generation unit is greater than that of the high-efficiency boiler.

The process temperature must be higher than 200°C before the co-generation equipment has better exergetic efficiency than does a combined-cycle power plant. In fact, Table 3.1 is still too optimistic for some co-generation installations because as the quality of the process increases, the electric efficiency α_E may decrease for the same generated heat quantity α_Q .

This means that the energy saving for co-generation equipment must be found on the heat side. For heat generated at temperatures lower than 200°C, co-generation may save primary energy, but the saving must come entirely from the heat side, i.e. the boiler; moreover, the heat side will have to overcompensate for the electric side.

3.4.3. *Energy saving in co-generation compared to separated generation*

In the example shown in Figure 3.8, the use of co-generation to generate a quantity of electricity $\dot{E} = \alpha_E$ and of heat $\dot{Q} = \alpha_Q$ results in a primary fuel saving (PFS), which is sometimes also called primary power saving (PPS):

$$EEP = \frac{\alpha_E}{\eta_E} + \frac{\alpha_Q}{\eta_Q} - 1 \qquad [3.7]$$

The first two terms $\frac{\alpha_E}{\eta_E} + \frac{\alpha_Q}{\eta_Q}$ represent the fuel consumption for separated generation. For co-generation, one unit has been taken as fuel consumption.[15]

The relative energy saving, compared with the separated generation is established by dividing equation [3.7] by the fuel consumption of the separated generation. This gives:

15. In fact, equation [3.7] represents the per unit fuel saving in a co-generation installation; it is therefore the relative fuel saving compared to fuel input in the co-generation installation.

$$EEP\ REL = 1 - \frac{1}{\dfrac{\alpha_E}{\eta_E} + \dfrac{\alpha_Q}{\eta_Q}} \qquad [3.8]$$

This relative energy saving, compared to separated generation, with η_E =0.55 and η_Q=0.9, is often defined by the quality index (QI) of a co-generation installation:

$$QI = 1 - \frac{1}{1.82\alpha_E + 1.11\alpha_Q} \qquad [3.9]$$

If QI > 5%, the co-generation installation receives the quality co-generation label, which means that the co-generation installation uses 5% less fuel than the separated generation with a CC and a HE boiler.[16] For the example in Figure 3.8, QI = 16%.

The relative primary energy saving given by the formula in equation [3.8] strongly depends on the value of each of the different efficiencies. The influence of the α_E/η_E ratio has to be emphasized. Table 3.2 gives some values for REL PFS (relative primary fuel saving) for η_Q=0.9, with η_E and α_E variable.

η_E / α_E	**0.30**	**0.40**	**0.50**	**0.60**
0.25	33%	23%	14%	8%
0.30	38%	27%	17%	10%
0.35	42%	30%	20%	12%
0.40	46%	33%	23%	14%
0.45	49%	36%	26%	16%

Table 3.2. *Relative primary fuel saving (REL PFS) expressed in %, compared to separated generation with an electric efficiency η_E and a thermal efficiency η_Q = 0.9. The value of α_Q varies according to α_Q=0.85-α_E*

16. This 5% margin is defined because the saved energy quantity is in practice always lower than the one forecast by the simplified relation based on the data in Figure 3.8.

The figures in Table 3.2 show that a high value of α_E corresponds to significant fuel savings. On the other hand, as the η_E of separated generation increases, the REL PFS of a given co-generation installation decreases for a given value of α_E .

It is worth looking more closely at equation [3.8] for relative fuel saving. Too often the saving is estimated from representative values. However, this does not mean that equation [3.8] represents the effective energy saving for a co-generation installation actually in operation.

The relative economy saving, REL PFS, calculated with equation [3.8], is accurate only when all the heat delivered by the co-generation installation is efficiently used. Let us assume that only a fraction, $\xi < 1$, of the heat is effectively used and that the remaining is cooled; the expression of REL PFS will be defined by:

$$\underset{\text{(cooling of part of heat } 1-\xi)}{EEP\ REL} = 1 - \frac{1}{\dfrac{\alpha_E}{\eta_E} + \xi \dfrac{\alpha_Q}{\eta_Q}} \qquad [3.10]$$

Using only 75% of the heat generated (ξ=0.75), a quarter of the heat is lost. Equation [3.10] applied to the example of Figure 3.8 leads to an REL PFS of barely 5%.

The use of generated electricity does not generally raise any problems. In principle, generated electricity sent to the grid will always be used.

It is also important to note that in equations [3.8] and [3.9], the values obtained are constant only if the said efficiencies are constant. In the definition of quality index QI, it is assumed that the co-generation installation always operates at its nominal power and that the comparison is always made with a combined-cycle power plant and a boiler operating at nominal power.

Engines may work at partial load, since α_E and α_Q will only vary moderately. However, partial load operation of gas turbines must be avoided as much as possible. It is useful to know that α_Q changes with the temperature of the available heat.

The third point concerning equations [3.8] and [3.9] relates to the choice of separated generation reference. Opinions seem to diverge, in particular with regard to electricity generation. Some favor a combined-cycle power plant, others a nuclear plant, whilst others take the average efficiency of the electricity installation.

The only scientifically accurate reference plant for separated generation in equation [3.8] will be the plant which, at the time of investment in a co-generation installation, uses the most efficient electricity generation technology based on the same fuel.

For gas co-generation installations, the combined-cycle power plant, with a $\eta_E \cong 0.55$, presently represents the top technology.

For diesel co-generation units, it is better to make a comparison with standard oil plants, with a $\eta_E \cong 0.40$–0.45.

3.4.4. *The energy equivalence and the influence of a central park*

As described in section 3.4.1, the scheme of Figure 3.8 can only be used for comparison when the co-generation and the separated generation installations simultaneously generate the same final power $\dot{Q}$ and $\dot{E}$. A comparison based on power equivalent is then possible. For the sake of simplicity, this will be referred to as energetic equivalence between co-generation and separated generation[17].

In reality, the energy equivalence situation in Figure 3.8 only occurs when a co-generation installation is expected to continuously deliver a constant heat for almost a year (for example, in the processing industry). However, for co-generation installations used during limited periods of time, the simple representation in Figure 3.8 is not accurate because it does not take the electric generation system into account.

In developing the central electric generation system, on the one hand to replace old units, on the other hand, in anticipation of the growing demand for electricity, there are two possibilities open to the investor who will be guided by the peak demands made on the electricity generation system.

Either the central generation capacity is extended, generally with a combined-cycle power plant, or many co-generation units are installed, thus reducing the need for new combined-cycle power plants. Because of the co-generation installations,

17. The term "immediate and integrated equivalence according to final delivered energy" would be more accurate; strictly speaking, the fact that both types of installation deliver the same final energy is not enough; the energy must also be delivered at the same time.

combined-cycle power units may be avoided. The type of new investments in electric plants will depend on factors such as capital cost, fuel prices, CO_2 emission costs, and the most convenient fuel solution for the supply security. For the sake of simplicity, only an investment based on gas is considered here because gas co-generation has already been studied. These considerations may, however, be extended by analogy to other types of power plant.

If a new, combined-cycle power plant were built (without a large co-generation installation), it would be relatively low in the ranking order of electricity-generation systems because of its high efficiency and the CO_2 penalties on coal plants. Use would be very high, both in winter and summer, but its dependency on the price of gas would be taken into consideration. This new, combined-cycle power plant would replace the oldest plants as well as the less efficient gas and coal units. Thus, these old plants would become less used and thus the global system efficiency would consequently increase.

If co-generation units with limited use ($< 4{,}000$ h), operating during the peak hours for electricity demand, are installed, the new, combined-cycle power plants would be avoided. However, when the co-generation installations are not operating, the electricity will be generated by the aforementioned less efficient gas and coal plants, which is something that would have been avoided had the investment been in a combined-cycle power plant.

Consequently, co-generation installations offer a less efficient operation for electricity generation systems than would combined-cycle power plants. A typical example of this situation occurs when choosing large scale co-generation installations for internal heating in residential and tertiary areas.

In order to have an accurate comparison basis between the co-generation and the separated generation, a comparative simulation of a specific area is necessary. This simulation takes into account the electricity and heat-demand patterns, the dimensioning of the co-generation installations, the additional heating need and the total electricity generation system behavior. The final result of such a simulation will demonstrate that the energy saving with co-generation will generally be lower than Figure 3.8 would suggest.

For thermal applications with micro-co-generation installations in residential and tertiary areas, the primary energy saving is very low, which means that it may be difficult to classify such installations as representative of qualitative co-generation. The installation of micro-co-generation units for such applications should not rely on environmental or energy-saving justifications. Justification will only be found for private, economic reasons. We can assume that these economic reasons will have no, or a negligible, negative effect on the country's energy saving or environment.

Especially if the unsuperseded (and less efficient) units are of the coal type, the situation described above will have important consequences for CO_2 emissions. The difference in CO_2 emissions on the scenarios with and without co-generation in tertiary and residential areas is about zero; sometimes, a slight CO_2 reduction will emerge, sometimes, the co-generation scenario will generate a little bit more CO_2 than the combined-cycle power scenario. The impact on global CO_2 emissions of an electricity generation system with or without co-generation is proportional to the energy saving, at least if the combined generation and the separated generation have gas as their primary fuel.

The situation where co-generation overtakes nuclear electric generation and separated heat generation is completely different. In the long term, industrial co-generation will prevent or avoid the construction of new nuclear units. The co-generation scenario will save a lot of primary energy because the efficiency of the nuclear unit is relatively low (about 33%). However, at the moment, the situation is completely different because the mix of fuels is completely different (uranium compared to gas and/or oil). As far as CO_2 is concerned, the co-generation scenario will generate many more emissions than separated generation with nuclear electricity generation.

3.5. Energy advantage allocation

Apart from the above considerations, which state that co-generation is an intelligent way to generate heat (section 3.1.3), a question is raised: to which source must this energy saving be assigned: heat or electricity?

For the sake of simplicity, let us look again at the energy-equivalent scheme in Figure 3.8. Beginning with the figures shown in this diagram, it is possible to obtain the next energy balance in order to generate $\dot{Q} = 0.5$ heat units and $\dot{E} = 0.35$ electricity units.

Co-generation	**Separated**	
Fuel input: 1	TGV	Fuel input: 0.64
	Boiler	Fuel input: 0.56
	Total	Fuel input: 1.2

Table 3.3. *Energy assessment of co-generation and separated generation*

3.5.1. *Electricity advantage allocation, effective electric efficiency*

Let us assume that the owner of a co-generation installation considers that for the generation of a heat quantity $\dot{Q} = 0.5$ based on co-generation, the same quantity of fuel as before will be needed in a boiler, i.e. 0.56 units. According to this scheme, the co-generation installation requires only $1.0 - 0.56 = 0.44$ fuel units to generate an electricity quantity of $\dot{E} = 0.35$ units. By this reasoning, the co-generation installation needs 0.20 units less than a combined-cycle plant. Thus, the electricity becomes a very low cost sub-product. In other words, the effective electric output α_E^{eff} of the co-generation installation is 0.35/0.44 = 0.795 or

$$\alpha_E^{eff} = \frac{\alpha_E}{1 - \frac{\alpha_Q}{\eta_Q}} \qquad [3.11]$$

where α_Q / η_Q is the fuel the boiler would have needed in the case of separated generation to produce $\dot{Q}$ heat units. The 1 is the fuel input of the co-generation installation.

Equations [3.8] and [3.11] show that the co-generation installation must save energy (REL PFS > 0); this can be described as follows:

$$\alpha_E^{eff} > \eta_E; \text{ energy saving co-generation} \qquad [3.12]$$

According to the above reasoning, co-generation is considered a heat generator with a turbine added in order to generate electricity: the co-generation is a "smart" boiler.

If we assume that, globally speaking, the quantity of fuel consumed in order to meet a nation's heat demand is equal to $CO_{2\text{-eq}}$[18] emissions when a separated generation system is used, then the widespread introduction of co-generation plants would lead to a decrease in the amount of fuel required to supply that nation's electricity demand. The effective average efficiency of the electricity generation would rise. In other words, the implementation of co-generation would perceptibly reduce the $CO_{2\text{-eq}}$ emission of the electricity sector.

18. If it is the same amount of primary fuel, for example gas.

3.5.2. *Heat advantage allocation, thermal effective efficiency*

In this hypothesis, let us assume that an electricity producer invested heavily in co-generation installations at sites belonging to the main industrial-heat users. We assume that to produce electricity quantity $\dot{E} = 0.35$, the same amount of fuel is required by the co-generation installation as would be needed by a combined-cycle power plant, i.e. 0.64 units. To generate a heat quantity $\dot{Q} = 0.5$, the co-generation installation would require only 1–0.64 = 0.36 units instead of the 0.56 units needed by a separated production system.

Following this reasoning, a co-generation installation requires 0.20 units less than a common boiler, making heat, in effect, an almost free product.

The co-generation installation is characterized by an effective thermal efficiency $\alpha_Q^{eff} = 0.50/0.36 = 1.39$. The corresponding formula gives:

$$\alpha_Q^{eff} = \frac{\alpha_Q}{1 - \frac{\alpha_E}{\eta_E}} \qquad [3.13]$$

According to this philosophy, the co-generation installation is hypothetically considered to be a small electrical plant (combined-cycle) which will deliver an amount of high-quality heat with an efficiency which could exceed 1.

Equations [3.8], and [3.13] rely on the requirement that a co-generation installation is more economical with primary energy than separated production (REL PFS > 0); This may be written as:

$$\alpha_Q^{eff} \cdot > \cdot \eta_Q \cdot \text{; cogeneration saving energy} \qquad [3.14]$$

Let us take the example in section 3.5.1 and extrapolate it at national level. We know from section 3.5.2 that the same amount of primary fuel would be required for the same electricity demand, so the electricity sector would be responsible for the same quantity of $CO_{2\text{-eq}}$ emissions as before. However, the same quantity of heat would require less fuel. If $\alpha_Q^{eff} > 1$, the co-generation installations being considered would serve as kind of energy amplifier for the heat part; operating as a *II*. By this reasoning, the process industry will emit much less CO_2 to generate the same quantity of heat.

3.5.3. *Suitable allocation of the energy advantage, proper efficiency*

Sections 3.5.1 and 3.5.2 only consider two extreme points of view. The reality is somewhere in between. It is, however, not possible to assert only one statement for allocation of energy efficiency. The best option is to propose a "meaningful" allocation in each case.

The expression of formula [3.7] of the primary fuel saving PFS, for the fuel input to a particular co-generation installation, may also be described as

$$PBB = \frac{\alpha_E}{\eta_E} + \frac{\alpha_Q}{\eta_Q} - 1 = \left(\frac{\alpha_E}{\eta_E} - X\right) + \left(\frac{\alpha_Q}{\eta_Q} - Y\right) \qquad [3.15]$$

for X + Y = 1.

This expression suggests that the co-generation installation requires X fuel units ($X \leq 1$) in order to generate α_E electricity units. In the same way, Y units ($Y \leq 1$) would be required to generate α_Q heat units. In this way the co-generation installation is characterized by what might be called the appropriate efficiencies α_E^{app} and α_Q^{app}:

$$\alpha_E^{app} \equiv \frac{\alpha_E}{X} \qquad [3.16]$$

$$\alpha_Q^{app} \equiv \frac{\alpha_Q}{Y} \qquad [3.17]$$

For the allocation of energy efficiency in section 3.5.1, $X = 1 - \alpha_Q / \eta_Q = 0.44$, whereas $Y = \alpha_Q / \eta_Q = 0.56$. According to this allocation, the results are as follows:

$$\alpha_E^{app} = \alpha_E^{eff} \text{ and } \alpha_Q^{app} = \eta_Q$$

(allocation according to section 3.5.1, equation [3.11]).

For the application in section 3.5.2, Y = 1 – α_E / η_E = 0.36 and X = α_E / η_E = 0.64. These considerations results in:

$$\alpha_E^{app} = \eta_E \text{ in } \alpha_Q^{app} = \alpha_Q^{eff}$$

(allocation according to section 3.5.2, equation [3.13]).

Although, at first glance, it seems logical to consider the examples in sections 3.5.1 and 3.5.2 as extremes, and to request that 0.44 < X < 0.64 and 0.36 < Y < 0.56 (taking into account that X + Y = 1), this does not necessarily need to be the case. The numeric values are based on the efficiency η_E = 0.55 of a combined-cycle power plant and a heat efficiency η_Q = 0.90, which are used to make a comparison with a separated production plant. The co-generation installation itself does not take into account the reference given for the comparison; the values X and Y may thus vary on a wider scale than suggested above.

A judicious allocation of fuel in a co-generation installation may be made exergetically. Accordingly, a smart co-generation installation, which will allocate more fuel to the heat generation when its quality, or exergy increases, will be the preferred option. In other words, the co-generation system uses little fuel to generate low-temperature heat, but if high-temperature heat is required, the system will decide to allocate more fuel to heat production. Instead of simply supposing X = $\alpha_E/(\alpha_E + \alpha_Q)$ and Y = $\alpha_Q/(\alpha_E + \alpha_Q)$, it is better to say:

$$\text{X} = \frac{\alpha_E}{\alpha_E + \alpha_Q\left(1 - T_0 / T\right)} = \frac{\alpha_E}{\alpha_{cog}^{ex}} \qquad [3.18]$$

and

$$\text{Y} = \frac{\alpha_Q\left(1 - T_0 / T\right)}{\alpha_E + \alpha_Q\left(1 - T_0 / T\right)} = \frac{\alpha_Q\left(1 - T_0 / T\right)}{\alpha_{cog}^{ex}} \qquad [3.19]$$

The relative exergy factor $(1\text{-}T_0/T)$ for heat is recognized; for electricity, the relative exergy factor is equal to 1.

The table below gives some numeric values for X and Y, based on equations [3.18] and [3.19], with T_0 = 15°C = 288 K.

T (°C)	15	100	150	200	250	300	400	500
X	1	0.75	0.69	0.64	0.61	0.58	0.55	0.53
Y	0	0.25	0.31	0.36	0.39	0.42	0.45	0.47

Table 3.4. *Numeric values for X and Y according to equations [3.18] and [3.19] with $T_0 = 288$ K, $\alpha_E = 0.35$ and $\alpha_Q = 0.50$*

One extreme, where T = 15°C, indicates a production entirely dedicated to electricity with a simultaneous delivery of heat without value. In this case, and as long as T < 200°C, co-generation requires more fuel than a combined-cycle power plant to generate α_E electricity units. The fuel profit from co-generation must come entirely from the heat side (small values for Y).[19] Moreover, it is clear that co-generation always requires less fuel than a boiler, i.e. 0.56 units to generate α_Q heat units, which again demonstrates the ability of co-generation to generate heat economically.

3.6. The electrical aspects of co-generation installations

The co-generation installation delivers power through the coupling of the turbine or engine axis to an electric generator. For higher amounts of power (around > 100 kW$_e$), synchronous generators are used; for smaller amounts of power (about < 100 kW$_e$), asynchronous generators are generally used.

An asynchronous generator takes its magnetization current from the stator. Generally speaking, an asynchronous generator will be connected to the grid in order that its magnetizing current can be delivered by the network. It is also possible to operate an asynchronous generator while operating in isolation, but it must have a condenser bench that will deliver the magnetization current. Isolated asynchronous generators are small because otherwise the condenser benches would become too large and expensive. Even generators linked to the grid are limited to powers below 100 kVA, otherwise the inductive load would become too heavy for the grid and the condenser benches would be used for a correction $\cos\varphi$.

19. In fact, the PFS when T = 15°C must be calculated by adding a factor ξ before α_Q / η_Q in equation [3.7] – see also equation [3.10] – where $\xi = 0$ may be chosen.

These limiting factors explain why synchronous generators are chosen for certain power requirements. The rotation speed of an asynchronous generator is slightly higher than the rotating magnetic field generated by the stator and which rotates at the grid frequency or at a part of it, depending on the number of generator poles.

The synchronous generators are magnetized by means of a DC current flowing in the rotor coils. The rotation speed of a synchronous generator is exactly equal to a fraction of the grid frequency. A tetra polar generator with a rotor turning at 1,500 t/min is often used. A synchronous generator may operate in isolated mode without any problem; a synchronous display is required to operate on the grid.

In isolated mode it is, of course, impossible to deliver power to the grid. In the case of an insufficient local electricity demand, the co-generation installation should operate at partial load or be switched off.[20] An asynchronous generator, connected to the grid, always absorbs inductive reactive power. There is no specific adjustment of reactive power. In adjusting the magnetic field on the rotor of a synchronous generator, however, it is possible to locally manage reactive power in the grid.[21]

Large asynchronous generators in electric plants have adjustment apparatuses for reactive power. In a standard co-generation installation, this device is not generally provided. Such a co-generation installation may, however, be designed in order to use, or deliver, reactive power, but it will not have any possibility of dynamic regulation.

Generally speaking, the owners of co-generation installations wish to optimize their installation to generate active power, and the compensation of reactive power is not planned.

It is important that the transmission (or distribution) system operator gives at least its requirements in terms of static reactive powers which will be appropriate to integrate co-generation installations to the grid. These requirements will be different according to the installation localization in the grid.

In a country where non-adjustable and decentralized power excesses are exploited, the control of unexpected grid disturbances will be more difficult, with possible blackouts. The major blackout which took place in the Utrecht (Netherlands) area in June 1997 was caused by an excess of non-adjustable production units of the co-generation type.

---

20. The dissipation of the generated electricity in the resistances is not considered a sensible option and electricity storage is not possible either, at least not in reasonable amounts).
21. The reactive power is important for the network voltage support.

The implementation of co-generation installations distributed throughout the network has the advantage of enabling the central grid operator to foresee less reserve power. If a $1GW_e$ power plant suddenly stops, a reserve power of $1GW_e$ must be available. In contrast, the probability that 20 units of 50 MW_e are not available at the same time is low. Some 100 MW_e of reserve power should be enough in the same situation

In principle, a well-chosen dispersion of co-generation installations in the network helps the reduction of transmission line congestion. In order to achieve good implantation configurations, a grid operator may make the network connection conditions dependent on localization or give price signals. As a good co-generation installation must be administrated by the heat demand and, consequently, must be installed near a large consumer of heat, the location depends more on the heat-consumer location than on network considerations. Moreover, price signals may not be accepted, at least not from a social point of view.

With a well-chosen location, co-generation installation may also limit losses in distribution and transmission networks. Losses in the transmission grid are typically around 2–3%, whereas they are in the range of 5–7% in the distribution grid.

According to introductory books, co-generation enables the avoidance of 7–10% of transport losses. This is, however, not correct. The main losses are located in the distribution network and more than half of the total losses are due to zero load loss in transformers.

As most of the co-generation users want to use the distribution grid in order to get power in case of the unavailability or inefficiency of the co-generation installation, or even to deliver a surplus of electric power to the grid, these users require the availability of a sufficiently robust distribution grid where all transformers must be maintained and operational. Main losses in the distribution grids cannot be avoided by using co-generation.

Precise calculation of the reliability of the implementation configuration of the network would be required to identify the transformers that could be eliminated. Due to co-generation, some percentage of transport losses may be saved, but probably not more than 3–5%.

In the future, the possibility will exist for the network administrators to sign agreements with the operators of large co-generation installations in order to have support in case of incidents such as blackouts. The grid operator could have an influence on the operation of the co-generation installation by means of remote-control management techniques under agreement stipulated limits. This is technically possible; the question is whether a company will be tempted to let the operation of its co-generation installation be influenced by the grid operator.

3.7. Cooling by absorption and tri-generation

To save energy, it is important that a co-generation installation uses all heat that is produced.[22] Apart from the standard use of heat for heating purposes, heat may also be used as a main energy source for the generation of cold. This generation is obtained in "absorption cooling machines".

In a standard cooling machine[23] the cooling agent follows a two-phase cycle (from liquid to vapor conversion in the evaporator and inversely to the condenser) where the vapor at the output of the evaporator is compressed by the condenser pressure by means of a compressor. The cooling of a volume by extracting heat via an evaporator describes the method of a "compression cooling machine". A "compression heat pump", inversely, heats a volume by means of the condenser heat.

An absorption cooling machine (or pump) uses a secondary fluid in which vapor of the primary-cycle fluid is soluble, thereby enabling the replacement of the compressor stage by an increasing of the pressure by a pump. As a pump requires less input energy, such a cooling machine will consume less energy. However, in order to enable the cycle continuity, it is necessary to regenerate the primary vapor out of the solvent after the pump stage. This requires heat absorption, which explains the name of the absorption cooling machine.

It is possible to absorb the heat generated by a co-generation installation by using an absorption cooling machine. If at the same time, part of the heat is used in a normal calorific application; this is called tri-generation; the co-generation installation generated three secondary energy products: electricity, heat, and cold.

The application of tri-generation is possible in areas which require much cooling: areas where there is food, e.g. fruits and vegetables, supermarkets, slaughterhouses, etc. Also in the tertiary area, where heat is required in winter and cool in summer, co-generation combined with an absorption cooling machine might be the possible answer.

---

22. In some very specific, but rather subtle, cases it may be reasonable to partially discharge the generated heat. However, the main rule is that the cooling of the co-generation-generated heat must be avoided at any price.
23. Or in a heat pump which follows an identical thermodynamic process.

3.8. Estimation of the potential of co-generation

It is, first of all, useful to make a few comments.

In principle, each classic thermal-power plant where there is a simple steam purge and in which heat is used for a thermal application could be considered as a co-generation[24] installation.

This is not wrong, although it does not make sense to consider the electric power of the co-generation installation as a reference of the installed co-generation power.

In these cases, it is better to use installed thermal power. The most consistent criteria consists of offering the thermal power with the $\dot{Q}/\dot{E}$ ratio of the installation.

In order to avoid the use of inflated figures in national statistics, the electric power of a co-generation installation can, since 1994, be taken into account in the official statistics of Eurostat only when $\dot{Q}/\dot{E} > 0.25$.

3.8.1. *Potential of co-generation energy*

In order to be considered in the energy potential calculation, a co-generation installation must satisfy a few criteria.

The most important condition is a relative primary fuel saving (REL PFS in equation [3.8]) of at least 5%. This condition is based on a static comparison between a co-generation installation and a separated generation in the form of combined-cycle power plants of 53%, including the network losses and a 90% boiler. If this criterion is satisfied, the term "quality co-generation" is used.

This limit value of 5% is reasonable because it deals with full-load efficiencies whereas in practice these co-generation installations will also work at partial load. In this case, the immediate energy saving decreases; occasionally it even becomes negative.

In order to have an overall energy saving which, on average, is positive whatever the load, it is wise to foresee a certain margin in the selection criteria of the energy system. Moreover, as described in section 3.4.4, the structure of the central system

24. Let us take as an example an existing thermal plant with four groups (3 x 124 MW_e and 1 x 300 MW_e). Each of these groups may deliver 10 tons/h of steam (although not simultaneously) to a nearby factory. Is it justified to talk about 675 MW_e of co-generation?

plays an important role in the final energy balance. Especially for co-generation installations that do not operate throughout the year (but only during the peak points of national electricity consumption), a true energy saving (with respect to the separated generation) that is lower than the one given by equation [3.8] can be expected. The 5% margin explained at the beginning of this section is then reasonable to obtain a more or less correct evaluation.

To estimate the potential of co-generation, two methodologies (although not equivalent) are possible. These methodologies are based on the principle that co-generation must be controlled by the heat and that the heat demand is dominant. These two methodologies are used in practice.

According to the first methodology, the co-generation unit is chosen based on the maximum demand for heat (defined by the maximal area in the heat-monotonous diagram in Figure 3.7).

According to the second methodology, gas turbines are designed in order to operate during 4,000 h/a at full load; for the gas engines, 2,000 h/a have been proposed as a principle, while enabling them to operate at a partial load (up to 50%). This second design philosophy aims to maximize the co-generation "power", whereas the first methodology aims to reach a maximum of heat (of thermal "energy") generated by co-generation.

The potential, expressed in electric power according to the options 4,000 h or 2,000 h is always greater than or equal to the potential obtain according to the methodology using the "maximum area". This is considered to be the base potential; the 2,000 h/4,000 h option is strictly a maximum potential (and, in fact, is too optimistic).

Some investigations suggested considering urban heating as part of the potential of the residential and tertiary areas. This practice leads to significant long-term uncertainty because of the investment costs of the heat distribution network. Although we must be very cautious about the European examples, it must be noted that many heating projects in districts in the Netherlands are considered to be unprofitable and are integrated into the so-called "stranded costs" in the liberal market. Moreover, urban heating leads to a strange situation: it is more profitable (because of the investment depreciation) when the heat demand increases.

Applications of rational energy use, such as better housing insulation, which in itself represents a more efficient way to save energy than urban heating, will then make centralized heating even more expensive. We must also question whether the centralized heating idea is really within the realm of social expectations. Society is increasingly accepting the idea of decentralization as far as electricity is concerned.

3.8.2. *Economic potential of the co-generation market*

Starting from the basis of energy potential , it is possible to extract an economic potential. Economic potential is generally based on a minimum period in which to recover the costs, which is defined in advance, or on requirements regarding the net present value for the co-generation installations. In these estimates, the country subsidies must not be included; on the other hand, it may be useful to include the external costs. Experience shows that the economic potential is situated around $^{2}/_{3}$ of the energy potential. The so-called market potential takes into account even more elements, sometimes even psychological or less rational, to determine the true potential. Here, the market potential is estimated, by experience, at around $^{1}/_{2}$ of the energy potential.

3.8.3. *Co-generation installations already in existence*

From all the results of the previous sections, the existing installations must be deducted in order to obtain the remaining potential (at a given instant). It is also recommended that the existing co-generation capacity be estimated since it is often historical co-generation which would not now be considered as quality co-generation.

Finally, because of some inaccuracies in the co-generation potential estimation (for example, not taking into account thermal storage in buffer tanks), and to keep a reasonable security margin, we are advised to use both methodologies: the estimate based on the maximum area is accepted as a minimum estimate and the results obtained by the 2,000/4,000 h method as a maximum estimate. The average between both generally offers a reasonable estimate of the co-generation potential to be installed.

3.9. Influence of co-generation on the environment

The environmental impact of co-generation installations may be considered to be the undesirable product emissions of the "prime mover" (gas turbine, gas engine) with respect to the same emissions of heat and electricity separated generation installations.

3.10. Conclusions and perspectives

Co-generation, through the simultaneous generation of heat and electricity, is a very useful energy-conversion technique as it enables the saving of a large amount of primary energy as compared to the separated generation of electricity and heat.

The use of a high-quality co-generation is strongly recommended, but co-generation must not in any case be considered as *the* miracle solution to all energy problems. Co-generation does not make sense unless it is controlled by heat use, which consequently limits its potential to the heat demand (especially in industry, services and the commercial sector).

Moreover, because of the continuous improvements made in the standard means of generation, publicity on co-generation must be condemned, particularly when it concerns low-scale usages.

Indeed, in the residential sector, the main objective to be first considered is to insulate, as well as possible, housing, and often the co-generation, as presently considered, is not the best solution.

For the same reasons, the massive use of urban heating must be analyzed very carefully. Before launching a large-scale promotion or subsidy scheme for some areas, a systematic analysis must be made taking into account the response of the overall energy system.

The future of co-generation is rather uncertain in the short- and medium-term because in many countries the available potential has not been exploited. The generation means based on gas (turbines and gas engines) are very active, but they may be affected by the high gas prices. Although the choice of the urban heating must not be, as previously explained, decided by the "fashion effect", it may grow in some places in Europe where coal plants are again being introduced in the electric system.

The advent of some promising technologies, such as the integration of a Stirling engine in a boiler, or the advent of fuel cells, could increase the prospects of small-scale co-generation. However, although we recommend making a effort on the development of these technologies, in the future it is only their economical and industrial successes that will ensure their permanence.

3.11. Bibliography

[BAR 98] F. J. BARCLAY, *Combined Power and Process and Exergy Approach*, Professional Engineering Limited, London, 1998.

[CAT 97] J. A. CATON and W. D. TURNER, "Co-generation", in F. Kreith and R. E. West (eds.), *CRC Handbook of Energy Efficiency*, CRC Press, Boca Raton, FL, 1997, pp. 669–711.

[EDU 01] EDUCOGEN, *The European Educational Tool on Co-generation*, 2nd Edition, December 2001; available at www.cogen.org/projects/educogen.htm.

[SCH 05] K. W. SCHMITZ and G. SCHAUMANN, *Kraft-Wärme-Kopplung 3*, Auflage, Springer, Berlin, 2005.

Chapter 4

Hydrogen: An Energy Vector

4.1. Context

The idea of using hydrogen as an energy vector has kept scientists busy since the beginning of the 1980s, but the public and the media have only been interested in this topic since the beginning of this century.

This is not just due to chance or to the determined will of some lobby, worried about change, but results from the analysis of am energy and environmental situation, which worries industrial countries, as well as developing countries. How are these situations worrying?

4.1.1. *Geopolitics and the energy situation*

The following evidence characterizes the 20th century and the beginning of the present century: today 85% of world energy needs are still being supplied by fossil fuel, that is oil, gas and coal. Specialists all agree that reserves are now seriously depleted, and predict an oil peak[1] between 2020 and 2040, a gas peak 30 to 50 years later, and a coal peak in the next century.

To try to predict the most realistic situation for the middle of this century, specialists have published many prospective[2] studies which show that a consensus

---

Chapter written by Thierry ALLEAU.

1. An oil peak is when consumption is no longer compensated for by new discoveries.

2. Royal Dutch-Shell, *Energy needs, Choices and Possibilities. Scenarios to 2050*. Available at: www.shell.com.

has been reached. This can be summarized by two figures: by 2050, the world energy demand will have increased by a factor of between two and three, compared with levels in 2000. Fossil fuels will account for almost 63%, which means a relative reduction, but, in absolute terms, this will be an increase of the fossil contribution to the global energy production.

If we add to this the various global deregulations which already result from it and which will probably continue to increase, it is easy to understand that the problem of energy resources is urgent, and it is vital that we waste no time in finding possible solutions.

These solutions will necessarily come through one or two of the other primary energy sources at our disposal: sustainable energy and nuclear energy. Their contributions may be reduced by a strong economy and energy recovery policy.

Through their exploitation methods, these energies are converted into electricity, which is then easily converted into hydrogen, by means of water electrolysis.

4.1.2. *The environmental situation*

We cannot any longer ignore the fact that our environment is being affected by the growing concentration, of greenhouse gases in the atmosphere and which are the cause of climatic changes[3] and are responsible for atmospheric phenomena, which are not necessarily more frequent, but are certainly more violent.

These changes are now evidenced by many manifestations such as increases in sea levels, the disappearance of some animal species, and many other problems which we will not detail here.

Of all the greenhouse gasses, carbon dioxide[4] (CO_2), which is directly caused by the use of fossil fuels, is the most worrying because it is the most active.

The majority of the world leaders became aware of this situation early enough, and signed the Kyoto agreement, which became effective in February 2005. Unfortunately, this agreement is largely insufficient since they forecast emission reductions of only 5.2% by 2012 compared to 1990, which will not even stabilize

---

3. The surface temperature of the Earth has increased (by 2005) by 0.6°C since 1861 and the increase forecast for 2050, if nothing is done, is between 1.5 and 4°C depending upon the hypothesis and the models used.
4. The CO_2 concentration in the atmosphere was 380 ppm in 2005 compared to 280 ppm at the beginning of the 21st century.

the situation. Europe has decided to bring this rate up to 8%, which will still not be enough. To reverse the situation, a good objective would be as follows: reduce emissions by industrial countries by a factor of four from the beginning of 2005 until 2050. It is a very ambitious objective, and it will not be easy to achieve, not least because, for financial and logistical reasons, alternative solutions have not yet been developed, which means that we will need to continue to rely on fossil fuels for the greater part of this century.

There is only one solution: the confinement of CO_2. This means that the primary fuel, i.e. oil, gas, or coal, is converted, by reforming, or gasification, into a non-carbon fuel, and the resulting CO_2 is stored on site. The only non-carbon performing fuel obtained by this method is hydrogen.

4.2. Hydrogen: an energy vector for the future?

Whatever the available primary energies (fossil, renewable, or nuclear), they will generally be available to us in the form of electricity or hydrogen.

Whilst they may be less important, two other forms of energy must not be ignored: heat, i.e. solar, thermal, geothermal and biomass. Biomass of course leads to carbon components, but this carbon, captured by photosynthesis, is already included in the atmospheric cycle and is therefore not harmful.

Thus, electricity and hydrogen will become the two elements of a new virtuous couple which will progressively replace the present couple: electricity and fossil fuel. Liquid fuel is progressively being replaced by gas in the transition phase at the beginning of this century.

The electricity-hydrogen couple has a significant advantage compared to the present day couple in that it is completely reversible. In the present couple, whilst electricity can be obtained from fossil fuel, the inverse is not possible. In the future couple, it will be possible to go from electricity to hydrogen by means of water electrolysis. The inverse will also be possible through the fuel cell (see Chapter 5).

Hydrogen is far from new: Paracelse discovered it in the 16th century. Its public use has been spreading since the 19th century, in street lighting and domestic applications, and the 20th century saw its development in industrial applications, particularly in chemistry. It is important to consider its advantages and disadvantages.

Advantages:

– it is abundant on Earth in its elementary form (atomic), particularly in water and in hydrocarbons;

– it is the most energy-efficient fuel: 120 MJ/kg, twice the mass specific energy of natural gas;

– it is neither pollutant nor toxic;

– it is the lightest gas, which is an advantage for security; its diffusion rate the in air is four times faster than that of natural gas;

– it is possible to transport safely;

– it can be produced and converted in many different ways.

Disadvantages:

– it is light, its volume energy density is low, which is a problem for transportation and storage;

– it is five times more flammable than natural gas, but it is almost impossible for it to explode in the open air;

– the energy required for it to burn is low (20 μJ for 290 μJ with natural gas) and its flame is invisible, and very hot (2,000°C), which is a security risk in case of fire;

– its public image is generally negative: hydrogen is dangerous because it explodes!

The various properties listed above show that hydrogen use is not without danger, but we must accept the fact that there is no completely risk-free method of transferring and storing energy.

To use hydrogen safely we must first identify the potential risks and put in place procedures that will, as far as possible, eliminate the risk and consequences of accidents.

If used in accordance with agreed safety regulations, experts agree that hydrogen is no more dangerous than any other vector, particularly natural gas

4.2.1. *Today's uses*

As an energy vector the use of hydrogen is limited largely to the fueling of space craft, however significant quantities have for some time been used by the oil refinery and chemical industries. The chemical industry's main product is ammonia (NH_3), which is used mainly for the manufacture of fertilizers, and methanol and hydrogen peroxide which are used in the manufacture of glass.

World hydrogen production in 2000 was about 45 million T/year (or 500 billion of Nm^3/year[5]), 13% of which was in Europe. This represents about 1.5% of world energy production.

4.2.2. *Tomorrow's uses*

If we accept that hydrogen will become a widely distributed energy vector, we may assume that by 2050 it will represent about 20% of world energy transfers. This figure is based on a world consumption which will have at least doubled from 410 EJ to 820 EJ, which means a hydrogen equivalent of 164 EJ: 25 times more than in 2000, or more than 1,100 million tons of hydrogen to manufacture each year.

These gaps, and consequently the industrial impact on the hydrogen production industry, are so important that they do not allow anticipation of the future manufacturing processes, compared with those existing at the beginning of this century, since the environment and economic factors will play an increasing role.

4.3. How do we produce hydrogen?

In its atomic form hydrogen is plentiful; but in its molecular form, i.e. the form which interests us, it does not occur naturally.

We must remember that hydrogen is not, as it is sometimes, incorrectly referred to, an energy source, but an energy vector, i.e. the form in which the three primary energy sources available on Earth, fossil, renewable, and nuclear, may be converted into usable energy.

So-called hydrogen sources are natural chemical bodies, which occur abundantly in nature. The two forms are:

– fossil fuels i.e. oil, gas, coal, in the form C_xH_y. Biomass may be added to this list in the form $C_xH_yO_z$;

– water (H_2O).

All the hydrogen manufacturing processes to be described will therefore consist of breaking down these molecules, in one way or another. As these molecules are

5. Or 6.5 EJ (exajoules) per year, one exajoule is equivalent to 1,018 joules.

very stable, it is not easy to break them down, and the process uses a quantity of energy of which only a fraction will be transformed into hydrogen. For a process to be considered economically interesting, the efficiency of the operation, i.e. the energy converted into hydrogen divided by the energy needed for the process, must be high.

4.3.1. *Hydrogen production from fossil fuels*

It is interesting to note that this is the most commonly used process today; it currently represents over 90% of hydrogen production, through natural gas.

There are three main types of process: water-steam reforming also known as steamer reforming, partial oxidation (or a combination of both), and coal gasification.

Other processes are technically feasible, but have not been fully developed because of their poor performance; for example, plasma reforming, largely studied in Russia, through which the natural gas molecule is first excited in an electric discharge and then more easily.

4.3.1.1. *Steam reforming or reforming with water steam*[6]

Steam reforming is the process of transforming a light hydrocarbon, usually methane CH_4, a natural gas, into a gas which is a synthesis of: H_2, and CO with methane traces. This operation is carried out at high temperature (1,100 K to 1,200 K) and under moderate pressure (20 bars to 30 bars) with a catalyst, generally nickel-based.

The result of the two reactions, i.e. steam reforming itself and the conversion of CO[7], is:

$$CH_4 + 2H_2O \rightarrow CO_2 + 4H_2$$

This reaction requires a heat quantity of approximately 165 $kJ.mol^{-1}$: this energy is supplied by a fraction of the same fuel as the one transformed. The quantity of the generated CO_2 is between 16% and 20%. The following stages depend on the hydrogen pureness required.

We see that hydrogen conversion processes will have a direct impact on hydrogen costs, and consequently on the overall outcome.

6. Also called steam reforming.
7. Also called water-gas shift.

Hydrogen quality will be different depending on the way it is to be used: via Polymer Exchange Membrane (PEM) fuel cell[8] which cannot use more than 10 ppm of CO, or, via an SOFC (solid oxide fuel cell) which uses CO is a fuel. Furthermore, a thermal engine will have other requirements.

If very pure hydrogen is required, particularly for PEM fuel cells, the solution which might contain 0.3 to 3% vol. of CO, is processed by one of the two following methods:

– methanization ($CO + 3H_2 \rightarrow CH_4 + H_2O$) which produces hydrogen with a purity of 95% to 98%;

– selective adsorption, also called PSA[9] process, which produces a purity of 99.9%. This adsorption is operated on molecular sieve beds.

The following step will consist of separating, and possibly storing, the CO_2 (see section 4.3.1.4.1).

4.3.1.2. *Reforming by partial oxidation*

Partial oxidation may be carried out with fossil fuels, of varying weight, from natural gas to coal. This happens at high temperature (1,470 K to 1,770 K) and under high pressure (20 bars to 100 bars), with oxygen as the oxidant and a temperature moderator (water steam).

Similarly, partial oxidation leads to a gas synthesis, beginning with natural gas, and which is described as follows:

$$CH_4 + H_2O + 1/2\ O_2 \rightarrow CO_2 + 3H_2$$

By contrast, this reaction generally works without a catalyst and produces a heat that must be dispersed and which is of the order of 77 $kJ.mol^{-1}$. The following purification and separation steps for CO_2 are the same as previously described.

4.3.1.3. *Auto-thermal reforming*

This process, a combination of both the processes described above, is probably the most interesting. Its major advantage is that it compensates, again in a fixed-bed reactor, the endothermal reactions of steam reforming by exothermal reactions of partial oxidation. In this process, natural gas and oxygen are mixed with water vapor before being preheated; they are then released towards the reactor at a pressure of between 20 bars to 60 bars, and a temperature of between 1,170 K and 1,270 K, with

8. For details see Chapter 5, section 5.2.4.2.
9. Pressure Swing Adsorption.

a nickel-based catalyst for the production of synthesis gas. The composition of the gas obtained is typically:

68% H_2, 20% CO, 10% CO_2, and traces of CH_4 and N_2

The following steps are the same as above:

– Performance: the energy efficiency of these processes depends on their size and technology, but it will be in the range of 70–80%, which is quite satisfactory.

Many companies in the world offer products in this field, e.g. N-GHY in France.

4.3.1.4. *Coal gasification*

Coal gasification is the process of transforming carbon, partly or wholly, into a family of gases: carbon monoxide, carbon dioxide, hydrogen, methane, and nitrogen, some of which are flammable. It is achieved by burning coal underground, in situ.

Although this technique is well known for its efficiency, it requires many different disciplines, such as chemical kinetics, thermal exchange, gas flow, hydrology, and geophysics.

The technology developed by Professor Yu Li, at the Chinese University of Mineral Industries, enables the successful extraction of high temperature water vapor and low cost hydrogen. According to Professor Yu Li, one ton of coal may generate 1,300 to 1,500 m^3 of hydrogen at a much lower cost than natural gas. The gasification should enable China to ecologically exploit its coal reserves, provided the CO_2 is separated from the other gases and captured at the end of the process.

4.3.1.4.1. CO_2 sequestration

It is obvious that the transformation of fossil fuels into hydrogen depends on the CO_2 treatment: if the entire operation consisted only of making successive transformations and then finally releasing the CO_2 produced, the environmental benefit when compared with the benefit of direct use of fossil fuel would be negative or, at best, barely positive.

The idea is to convert fossil fuel in sufficiently large units; these units must be big enough for hydrogen separation and capture to be economically profitable; the hydrogen would then need to be stored, possibly following transportation; the combination of these three steps is what we refer to as sequestration.

4.3.1.4.2. CO_2 captures

For separating CO_2 from synthesis gas there are currently many options which are currently being evaluated, both technically and economically:

– amines or potassium carbonate solvents;

– adsorption on molecular sieves.

The drawback with these two processes is that the products must be retreated in order to extract the CO_2.

They require the use of separate thin membranes which have computerized porosity, and which allow the flow of small molecules of hydrogen, whilst retaining CO_2.

Many types of membrane, e.g. inorganic, metallic, ceramic, can be used, depending on the pressure, temperature, and purity conditions. This method seems to be the most promising.

4.3.1.4.3. CO_2 transportation

CO_2 is a gas where its physical properties make it relatively easy to package and transport. It is a heavy gas which can be easily liquefied at an ambient temperature, thus increasing its commercial possibilities. It can also be compressed for long distance transport in pipelines.

4.3.1.4.4. CO_2 storage

It is often claimed by those who are opposed to sequestration that one of the drawbacks of the system is that it is very difficult to ensure that the containers would remain waterproof for long periods.

In fact, the objective is not to store CO_2 indefinitely, but just long enough for its re-emission, or natural leakage, not to add to the greenhouse effect. This period of time would only be a few centuries, which is, in fact, a relatively short time. To be cautious, storage for thousands rather than hundreds of years is currently being researched. Many sites are theoretically suitable and are, or have been, investigated:

– Deep sea storage: in order to store CO_2, efficiently, it must be in liquid form, and storage must be at a pressure higher than 300 bars, well below 3,000 m, and at a pressure of at least 300 bars. We must then prove that its presence does not affect the sea ecosystem or that an air lift phenomenon does not bring it suddenly to sea level as happened in Lake Nyosin Africa in 1986, causing the death of nearly 1,800 people. For this reason, environmentalists are not enthusiastic about this solution.

– Storage in depleted or underexploited fossil fuel beds: this option is particularly interesting. Quite apart from the fact that carbon would be returned to its natural, waterproof container, its presence would improve production *in situ* by providing pressure when the site is still being exploited.

– Storage in deep salty aquifers: this solution shows the highest potential in terms of storage capacity. The trapping mechanism here is the dissolution of gas in water. The current weakness of such a solution is the lack of information on the behavior of these sites, but the carbonation should not be harmful insofar as this kind of aquifer due of its salinity is of no interest for other applications.

4.3.1.4.5. Social and economic acceptability of CO_2 storage

In the economic field, the cost of storage is not very well known, but generally estimated. It was between €20 and €50 per ton of CO_2 in 2005.

In the social field, demonstration operations are planned and in progress. Whatever the context, whether at national level, e.g. studies by the Institut Français du Petrole (IFP), experiments in the North Sea by Statoil in the Sleipner field, and the current Total project at Lacq, at European level, e.g. the 6th Framework Program project, or at global level, e.g. incentives in the context of new CO_2 quota markets, these investigations should provide answers by 2010–2015.

4.3.2. *Producing hydrogen from water*

Besides the thermal method itself, which requires temperatures of over 3,000 K, and is therefore not very realistic, there are other ways of breaking down water molecules. Three methods are currently being developed: thermochemistry, photoelectrolysis, and electrolysis.

4.3.2.1. *Thermochemistry*

The principle is to break down water at a temperature lower than its direct break down temperature by means of chemical substances which, through a series of combination reactions with water, will lead to hydrogen and oxygen production, while recycling the initial chemical substances.

Thousands of cycles using this principle have been identified for decades. Each has been investigated from the point of view of technology, performance, and safety. As an example, some information is given below on the sulfur cycle, which currently seems to be the most interesting.

This cycle has been retained for many reasons, especially its compatibility with realistic temperature levels. The chemical reaction chain is as follows:

$$I_2 + SO_2 + 2H_2O \rightarrow 2HI + H_2SO_4 \quad (1)$$

$$2HI \rightarrow H_2 + I_2 \quad (2)$$

$$H_2SO_4 \rightarrow SO_3 + H_2O \quad (3)$$

$$SO_3 \rightarrow SO_2 + 1/2\ O_2 \quad (4)$$

The cycle may be explained as follows: water vapor is in contact with SO_2 and iodine; this reaction (1), shown by Bunsen, happens at 450 K. The HI acid produced is distilled (2) around 550 K to generate H_2 (it is the hydrogen source) and iodine which will supply the reaction (1). The sulfuric acid H_2SO_4 of reaction (1) is broken down (3) around 900 K to give water vapor and SO_3; it is then broken down (4) around 1,200 K, in SO_2 necessary to reaction (1) and in oxygen, which is emitted.

By the end, the four reactions have thus generated hydrogen and oxygen and regenerated all the other chemical substances. All these reactions are globally endothermal.

We see that reaction (4) is the most rigorous in terms of temperature: in the order of 1,200 K which may be available, either with a high temperature nuclear reactor, gas (helium) cooled, or at the focus of a solar concentrator.

The USA, Japan, and France are currently working together on this method of hydrogen production, which is coupled to a Generation IV nuclear reactor. The global efficiency of this production method could be around 50%.

4.3.2.2. *Photoelectrolysis*[10]

Photoelectrolysis of water is the process by which solar light is used in order to dissociate the water molecule; its implementation is operated by lighting of a semi-conductor photocatalyst, which supplies electric voltage at a level high enough to break down the water molecule. For this purpose, specific semi-conductors are used, such as TiO_2 and AsGa.

Current research deals with light absorption improvement by different processes, e.g. modification of semi-conductor structure, coupling with photosensitive structures, separation of hydrogen and oxygen production zones, etc., and laboratory results showing efficiencies in the order of 12% are very encouraging.

10. Otherwise called a photo-electrochemical or PEC cell.

However, this process is in competition with an integrated system that directly couples a photocell with an electrolysis cell; the results of this process are currently higher: 16% efficiency obtained by the NREL (National Renewable Energy Laboratory)[11] with a AsGa/GaInP$_2$ photocell coupled to an alkaline electrolysis cell.

4.3.2.3. *Electrolysis*

The breakdown of water by electrolysis is written thus:

$$H_2O \rightarrow H_2 + 1/2O_2$$

with a water dissociation enthalpy of $\Delta H = 285$ kJ/mole.

An electrolysis cell is made of two electrodes separated by an ionic conductor electrolyte (liquid or solid). At the terminal of the electrodes, an electric current is supplied by a direct current generator; this voltage is theoretically equal to 1,481 V at 298 K, but in practice is between 1.7 and 2.1 V. These values correspond to electrolysis efficiencies of 70% to 85%.

Three different processes may used for the electrolysis of water: alkaline electrolysis, PEM-type acid electrolysis, and the so-called high-temperature electrolysis.

Alkaline electrolysis

In this technology, the electrolyte (liquid) transports the OH^- ions. It is made of an aqueous solution of potassium hydroxide, the concentration of which varies according to the temperature, from 25% (in mass) at 358K to 40% at 433K.

This is the most commonly used technology and there are already many suppliers[12] in the market.

PEM-type acid electrolysis

This type of electrolysis, called acid electorlysis, differs from the previous type by a solid electrolyte with polymer membrane, PEM[13] which is a proton[14] conductor.

This type of electrolyzer is used far less frequently than the alkaline electrolysis, but its technology benefits from the many developments currently being made to fuel cells of the same type and could replace alkaline electrolysis in the near future.

---

11. At the University of Colorado, USA.
12. Available at www.afh2.org.
13. PEM: proton exchange membrane.
14. Proton: it is the H+ ion.

Its advantages are its compactness, design simplicity, operation, absence of corrosion (potassium hydroxide is very caustic), absence of the necessary electrolyte liquid circulation, slightly higher performances, and the possibility of operating this kind of electrolyzer under high pressure (many hundreds of bars) which simplifies the problem of hydrogen compression for its storage.

Its drawbacks are the high cost of the electrolyte membrane and the need for expensive catalysts for the electrodes.

Its technology is well established; UK and US nuclear submarines have been using such equipment to supply onboard oxygen for decades, and many companies sell this process, in particular the Japanese Mitsubishi Corp. and Proton Energy Systems Inc.

High-temperature electrolysis (HTE)

This electrolysis is identified by the use of a solid, ceramic electrolyte, identical to the one operating in SOFC fuel cells whose ionic conductivity (ions O^{2-}) is satisfactory beyond 1,000 K.

Water is converted into steam and the temperature is maintained by an external heat source. The thermal level of the source comes, as it does for the thermochemical production described above, from a high-temperature nuclear reactor (Generation IV HTR (high-temperature reactor)) or from a solar concentrator. The electric energy comes from a thermodynamic cycle (Brayton cycle gas turbine), supplied by the same source.

This technology has been at the research stage since 1975 in many laboratories, such as CEA (Commissariat à l'Energie Atomique) in France, and is profiting from the developments in progress on SOFC-type fuel cells.

Its efficiency is high (about 80%) because of the temperature, which, when coupled with a nuclear reactor, could achieve a global efficiency of the conversion chain of about 50%, as for thermochemistry.

4.4. Hydrogen transportation

The transportation of hydrogen from its place of manufacture to its destination is an important element of the hydrogen chain.

As hydrogen is manageable in its two forms, liquid and gas, two transportation methods are possible.

4.4.1. *Transportation in liquid form*

To be transported in liquid form, it must first be brought to a temperature of 20.3 K. At this temperature, its volume density is 71.1 kg/m^3. There are many ways to carry out this operation, but they all use a significant amount of energy. The energy used for liquefaction is about 3.9 kWh/kg for hydrogen under atmospheric pressure and the efficiency of the liquefaction operation depends on various factors, such as the liquifactor size. Thus, the efficiency varies from 10% for the small units to about 50% for equipment of many tens of T per day.

Once liquefied, hydrogen has the advantage of a volume density high enough[15] to be transported by road or sea. Thus, many tankers travel the highways with a load of 3.5 T of liquid hydrogen for a total weight of 40 T. Many projects are planned in which tanker ships will transport 14,000 T of hydrogen in spherical tanks for 6,000 nautical miles; these ships will be powered by the evaporation losses of the liquid hydrogen.

4.4.2. *Transportation in gas form*

There are two ways to transport hydrogen in its gas form, either in bottles or in pipelines; these two methods are not new and their technology is well established.

In the first case, gas suppliers usually transport hydrogen in metal cylinders, at high pressure up to 200 bars. The drawback is that the cylinders are very heavy.

Transportation by pipeline has been in operation since 1938 in the chemical field of the Ruhr. Today, Western Europe alone owns more than 1,500 km of this type of pipeline, many used by Air Liquide. The pipelines have a diameter of between 10 and 300 mm, and operate under pressure of between 3.4 and 100 bars. The pipelines are made of standard steel and have not, to date, suffered any problems. The energy used by the high-pressure pumps, placed along the network to guarantee a constant flow, is 1.4% of the energy power of the transported hydrogen, on 150 km. It is the cheapest of the transportation methods and it will probably be used more extensively in the next decades.

In addition to the transport of pure hydrogen described above, we must consider, as a transition phase which may last quite a long time, the transportation of hydrogen mixed with natural gas. Indeed, if we take into account the two following points, we can see how easy this would be:

---

15. From 70 to 79 kg/m^3.

– the excellent chemical compatibility between the natural gas (methane) and hydrogen, which are miscible;

– the existing and well-developed natural gas network, e.g. more than 30,000 km in France.

We can imagine the interest of this solution:

– in each place, the manufactured hydrogen, particularly that of renewable origin through wind or photovoltaic stations, could be poured into the natural gas network, without additional infrastructure cost for transportation;

– at each point on the grid, the hydrogen could either be used directly in its mixed form without any problems for the majority of users, or recovered by filtering the solution, for specific uses or to supply fuel cells.

4.5. Distribution

When hydrogen is used for applications other than chemical, the problem of its distribution takes on a greater importance. Because of their nature, distribution methods can be split into two categories: road transport and others. The second category includes domestic users. We tend to think of domestic users being supplied via a network, as is the case with natural gas today. We are aware of the difficulties with the distribution of liquid hydrogen or bottled gas to individuals. The only new advance is that of service stations supplying hydrogen to certain types of vehicles.

Due to the technology of the vehicles manufactured and currently being tested, two forms of hydrogen must be considered: liquid hydrogen used mainly by BMW and General Motors for HydroGen 3 prototypes, and pressurized hydrogen, more commonly at 350 bars, or at 700 bars, used by the majority of the more recent car prototypes.

4.5.1. *Distribution of liquid hydrogen*

Handling liquid hydrogen is difficult because of its very low temperature (-253°C) and of the flammability of its vapor. However, its handling is well controlled, due, in particular, to its development in the space industry, but it needs to be adapted for the automobile sector.

In principle, a liquid hydrogen distribution station is a large liquid hydrogen tank from which the required quantities are extracted by means of a slight overpressure. In order to reduce transfer times and to limit parasitic consumption of liquid hydrogen, the tank which is to be filled must not be at ambient temperature This, in effect, means that the tank should not be completely emptied, and must be cooled to the required temperature before filling.

4.5.2. *Distribution of hydrogen*

Because of the technology under progress, the gaseous transfer is made either about 350 bars (35 MPa), or around 700 bars (70 MPa). This technology of hydrogen under pressure is perfectly controlled by the industry of the compressed gases up to about 250 bars and has needed many developments to be adapted to higher pressures.

The transfer is made via a reserve gas bypass at a pressure about ten bars higher than the pressure in the tank. The bypass tank is supplied through a compressor which regulates the bypass pressure and is in turn supplied by another tank under an intermediate pressure.

The remaining problem is the transfer time: the increase in pressure in the tank to be filled causes a temperature increase which can adversely affect the strength of the tank, which has a composite structure, if not permanently controlled. Cooling due to reduction of pressure in the filling nozzle unfortunately does not compensate for the rise in temperature.

During the filling process the internal temperature of the tank is constantly monitored, which directly affects the speed of the gas flow. In this way a full load of 5 kg of hydrogen can be transferred in less than five minutes.

At the end of 2007, more than 160 of these service stations were in existence across the world – a big enough sample to give an accurate idea of how the equipment performs.

Many countries have started to develop hydrogen highway service station equipment: Norway, Canada, Germany, and the USA with California's equipment project "California Hydrogen Highway" which started in 2004 and planned 100 stations by 2010.

Figure 4.1. *Total service station, built in Berlin (2006)*

Figure 4.2. *Air Liquid service station used at the European Hydrogen Energy Conference (EHEC) 2005 (Saragosse)*

4.6. Hydrogen storage

There are many different types of storage tanks, depending on requirements; large fixed tanks, small fixed tanks, transportation tanks, mobile tanks for vehicles, and tanks for portable equipment.

4.6.1. *Large fixed tanks*

The existing technologies are mainly adapted from those which have long been used for storing natural gas, but they are technically more difficult and more expensive because of the small molar mass of hydrogen. Two types have already been implemented.

4.6.1.1. *Underground storage of pressurized hydrogen gas.*

This is possible in many types of cavities, such as aquifer layers, salty cavities (some of which are more than 300,000 m^3), and old natural gas caverns. For instance, the natural cavity under the city of Kiev was used to store the city's gas with about 65% of hydrogen, until the end of the last century.

4.6.1.2. *Storage in cryogenic form*

This is mainly used in the space industry, which is why NASA has equipped its Cape Kennedy site with four 3,800 m^3 tanks and ESA (European Space Agency) has four 360 m^3 tanks at Kourou. The evaporating hydrogen losses in this type of tank are about 0.1% of mass per day.

4.6.2. *Small fixed tanks*

This type of tank is mainly used in distribution sites, such as service stations, and by the end user. The pressurized gas and cryogenic technology used for this type of tank is well established. The tanks are available from many suppliers.

4.6.2.1. *Storage systems for transporting gas*

These storage systems are used to transport liquid or hydrogen gas from a place of production to a user.

They are spherical tanks (cryogenic hydrogen) or cylinder-spherical tanks transported on barges or trucks.

The metallic bottles adapted to the transportation of hydrogen gas are generally pumped up to 200 bars and enable the trucks to transport from 2,400 to 3,600 Nm^3[16].

16. Nm^3: measuring unit used for gases. It is the volume used in normal conditions, namely 0°C and pressure of 760 mmHg or 14,503 PSI.

Figure 4.3. *Storage bottles for road transportation*

4.6.2.1.1. Mobile tanks for space, air, sea, and land vehicles

Land vehicles currently using hydrogen are mainly light vehicles. The majority of current prototypes use fuel cells, but a growing proportion have a thermal engine. These vehicles load hydrogen (about 5 kg) either in liquid form (cryogenic) or in its pressurized, gaseous form.[17]

Figure 4.4. *Liquid hydrogen storage in a BMW vehicle*

17. At the end on the 1990s, some fuel cell vehicle prototypes, mainly from Japan, were made with hydride storage, but this solution seems to have been abandoned for vehicles because of the mass density (from 1.5 to 2%), which is at too low a temperature to be easy to control (150°C would be the maximum).

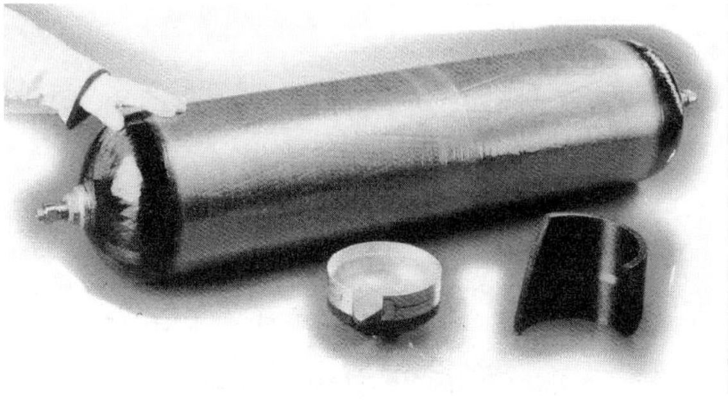

Figure 4.5. *Composite tank storing 700 bars*

In the first case, the tank is quite bulky and takes up, in current prototypes, part of the trunk (as in Figure 4.4). Between 1 and 2% of mass is lost per day due thermal losses, consequently the tank will not remain full for more than a week or two, depending on the quality of the insulation.

In Figure 4.5 the cylinder's spherical form enables better integration into the vehicle: a set of three or four bottles is generally housed in the chassis.

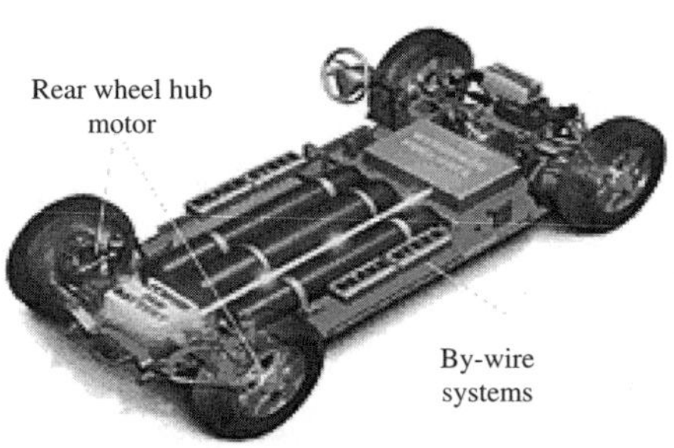

Figure 4.6. *700 bar hydrogen bottles integrated into the chassis of the Sequel from General Motors*

These bottles are of a composite structure: a waterproof casing of aluminum or thermoplastic[18] alloy, the mechanical behavior of which is assured by a carbon fiber coil, often with an additional glass fiber coil, which is cheaper but does not perform to the same level. This technology is very light and it can reach a mass density of 7–

18. As the high-density polyethylene.

8% in commercialized structures.[19] The Atomic Energy Commission, associated with the French SME Ullit, gave the first demonstrations at 700 bars[20] at the beginning of the 1990s. Since then, other companies have also started to develop the technology.[21]

Submarines and surface ships are gradually being equipped with fuel cells that are usually supplied with pressurized hydrogen for surface ships and with hydrogen stored in hydrides for submarines, whether they are military or civilian. This is why the Class 212 military submarines built by the HDW German shipyards are equipped with a 300 kW fuel cell supplied by hydrogen stored in a FeTi alloy.[22] The tanks weigh 160 T for an on board supply of 20,000 Nm^3, which enables many weeks of complete immersion.

Aircraft vehicles that use hydrogen are still at the prototype stage, but may be developed further. Demonstrators have been built for three categories:

– Surveillance drones with electric motors supplied either by photocells during the day or with the hydrogen generated during the day by electrolysis. The hydrogen is then stored in a bottle under medium pressure.

– Jumbo jets, Russian and European projects (such as the Cryoplane project, on an Airbus frame), which consist of loading liquid hydrogen to supply the reactors.

– Space rockets with on-board liquid hydrogen, either for propulsion (all the present day rockets have a cryogenic stage), or to supply fuel cells, as in the Apollo projects and space shuttles.

4.6.2.1.2. Storage for portable applications

Portable applications for hydrogen are directly associated with fuel cells and the stored energies must be higher than those obtained with the best lithium ion rechargeable batteries, i.e. around 150 Wh/kg, by a factor of at least three to four.

Many of the companies that began to research these technologies do not seem to have made their final choice of the many solutions[23] proposed in 2006 in the three areas of portable equipment: mobile phone supply or PDA (range 100 mW), portable computers and cameras (range 20 W), and motor generator sets (range 200 W to 1 kW):

– small bottles filled with pressurized hydrogen;

---

19. The mass density performances have reached 13% on laboratory prototypes.
20. With a safety coefficient of 2.5.
21. Mainly the Canadian Dynetek and the American Quantum technologies.
22. This alloy, based on FeTi hydride, has a storage mass density approaching 2%.
23. In addition to the methanol solution retained by some industries.

– small bottles filled with a saturated alcohol;

– bottles filled with sodium borohydride, which breaks up in the presence of water, and a catalyst[24] to give hydrogen.

4.7. Applications of hydrogen as energy vector

Beyond the field of chemistry, there are many possible uses for hydrogen, most of them already being developed and nearing commercialization in 2006. They fall into three major categories: mobile, transportation and stationary. These applications are, in the main, associated with the fuel cell, the principles, operation, and performance of which are detailed in Chapter 5.

4.7.1. *Mobile*

In this category, hydrogen acts as fuel for the fuel cell which will replace the present electrochemical battery. Its performance is limited both by stored energy (around 130 to 150 Wh/kg for lithium batteries) and by charging time. The category of mobiles is also divided into three groups according to the required electrical energy:

– Mobile phone, PDA, MP3 player: because of the power requirement, the energy source for these devices is a hybrid combination of a battery (generally lithium) and its charger which is a fuel cell (with its hydrogen reserve), integrated or pluggable, depending on the model.[25] The power required is in the range 100–1,000 mW for a few days' running time, which represents energy of a few tens of watt-hours. The use of such a hybrid source would increase the running time of these devices by a factor of three to five and would also enable almost instant recharging.

Many storage methods are used for hydrogen, and they have been described. This is an important market, and many companies,[26] most of which are Japanese, have developed this technology and exhibited prototypes, the performance of which

24. Solution industrially developed by the American Millennium Cell company.

25. The fuel cells that equip portables are either directly supplied with hydrogen (PEM-type cell) or supplied with methanol (DMFC (direct methanol fuel cell)-type cell): in 2006 both technologies shared this market on a 50:50 basis.

26. In 2006, the companies concerned were the following: Toshiba, KDDI, Hitachi, Seiko, NTT DoCoMo, Casio, Cosel/Saito, Sony (all from Japan), CEA (France), Angstrom (Canada), and Medis Tech (US/Israel).

is extremely impressive. Many should be available on the market by 2008. This is probably the first mass market for hydrogen energy and fuel cells.

Figure 4.7. *Mobile phone and its Medis Tech*[27] *charger (2006)*

– Portable computers, cameras and camcorders, military equipment: the hybrid principle is the same as described above, but the power requirements are 10 to 50 times higher. This market attracts different companies, but the numbers involved are about the same as for the previous category,[28] and, as before, the majority are Japanese. The first items appeared on the market in 2008.

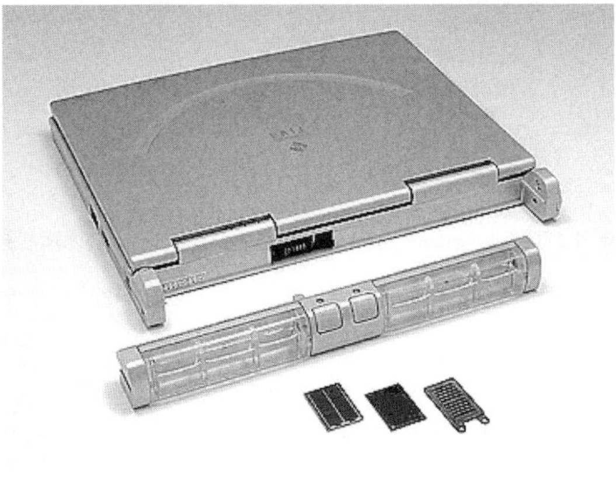

Figure 4.8. *Casio portable computer, with a fuel cell connection*

27. Medis Tech, an American and Israeli company.

28. Sanyo, NEC, Casio, Matsushita, and Canon from Japan; MTI and SandPiper Tech from the USA; LG Chemical and Samsung from Korea; and Paxitech from France. Different to the previous field, 40% have chosen hydrogen storage, the others methanol storage.

– Portable energy sources: this range has a requirement for hundreds to thousands of Watts. Some companies favor the use of methanol cells whilst others prefer PEM cells supplied by hydrogen either under pressure or in hydride tanks. Commercial production has been started by the British Voller Energy Ltd (hydrogen stored in a hydride), the Canadian Coleman Powermate (AirGen^{TM}), as well as the French Axane (a subsidiary of Air Liquide) with pressurized hydrogen supply.

Figure 4.9. *2 kWe transportable generator from Axane Roller Pac (2005)*

4.7.2. *Transportation modes and means*

There are many uses for hydrogen in transportation. Some have already been implemented in the areas of road transportation, sea and underwater transportation, and air transportation.

4.7.2.1. *Road transportation*

This is the flagship application for hydrogen energy, the most difficult to make economically viable, and the most ambitious in the environmental field bearing in mind its potential long-term impact.

Almost all of the automobile manufacturers in the world have tackled this challenge: more than 600 demonstration models made by 23 car manufacturers were running by at the end of 2005, 11 years after the first DaimlerChrysler (Necar 1) prototype in 1994.

It would take too long to detail all the programs in progress,[29] but the two main types from the years 2005–2006 are briefly outlined.

– Hydrogen and thermal engine vehicles: this type of vehicle is becoming increasingly attractive to many car manufacturers because the deadlines for converting to fuel cell vehicles are delayed from year to year and are now set at 2020. Of course, the performance of this kind of vehicle, operating with only a simple thermal engine (i.e. those developed by BMW, Mitsubishi and Mazda) or with a rotating engine (i.e. RX-8 and Ford) is not brilliant, but in a thermoelectric hybrid concept, their performance is quite interesting. For this reason the State of California has asked Toyota to make modifications to their Prius so that it can be adapted for hydrogen. A few dozen where running on Californian highways in January 2006 (see Figure 4.10).

Figure 4.10. *The hydrogen Toyota Prius (2006)*

– Hydrogen and fuel cell vehicles: this group has the largest number of prototypes. In 2006, it included: hydrogen storage at 350 or 700 bars in composite tanks housed in the vehicle chassis, PEM-type fuel cells of nearly 70 kWe, hybrid structure with energy storage in a battery (NiMH or Li-ion) or a supercapacitor, and a central electric motor or wheel motors.

29. All these programs are listed on the website site of the Association Française de l'Hydrogène, at www.afh2.org.

The same range is found in fuel cell buses, about 100 of which were circulating on the roads in 2006.

In addition, there has been active development of engines[30] for two-wheeled vehicles, i.e. motorbikes, scooters, motorcycles, (see Figure 4.11). They carry a bottle of pressurized hydrogen coupled to a fuel cell which supplies an electric motor.

Figure 4.11. *Masterflex hydrogen fuel cell motor cycle with a 120 km range (Germany, 2005)*

Within this category, progress is also being made with the development of the APU (auxiliary power unit) which would be useful for the provision of on-board power for air conditioning and refrigeration in vehicles.

4.7.2.2. *Sea and underwater transportation*

Numerous demonstrations have shown the desirability of a clean and silent energy source for commercial craft such as river taxis and cruiser ships equipped with a fuel cell supplied by hydrogen stored in a pressurized bottle.

For underwater engines, the Japanese have developed the Urashima 600 for civilian use, and the German shipyards HDW for military use. Since 2005, many foreign navies have been equipped with these duel-fuel (diesel and fuel cell) engines. The fuel cell is supplied with hydrogen which is stored in hydrid tanks placed outside the hull (see Figure 4.12).

---

30. The major manufacturers who develop engines for two-wheel vehicles are Piaggio, Bombardier, Honda, Intelligent Energy, Masterflex, Palcan, Vectrix and Yamaha.

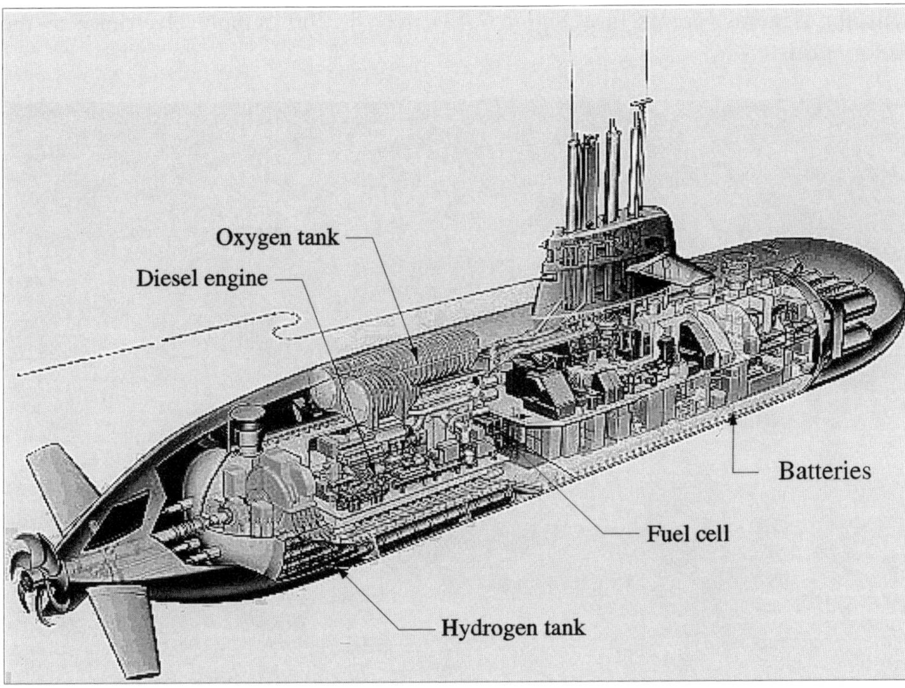

Figure 4.12. *Underwater engine U212 with fuel cell, operating since 2005*

4.7.2.3. *Air and space transportation*

Hydrogen has been used in this area for many years, particularly for rocket propulsion. Advanced projects for the propulsion of drones and commercial aircraft, e.g. the European project Cryoplane, driven by Airbus, have demonstrated the feasibility of an air carrier equipped with liquid hydrogen tanks.

4.7.3. *Stationary applications*

Although fuel cells are being used for more applications in this area, few of those in operation in 2006 were using hydrogen as their primary energy source. Almost all were using traditional fossil fuels, in particular natural gas and kerosene, to supply a reformer in front of the fuel cell. This is only because hydrogen distribution is not yet available.

However, one example that can be given and which should be developed is that of America's Dow chemical factory which uses its hydrogen leakages to supply fuel cells for General Motors (see Figure 4.13), who in turn supply electricity to the factory grid.

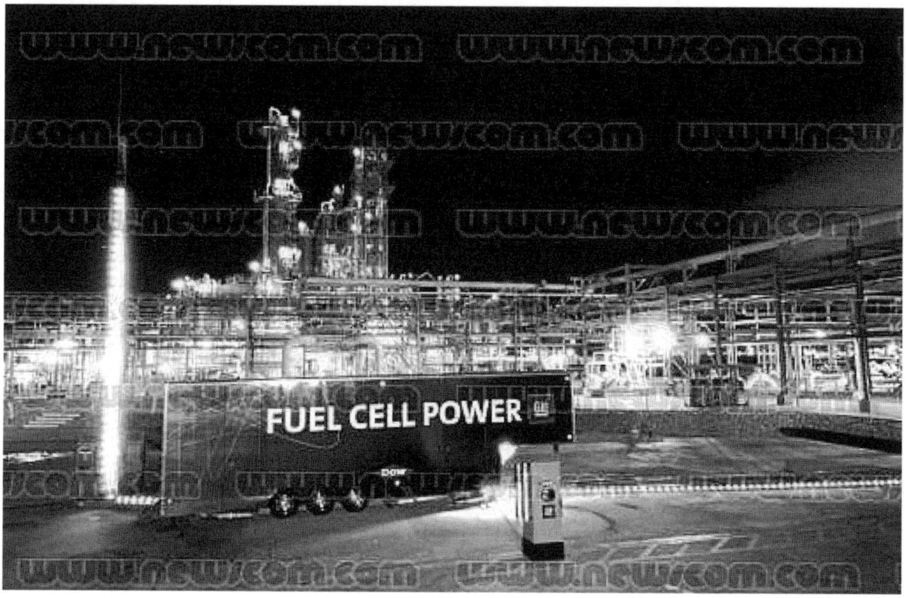

Figure 4.13. *General Motors cell supplied by hydrogen leakages in a Dow factory (2005)*

4.8. Risks, standards, regulation and acceptability

4.8.1. *Flammability and explosiveness*

The hydrogen properties given at the beginning of this chapter show that this gas is highly flammable. It must therefore be used cautiously.

In order for combustion to take place, two conditions must be satisfied: the presence of a combustive (oxygen in the majority of situations) and inflammation energy in the immediate vicinity (spark, heat source).

This combustion may turn into an explosion if it happens in a closed environment. This means that there must be no risk of hydrogen leakage in a confined space. This is a most important precaution.

The level of explosion will mainly depend on the hydrogen concentration in the mass of the enclosed air.

The deflagration occurs with hydrogen concentrations between 4% and 18%, and detonation will be reached in the range of 18% to 59%; beyond this concentration, we are back to a deflagration regime. Depending on local conditions, the explosion may turn from deflagration to detonation.

4.8.1.1. *Safety, transport and storage*

Although hydrogen has been used for many years in the chemical industry and storage and distribution techniques are today well controlled, it is important to consider the risks of hydrogen leakage because of the highly flammable and potentially explosive properties of this gas.

Hydrogen molecules are very small and, therefore, have a tendency to escape. Leakage problems may occur at any point in a system using this gas. In an open-space environment, hydrogen disperses immediately and does not reach a concentration which could lead to an explosion, except in cases of very large leakages which are well beyond most usual situations. Normally the question of combustion is a theoretical concern for the purposes of a security analysis.

This dispersion ability is such that an investigation has shown that in a garage, which is, after all, a closed space, a few small ventilation holes at regular intervals are enough to prevent the development of a hydrogen cloud which would be an explosion risk.

In the same field, many investigations and experimental evaluations of a storage tank at 700 bars have been adapted to automobile application, and have shown that the security levels are similar, or even better, than those of standard gas tanks. However, many investigations are still in progress around the world on the dispersion phenomenon and apparatus adapted for better detection and ventilation in order to improve even further the security of hydrogen use.

4.8.1.2. *Standards and regulation*

Specific regulations for and standardization of the use of hydrogen are being prepared at the international level.

4.8.2. *Standardization*

An ISO (International Organization for Standardization) technical committee, TC 197 (hydrogen technologies) was created in 1990 in order to write the standards in the field of systems and devices for the distribution, the storage, the transportation, the measurement, and the use of hydrogen. This committee has seven working groups.

Another technical committee, IEC (International Electrotechnical Commission) TC 105 (fuel cell technologies), was created in 1996 in order to develop standards in the field of fuel cell technologies for mobile and stationary applications. The committee has six working groups.

4.8.3. *Regulation*

At the regulation level, stationary applications, mobile applications, and portable applications are differentiated between:

– Stationary applications: guidelines exist at European level, they have been incorporated into French law and concern:

- worker protection,
- equipment protection, and
- environmental and population protection

(the references and details of these guidelines are available on of the Hydrogen Memo [AFH 3w]).

– Mobile applications: the use of hydrogen in these applications require the establishment of specific regulations in order to guarantee a high level of security. The European project EIHP (European Integrated Hydrogen Project), which ended in 2004, proposed European regulation on hydrogen vehicles and on the distribution infrastructure to replace the various different national regulations. This is currently being promoted by the European Commission.

– Portable applications: at the moment there are many European guidelines that are partly applicable. But it is obvious that, as the use of fuel cells and hydrogen becomes more widespread in the future, standardized regulations will have to be adopted. The Americans and the Japanese, as the main players in this area, are doing their best to implement such regulations, which should be up and running in approximately 2010. France is not very active in this area.

4.8.4. *Hydrogen acceptability*

Existing and successful technology, and standards and regulations dedicated to this technology, are not enough if the public does not accept it. Before making the technology widely available we must:

– identify the criteria that define the social acceptability, and pay particular attention to the perceived risks;

– study the potential users' perception of the technology in order to educate them as to its potential uses; and

– develop the information on this technology.

There are already some public projects in progress in France, which will promote social acceptability of the technology. Bearing in mind the results of the initiatives already implemented in other countries that are more advanced countries than France in relation to this technology, such as Germany, it is realistic to expect that the discussion on acceptability should not raise any particular problem in the near future. However we must remember that hydrogen has always suffered, from a heavy handicap: the attitude of the public being that hydrogen is dangerous because it explodes. This view results from some misunderstood events, the most well-known ones being:

– The H-bomb: this comparison is obviously not serious, but, it does exist.

– The explosion of the space shuttle Challenger in 1986: whilst it is true that there was liquid hydrogen on board, it was a very large quantity and does not compare with the quantities generally used, and it was in the presence of liquid oxygen which would only occur in the situation of a space shuttle.

– The 1937 accident of the German airship Hindenburg at Lakehurst which killed 34 passengers out of 97. This catastrophe[31] was finally explained following a long investigation by American investigators. They demonstrated that it was the airship's waterproof varnish that ignited and that hydrogen did not have anything to do with the accident. It should also be noted that the balloon was carrying 200,000 m^3 of hydrogen and, in spite of the fire, no explosion occurred.

The French public seems to have forgotten that gas distributed in French cities until the 1970s was 50% hydrogen and 50% CO. Accidents were more often due to CO than to explosions.

4.9. A hydrogen economy

In view of the number of industrialists investing in the field of hydrogen and fuel cells, we can think in terms of the birth of hydrogen economy. Of course, it was not yet ready to fly on its own at the beginning of 2006, but it should take off very soon.

31. http://www.hydrogenus.com/advocate/ad22zepp.htm.

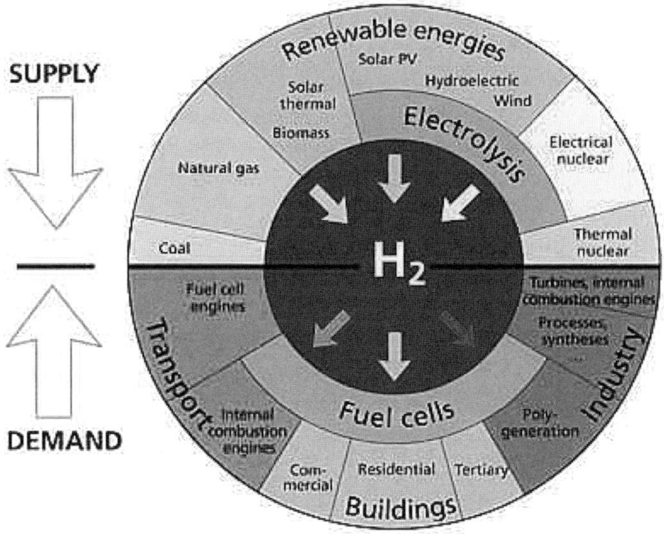

Figure 4.14. *Supply and demand in the hydrogen field (European Commission documents)*

An interesting indicator of the growing trend is the number of accessible internet sites, more than 300 worldwide, dedicated to commercial companies which have invested in hydrogen and fuel cells.

These companies are involved in all areas associated with the production and use of hydrogen as described in Figure 4.14, above, distributed by the European Commission.

It is generally estimated that the total global investment, public and private, in hydrogen and fuel cells[32] by 2005 was in the order of one billion US dollars per year.

In the case of France, which represents about 10% of this effort,[33] almost all of the public and private institutions involved in the area of energy (see section 4.9.1) are focusing on hydrogen.

---

32. It is extremely difficult to separate hydrogen and fuel cell investment because of the imbrication of the works and of the grouping of the two expenses in public budgets.
33. *Hydrogen and Fuel Cells in France*, EHEC website 2005.

These players are absolutely confident of hydrogen's emergence as the dominant fuel source. Industrialists in particular are faced with the challenge of a changing future. All are pioneers who are not sure that they will find El Dorado, but they dare not believe that it does not exist.

Some of the more confident institutions, such as fuel cell manufacturers (Ballard), automobile manufacturers (in particular, General Motors, DaimlerChrysler, Toyota, Honda), and computer and mobile phone manufacturers (Toshiba, Seiko, Sony, MTI, Samsung, NEC, etc.) risked investing significant amounts, whilst some others, more conservative or more fainthearted, felt it was sufficient to invest a small amount just to see, just as in a game of poker. Time will tell which approach was right.

Meanwhile, the public authorities worry about the future of hydrogen in the decades to come, from its manufacture to its use.

Most of the industrialized countries have worked on hydrogen technologies and, in general, have drawn up roadmaps until around 2050. Europe initiated this exercise in the framework of the 6th framework program through the HyWays project and gave its conclusions in 2007; the French part of this project, called HyFrance, gave its conclusions at the beginning of 2008.

4.9.1. *The cost of hydrogen*

Hydrogen is an industrial product which can be bought from many suppliers. Its price depends on the price of the raw material (the price of natural gas can be rather volatile), the quantity required, and the desired form (pressurized or cryogenic).

The following table gives orders of magnitude, in € per kilogram and per kWh, as at 2005, as compared to the selling price, excluding the French tax on oil products (TIPP), of gas from the refinery and at the pump, and the gross price of natural gas.

The last row of the table gives a price for hydrogen energy, which would be sold today at a service station pump, under 350 bars, and made from natural gas with capture of CO_2: this last figure comes from a technical economics study made by the AFH_2 [AFH 3w] in 2004.

We can see that the tax-free price would be equivalent to the present day price of gas at the station – thus at a competitive price – if the authorities would temporarily agree, not to charge TIPP for a period whilst this fuel is being launched.

Product	€/kg	€/kWh
Gasoline out of refinery (0.38 €/liter)	1.4	4.1
Gasoline at station pump (1.2 €/liter)	4.4	12.9
Natural gas (in bulk at US$10/million BTU)	1.2	3.7
Sale prices H_2 100 bars, in bulk	1.5	4.7
Sale prices H_2 100 bars, retail	4.6	15.7
Sale prices H_2 liquid, retail	8.3	25
Sale prices H_2 at the pump (excluding TIPP)	4.1	12.4

Table 4.1. *Hydrogen costs*

4.10. The hydrogen players

It would not be possible to list in this book all the parties involved in the development of hydrogen, from the fields of research, industry, and communication; however, amongst the major players, the following should be mentioned.

4.10.1. *The public players*

The public players are government structures that are in charge of the politics and financing of the research and development operations surrounding hydrogen.

In effect, all industrialized countries have included hydrogen technologies, generally coupled to those of fuel cells, in their research and development programs: for example, the DOE (Department of Energy) in the USA, the METI (Ministry of Economy, Trade and Industries) in Japan, the MNES (Ministry of Non-conventional Energy Sources) in India, MST (Ministry of Science and Technology) in China, and the MFE (Ministry of Finance and Economy) in South Korea.

Europe, through the European Commission, also manages hydrogen programs, at the research level (Community Research[34]) and also at the application level.

34. Via the SP1 – Priority 6–1, sustainable energy systems.

In France, most of the budgets for hydrogen are managed by two new organizations created in 2005:

– the ANR[35] (National Research Agency) which implemented a specific hydrogen program named PAN-H (National Action Plan on Hydrogen); and

– the AII (Industrial Innovation Agency).

These are the main organizations to apply to for public funding; these agencies work by means of the submission of proposals. To these we can add other organizations which also participate in the financing of hydrogen projects: regional councils, general councils, Ademe, Predit (Research and Innovation Program for Land Transportation), Oseo Anvar, etc.

4.10.2. *The main players in the research field*

There is no global research center which does not have teams working on hydrogen and fuel cells.

In France, the most important contributors are:

– the CEA (Atomic Energy Commission) which dedicates significant resources (more than 200 people) to hydrogen, its production (nuclear and biomass), its storage (high pressure), its security, and its conversion via fuel cells (PEMFC (polymer exchange membrane fuel cell) and SOFC);

– the CNRS [CNRS 3w] and the universities with about 30 laboratories in the areas of production, storage, fuel cells, materials, energy policy, socio-economics;

– EDF on fuel cells;

– Gaz de France on hydrogen and fuel cells;

– the IFP (French Petroleum Institute) for production, particularly in the area of fossil production with CO_2 sequestration, and conversion with thermal engines;

– higher education engineering schools: Mines of Paris, Sup-Elec, Grenoble INP, etc.;

– Ineris[37] for the study of risks, standards and regulation.

---

35. www.agence-nationale-recherche.fr/.
37. National Institute of the Industrial Environment and the Risks.

4.10.3. *The industrial actors*

As stated above, unlike hydrogen chemistry, it is not yet possible to provide hydrogen energy profitably.

This is why most of the industrial participators are big companies that are prepared to invest in the technology with no prospect of making a profit for some years.

This is also true in the case of big gas producers, such as Air Liquide and Linde, oil companies, such as Shell, BP, Esso, Total, etc., automobile manufacturers, and companies associated with energy, such as EDF, Gaz de France, and Areva in France.

In France, these big companies either have dedicated teams, as in the case of EDF, Gaz de France, Total, PSA, Renault, Snecma, etc., or they have started up subsidiaries, for example Helion from Areva, and Axane from Air Liquide, both of which are dedicated to hydrogen fuel cells).

Alongside these big groups, some start-ups or small and medium enterprises, bravely launched themselves into the adventure: PaxiTech in mini fuel cells; Ceth in the production of ultra-pure hydrogen; N-GHY in the reforming production; Ullit in high-pressure storage composite tanks; Alca-Torda in marine applications. Alphea Hydrogen, a competence center of excellence on hydrogen, which carries out investigations on behalf of its members can also be included in the list.

4.10.4. *Participators in communication: associations, periodicals, books, conferences, exhibitions*

The development of hydrogen energy has seen the emergence of many communication coordinators, some of which are identified in the following sections:

– the national associations in most European countries, including France, belong to a European association, the EHA (European Hydrogen Association);

– periodicals on hydrogen and the many issues in relation to fuel cells, and which also deal with hydrogen;

– websites;

– books;

– many conferences on the subject, such as the EHEC and the WHEC (World Hydrogen Energy Conference) which take place alternately every two years;

– periodic exhibitions, the most important of which in Europe is the Hanover Fair, which generally is in April each year. In France, the Pollutec symposium (alternately in Paris and Lyon) also deals with the subject, and has many stands and presentations;

– hydrogen and fuel cell demonstration kit manufacturers, the most famous of which is the German company h-tec.

4.11. Conclusions and perspectives

Global warming, with its inevitable damage to the flora and fauna and to mankind, is starting to be taken seriously by the governments of all industrialized countries.

Various financial measures, such as the introduction of quotas, markets for CO_2, high rates of repurchase of renewable energies, etc., and regulations, such as the Kyoto agreements, and other more restrictive agreements, such as those proposed by Europe, and France in particular, are heading in the right direction but are by no means sufficient.

The best solution to completely eliminate climate change would be to rapidly go from our world energy system, based on 85% of energy from fossil fuels to alternative energies, in particular, renewables. Unfortunately, this solution is a medium-term utopia, i.e. before the next century.

Global warming, which is the direct consequence of using fossil fuels, would reach unacceptable levels if we were content just to wait for it to happen. Moreover, technical and economical realities oblige us to continue using fossil fuels in the twenty-first century. How can we resolve this dilemma? It is possible to do so by:

– implementing more quickly the use of non-CO_2-emitting energies;

– continuing with the use of fossil energies, but converting them at the source and confining the CO_2;

– decreasing our energy consumption by increasing energy efficiencies and being more aware of waste and thermal emissions.

These three routes underline the necessity of developing a new energy vector which does not contain carbon, unless it comes from biomass, transportable and easy to store; hydrogen obviously meets these requirements. Moreover, it is the perfect energy vector.

Being the best possible energy choice does not mean that hydrogen is without problems; however, recent improvements have shown that there are solutions to its disadvantages that are economically and sociably acceptable.

The main worldwide players in the energy field acknowledge that hydrogen will have a brilliant future even if, although they point out the various obstacles which must be overcome. They have thus created a new technical-economic concept: the hydrogen economy.

In this chapter we have described the huge improvements which have taken place over the last 20 years enabling the emergence of a hydrogen economy in many fields, particularly the field of portable equipment, which will use hydrogen before the end of this decade, and the field of land transport, due, first, to hydrogen-fuelled thermal motors and, then, to fuel cells.

Despite improvements, the protection of our environment is not ensured. We can henceforth assume one thing: hydrogen will one day be part of our environment; the only, and important question remains: when?

Speed is vital and only politicians at the world level can promote it by a voluntarism which must lead to new financial and regulation incentives and measures.

4.12. Bibliography

There are hundreds of works of reference material accessible on the Internet, from traditional research engines, and available on the AFH_2 site. Below is a list of some of these reference materials.

Books

[AFH 06] AFH_2, *L'hydrogène*, Editions John Libbey, Eurotext, 2006.

[BOU 06] S. BOUCHER, *La révolution de l'hydrogène*, Editions le Félin, Paris, 2006.

[RIF 02] J. RIFKIN, *L'économie hydrogène*, Editions de la Découverte, 2002.

Internet sites

[AFH] Association française de l'hydrogène, "Veille technologique et memento de l'hydrogène", www.afh2.org/

[ALP] Association Alphea, www.alphea.com/

[EHEC] European Hydrogen Energy Conference, www.ehec.info/

[FCP] "The latest fuel cell news in Japan" (in English), www.fcpat-japan.com/

[H2F] "The Hydrogen and Fuel Cell Investor" (in English), www.h2fc.com/news.html

[HFC] "Hydrogen & Fuel Cell Letter", www.hfcletter.com/

[HYW] *HYweb* (in English), www.hyweb.de/english/

Events

[HAN] Foire de Hanovre, www.fair-pr.com/

[POL] Salon Pollutec, www.pollutec.com/

Manufacturers

Kit manufacturers

[CET] CETh (French company), www.ceth.fr/

[HTE] h-tec (Germany), www.h-tec.com/

French industrials (main activity associated with hydrogen and/or fuel cells)

[AXA] Axane, www.axanefuelcell.com/

[CET] CETh, www.ceth.fr/

[HEL] Hélion, www.helion-fuelcells.com/

[NGH] N-GHY, www.n-ghy.com/

[PAX] PaxiTech (French company), www.paxitech.com/

Chapter 5

Fuel Cells

5.1. Introduction

The fuel cell, or more precisely, fuel cells have interested those involved in energy technology for a decade. The constant increase in the price of so-called fossil energy (oil, natural gas), the legal reduction of the greenhouse-gas emissions, and the growing dominance of electricity as an energy vector constraint, have led these actors to research new energy chains. Those chains combine different energy sources (solar, wind, geothermal, coal) with better performing energy converters which are better adapted to these sources and to the dominant energy vector: electricity.

The fuel cell, which directly transforms a chemical energy contained in a fuel into electric energy, belongs to this challenge.

The main application of fuel cells is the direct transformation chain of hydrocarbon or fossil energy into electricity, thanks to its high conversion efficiency. Furthermore, the fuel cell is a key component of a hydrogen chain used as an electricity buffer. In an intermittent renewable energy system, electricity is transformed into hydrogen in order to compensate intermittent generation or temporary overproduction. Then this hydrogen is converted in electricity to satisfy the demand of the load. If this notion is extended to the whole storage and energy distribution (we then speak about hydrogen civilisation), it could lead to major economic growth.

Chapter written by Pierre BAURENS, Pierre SERRE-COMBE and Jean-Philipe POIROT-CROUVEZIER.

The fuel cell is the central element of these new energy chains. It appears in many reactor technologies (SOFC (solid oxide fuel cell), PEMFC (polymer exchange membrane fuel cell), etc.); the choice of which fuel cell depends on the upstream (fuel type) and on the downstream of the chain in which it is included (load dynamic, co-generation, etc.).

The apparent simplicity of the conversion mechanism is misleading where the design and the optimization of the fuel cell systems is concerned. This simplicity hides the complexity of the global energy system. This complexity is connected, on the one hand, to the complexity of the phenomena in the cell operation and, on the other hand, to the fuel conditioning before filling the cell.

5.2. Operation principles in different cell types

A fuel cell is an energy converter. It transforms chemical energy into electric energy. The chemical energy is brought by the fuel and by the oxidant that are introduced in the cell. In a reactor, the chemical energy used results from the chemical reaction (combustion) of all or part of the fuel and the oxydant. The product of this reaction is the third element that must be evacuated. Internal electrochemical reactions turn this combustion energy into electric energy (voltage and current) and heat. The latter comes from the uncompensated heat which depends on the entropy variations of the chemical reaction. It also comes from irreversible sources generated by physical mechanisms which happen in the internal electrochemical processes and in the transfer and transportation of chemical components and associated electric charges.

5.2.1. *Definitions*

In order to achieve sufficient output power, fuel cells are usually assembled in a stack. Each fuel cell is an elementary electrochemical cell (Figure 5.1). These cells traditionally contain a membrane-electrode assembly (also called MEA) and (half) bipolar plates on both sides. Inside the MEA, there is an electrolyte between two electrodes. Electrochemical processes which absorb or generate electrons and ions, use the fuel and the oxidant and supply the combustion product, take place in the electrodes. The electrolyte enables the transportation of ions by electric migrations from one electrode to the other because of the potential electric difference generated by the electrochemical processes that develop on these electrodes. Generally, the main functions of the bipolar plates are to conduct electrons from the cathode of the cell to the anode of its neighbor, to ensure gas-tightness between the corresponding gaseous media, to transport the reactants and the products to as near as possible the reaction areas, and to evacuate the generated heat.

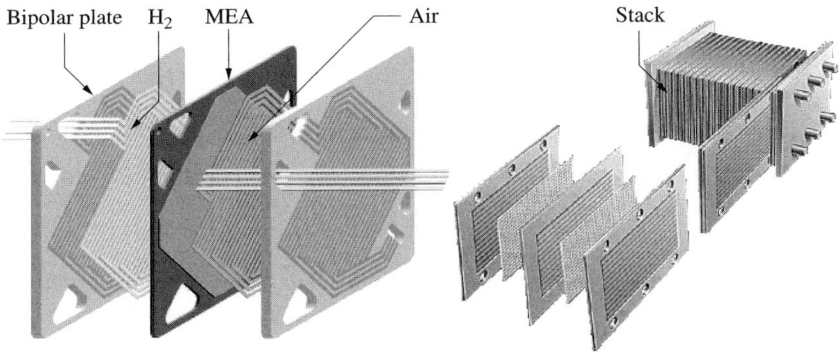

Figure 5.1. *Elementary cell, unit, and stack*

The different cell technologies are characterized by the type of electrochemical reactions realized on the electrodes and also by the associated electrolyte.

5.2.2. *General principle of function: bases of electrochemistry and of the elementary electrochemical reactor*

Electrochemistry is devoted to the study of phenomena involved in the flow of electrical currents from one electronic conductor to one ionic conductor (both of which are charge carriers that are not naturally compatible) through a chemical reaction at their interface (Figure 5.2).

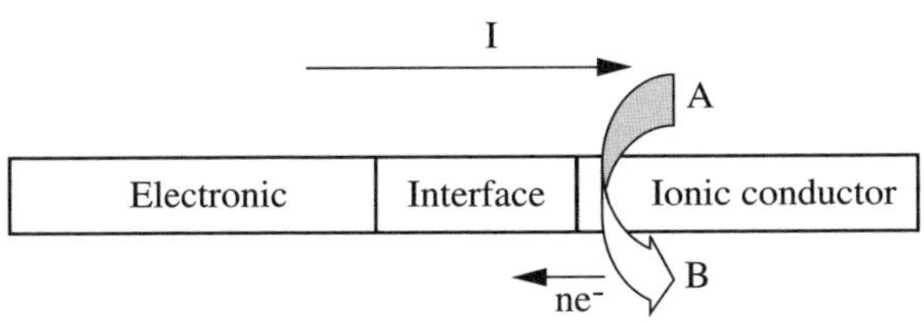

Figure 5.2. *Schematic principle of a chemical reaction*

An electrochemical system comprises a heterogenous assembly of ionic and electronic conducting media (or phases) generally associated with media or phases of non-electric conductors.

An electronic conductor is a medium containing free electrons (metal, carbon, semi-conductor, etc.).

An ionic conductor (or electrolyte) is a medium containing mobile ions. They may be aqueous or non-aqueous solutions, polymers or solids.

If an electric field $\vec{\mathcal{E}}$ is applied to this electrolyte, these ions will move because of the force: $\vec{F} = ze_0\vec{\mathcal{E}}$, where z is the ion charge number and e_0 the electron charge.

Such an electric field may be generated by placing two electronic conductors in the electrolyte and applying a potential electric difference to these *electrodes* as, for example, where the electrolyte is an aqueous solution of HCl in which there are platinum electrodes. This set represents an electrochemical cell (Figure 5.3).

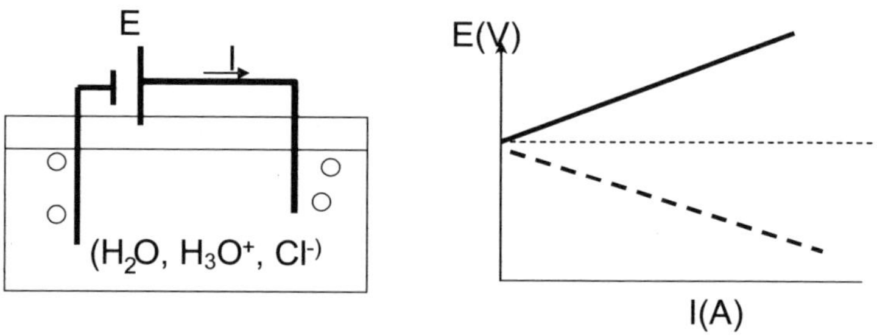

Figure 5.3. *Example of the electrolysis of a hydrochloric acid solution (E: potential difference; I: current flow; bold: operation in electrolysis mode; dotted: operation in cell mode)*

To let current flow through the cell, it is necessary to apply a minimum voltage. The current flow is accompanied of a chlorine emission on the positive electrode and a hydrogen emission on the negative electrode. The production flows are proportional to the current.

When looking in more detail, we notice that inside the solution the Cl^- ions go to the positive electrode where they are transformed into Cl_2 and that the H_3O^+ ions go to the negative electrode where they are transformed into hydrogen. The current flow due to the voltage between the electrodes is followed globally by a chemical transformation of HCl into H_2 and Cl_2. Each of the reaction products is generated on one electrode by a "half-reaction" along with the electron transfer from the electrolyte (ionic conductor) to the electrode (electronic conductor).

$$
\begin{aligned}
&2H_3O^+ + 2e^- \rightarrow H_2 + 2H_2O \\
&2Cl^- \rightarrow Cl_2 + 2e^- \\
&\overline{\qquad\qquad\qquad\qquad} \\
&2HCl \rightarrow H_2 + Cl_2
\end{aligned}
\qquad [5.1]
$$

This example corresponds to the transformation of electric energy into chemical energy. This electrochemical cell works in the electrolysis mode.

If during this type of operation the electric supply is disconnected from the electrodes and replaced by a resistance, we notice the existence of an electric voltage at the resistance terminals and the flow of the current which rapidly deminishes to zero. After the resistance is disconnected, a voltage at zero current is observed and maintained during a certain time. This voltage corresponds appreciatively to the minimal voltage observed for the electrolysis to start.

The bubbling of hydrogen around the negative electrode and of Cl_2 near the positive electrode enables the maintaining of a current in the resistance. The measured voltage is then lower than at zero current. This voltage and this current are developed by the inverse half-reactions of the previous ones on each electrode. Part of the hydrogen and chlorine injected is burned to generate an electric current. Chemical energy is transformed into electrical energy. The cell operates as electric generator. When the reactants can be continuously supplied to the electrochemical cell, this electric generator is also called a fuel cell. In other cases, the electrochemical cell is called a battery.

5.2.2.1. *Global thermodynamic approach of a fuel cell establishing a voltage-current relation*

A fuel cell can be modeled as a thermodynamic system made of a electrochemical reactor which is limited by its input and output of oxidant, fuel, and reaction product, and is further limited by its input and output of electrons (electric connections), and yet further limited by its input and output of heat (because of the fluid in the cooling circuit). The first and second principles of thermodynamics, when applied to this open system, enable us to write that, under a stationary regime, the net balance of the electrochemical power contained in the flow of the different chemical components (electrically charged[1]and not charged) entering and leaving the reactor is compensated for by the power dissipated by the irreversibility internal sources in the reactor.

1. The electrons entering or leaving the reactor are considered a charged chemical species.

The energy balance is written:

$$0 = -\left(\sum_k T\frac{dS_k}{dt}\right) + \left(\sum_{input} \mu_i^* N_i - \sum_{output} \mu_j^* N_j\right) \qquad [5.2]$$

where μ_i^* is the electrochemical potential of the species i, N_i its molar flow and $\sum_k T\frac{dS_k}{dt}$ the sum of the internal energy dissipations due to irreversible mechanisms (index k) during the process.

The molar potential electrochemical energy of the species i in the considered solution (or system) is equal to the species i chemical potential augmented of its electric potential energy in the case of an electrically charged species (electrons, ions).

$$\mu_i^* = \mu_i + z_i F\phi = \left(\mu_i^0 + RT\ln a_i\right) + z_i F\phi \qquad [5.3]$$

F is the charge of a mole of electron: 96,485 C; ϕ is the electric potential of the specie; R is the ideal gas constant; and T the temperature.

The chemical potential of the species i is also equal to its own chemical potential in the reference conditions[2] increased by the chemical potential due to the solution (multi-constituent gas, for example) which is expressed through the chemical activity (a_i) of the constituent i with respect to this reference.

The molar flows burnt or produced in the reactor by the electrochemical reaction are connected to each other by the half-reaction equations at the electrodes and they are both proportional to the current flow (electron species).

For an electrode equation, we write:

$$\sum_i \alpha_i A_i \rightarrow \sum_j \beta_j B_j + ne^- \qquad [5.4]$$

$$\frac{1}{\beta_j}\frac{dn_j}{dt} = -\frac{1}{\alpha_i}\frac{dn_i}{dt} = \frac{1}{n}\frac{I}{F} = \frac{1}{n}J \qquad [5.5]$$

2. The pure body at the temperature and pressure conditions and in the physical state (phase) of the system or considered solution.

with I the current supplied by the cell, n_i and n_j the numbers of specie moles burnt or generated by the reaction.

The species output molar flow rates (noted N) are obtained from the species per species material balances achieved in the reactor which, under a stationary regime, is:

$$N_i^{output} = N_i^{input} \pm \frac{\varepsilon_i}{n} J \qquad (\varepsilon_i = \alpha_i, \beta_j, ...) \qquad [5.6]$$

The equation between the electric voltage E at the cell terminals and the current generated by the cell is then:

$$E = (\varphi_c - \varphi_a) = \frac{1}{nF}\left[\sum_i \alpha_i \mu_i - \sum_j \beta_i \mu_j\right] - \frac{1}{nF}\left[\mu_e^c(T) - \mu_e^a(T)\right] - \frac{1}{nFJ}\sum_k T\frac{dS_k}{dt} \qquad [5.7]$$

The subscript e relates to electrodes.

Considering that the electron chemical potentials are identical in each electrode, which is generally the case (identical material, similar temperature), the fuel cell electric voltage contains three terms. The first term, positive, is proportional to the global chemical reaction affinity. The second term represents the thermodynamic state of the reactants and products solution in the input and output gases of the reactor. Finally, a negative term represents the internal irreversible dissipation generated by the reactor.

$$E = (\varphi_c - \varphi_a) = \frac{1}{nF}\left[A_{reaction}(T,P)_{ref.}\right] + \frac{RT}{nF}\ln\frac{\Pi(a_i)^{\alpha_i}}{\Pi(a_j)^{\beta_j}} - \frac{1}{nFJ}\sum_k T\frac{dS_k}{dt} \qquad [5.8]$$

The first two terms depend only on the achieved global chemical reaction, on the temperature and pressure of the reactor, and on the solution conditions of the reactive species introduced or extracted from the reactor. They formulate the chemical energy released in the reactor. They do not depend on the generated current. They represent the original electric potential, usually called reversible voltage of the cell.

The third term represents the heat dissipation of part of this original electric potential. It corresponds to the irreversible friction "losses" generated in the cell and in the reactor by the species motion (electrically charged or not). It only exists when current is supplied. Its value depends on the reactor design, the used materials, and the supplied current.

The above equation may also be written as:

$$E = (\varphi_c - \varphi_a) = E_{rev}(T, P, a_i, a_j) - \sum_k \eta_k \qquad [5.9]$$

where E_{rev} is the reversible voltage and η_k are over voltages due to the different irreversibilities.

5.2.2.2. *Equation of the zero current-generated electric potential*

At zero current, the cell voltage is equal to the first two terms of equation [5.8] which correspond to the potential chemical energy released by the reaction in its own conditions.

The global reaction affinity, as is every thermodynamic function, is calculated from a measured value in standard reference conditions, corrected for the temperature and pressure effects, according to the following equation:

$$\begin{aligned}\left[A_{reaction}(T,P)_{réf.}\right] &= \frac{T}{T_0}\left[A_{reaction}(T_0,P_0)\right]_{ref.} \\ &+ \int_{p0}^{p}\left[\sum_i \alpha_i v_i - \sum_j \beta_j v_j\right]_{ref.} dP - T\int_{T0}^{T}\left(\sum_j \alpha_i H_i - \sum_j \beta_j H_j\right)_{ref} \frac{dT'}{T'^2}\end{aligned} \qquad [5.10]$$

The reaction affinities values (those of the free enthalpy formation $A = -\Delta G$), established in the standard reference conditions, are tabulated for a given temperature, generally 298 K.

We remember that in the standard reference state, the reactants and the products are supposedly:

– pure;

– at a pressure of one atmosphere;

– at a given temperature T;

– in the physical state, or stable phase, at one atmosphere of the considered temperature.

Generally speaking, for a liquid fuel, the standard state is liquid if the considered temperature is lower than the boiling temperature.

In most cases, when there is a doubt about the physical state of the chemical body under these conditions, tables give corresponding values. This is the case particularly for water.[3] As an example, the original electric potential of the H_2/Air cell is given in the following points:

– Electric potential of a H2/Air cell in the standard state conditions.

When the combustible is hydrogen and the oxidizer is oxygen, which is the case of the H_2/Air fuel cell, the global reaction corresponds to the hydrogen combustion:

$$H_2 + \frac{1}{2} O_2 \rightarrow H_2O \qquad [5.11]$$

The reaction affinity corresponds to the water formation affinity:

$$A^{standard}_{reaction} = \frac{1}{2F}\left[\frac{1}{2}\mu_{O2}(T_0, P_0) + \mu_{H2}(T_0, P_0) - \mu_{H2O}(T_0, P_0)\right] = -\Delta G^f_{H_2O}(T_0, P_0) \qquad [5.12]$$

In the case where the produced water is in liquid state in the reactor output gases, the value of the standard affinity rates 237 kJ at 298 K. The original electric potential of the H_2/Air fuel cell under these conditions in then:

$$E_{rev}{}^{0}_{298} = \frac{1}{2F} A^{standard}_{reaction,}\text{(298 K, 1 atm., liquid water)} = \frac{23710}{2 \times 96480} = 1.228\, V \qquad [5.13]$$

In the case of water in the form of steam, the affinity value is 229 kJ and the standard reversible voltage at 298 K is 1.19 V.

---

3. Otherwise the calculation is possible with equation [5.10].

– Original electric potential in the chosen reference conditions.

At first approximation, the development of the equation gives:

$$E_{rev}^{H_2O,gas} = 1.186 - 0.4 \cdot 10^{-3}(T - 298) + \frac{RT}{2F}\ln\left[\frac{P}{10^5}\right]$$
$$E_{rev}^{H_2O,liquid} = 1.228 - 10^{-3}(T - 298) + \frac{RT}{2F}\ln\left[\frac{P}{10^5}\right] \quad [5.14]$$

– Original electric potential in the reactor-imposed conditions in the case of water produced in the steam phase:

$$E_{I=0} = E_{rev}^{H_2Ogas}(T,P) + \frac{RT}{2F}\ln\left(\frac{x_{H2}^{1}x_{O2}^{1/2}}{x_{H2O}^{1}}\right) =$$
$$1.186 - 0.4 \cdot 10^{-3}(T - 298) + \frac{RT}{2F}\ln\left(\frac{P_{H2}^{1}P_{O2}^{1/2}}{P_{H2O}^{1}}\frac{1}{10^5}\right) \quad [5.15]$$

5.2.2.3. *Non-zero current electric potential and current response of a fuel cell*

To write the non-zero current electric potential, the irreversible entropy sources in the system must be identified. These sources are due to the frictions generated by: the transport of many species, both those electrically charged and those uncharged, inside the reactor; the transfers at the interfaces whether or not they are electrified; the heat transports and transfers. These sources are located in different components (electrodes, electrolytes, gas compartment, exchanger, etc.). The generated mechanisms of these dissipations are: the uncharged species diffusion; the electric conduction (joule effect); the chemical reaction; etc. The thermodynamics of the irreversible processes shows that these source terms can be written as a product of a thermodynamic force and a flux of the thermodynamic flow due to this force.

$$T\frac{dS_k}{dt} = T(F_k\frac{dX_k}{dt}) \quad [5.16]$$

Table 5.1 shows the necessary elements to write these energy dissipation sources.

Process k	**Thermodynamic force F_k**	**Thermodynamic variable identification: X_k**
Thermal conduction	$\nabla \frac{1}{T}$	heat: Q
Diffusion	$-\nabla \frac{\mu_k}{T}$	Moles number: n_j
Electric conduction	$-\nabla \frac{\varphi}{T}$	Electric charge: q_e
Chemical reactions	$\frac{A_j}{T}$	Advancement degree of the reaction: ξ_j

Table 5.1. *Flux and forces in the mechanism of energy dissipation*

Generally speaking, the equation of the dissipation in an electrochemical system will contain the following generic terms:

$$\Sigma T \frac{dS_k}{dt} = \Sigma T \left(\frac{A_j}{T} \right) \frac{d\xi_j}{dt} - \Sigma T \left(\nabla \frac{\mu_k}{T} \right) \frac{dn_k}{dt} + \sum T \left(\nabla \frac{1}{T} \right) \frac{dq}{dt} - \sum T \left(\nabla \frac{\varphi_k}{T} \right) \frac{dq_k}{dt} \quad [5.17]$$

To represent the electric voltage drop due to one or another dissipation mechanism, we introduce the idea of mechanism-specific overvoltage, with the global overvoltage being equal to:

$$\eta = U - E_{rev} = \Sigma \eta_k = \frac{1}{FJ} \Sigma T \frac{dS_k}{dt} \quad [5.18]$$

The table aims to express the overvoltage form due to the major dissipation mechanisms in the fuel cells:

– diffusion overvoltage due to the transport of electro-active species to the electrochemical reaction location:

$$\eta_{diff} = -K\sum T\nabla(R\ln a_i) \approx -\frac{K}{e}T\sum R\ln(\frac{a_i(e)}{a_i(0)}) \\ = K_s T - K_n T\sum R\ln a_i \quad [5.19]$$

– ohmic overvoltage due to the ion transport through the electrolyte or in the electrode:

$$\eta_{ohmic} = -\sum \Re_k I \quad [5.20]$$

– reactional (or activation) overvoltage due to the electrochemical reaction on the electrode:

$$\eta_{reaction} = a + bT\ln i \quad [5.21]$$

The resulting form of the voltage-current law of an electrochemical cell is that presented in Figure 5.4. It shows that at low current, the dominant dissipative mechanisms are those related to the electric charges transfers, which means, strictly speaking, electrochemical reactions (activation overvoltage).

When the current becomes significant, the ohmic overvoltage issued from the joule effect related to the ions transport in the electrolyte is added. At very high current, a diffusion overvoltage may appear. It comes from the difficulty of supplying the reaction zone in electro-active species at the corresponding flow rate (mechanical energy dissipation in the internal transport to the reaction zone). This final overvoltage may drastically limit the supplied current (diffusion limit current).

In the cells where the generated water is liquid, a flooding overvoltage is added, which corresponds to the difficulty of evacuating the generated water far from the electrode, the limit being the complete flooding of the electrode, preventing the electro-active species from reaching the reaction sites.

The activation overvoltage may be important, especially at low temperature where the electrochemical phenomenon kinetic is slower. It depends on the electrode materials, their morphology, and their constitution.

There is a constant search for the best catalyst and the search for the best-suited morphology and constitution for the transport of all species implied in the electrode reaction.

The ohmic overvoltage depends on the electrolyte thickness (main source of ohmic dissipations) and on its ionic conductivity. The latter generally increases with temperature, but may also depend on other factors, such as the water content in the case of polymer electrolytes. The reduction of this overvoltage may be achieved by a reduction of the electrolyte thickness and by using an electrolyte with higher conduction in the operation conditions of the cell. This must be done in accordance with the other constraints on the electrolyte (manufacturing process, mechanical strength, length of life, and watertightness).

The diffusion overvoltages are generally related to local mechanisms in the electrode and depend on its morphology and constitution.

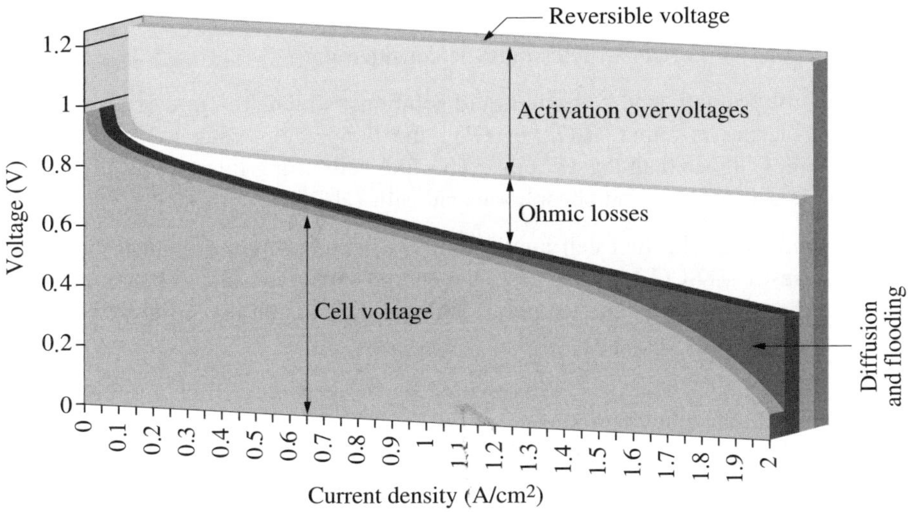

Figure 5.4. *Voltage drops for a cell under operation*

5.2.3. *Elements of a fuel cell*

5.2.3.1. *Fuel cell electrolytes*

The two major functions that a fuel cell electrolyte must fulfill are the transport of the ions from one electrode to the other and the electronic isolation of both electrodes.

An electrolyte is therefore a liquid or solid medium containing ionic-type charge carriers, the volume and mobility concentration of which are high enough to offer an acceptable conductivity for the cell and its application.

$$\sigma_{ionic} = \left[P_{ion}\right] u_{Pion} z_{Pion}^{2} F^{2} \qquad [5.22]$$

with $[P_{ion}]$ the concentration of the ionic charge carriers in the electrolyte and u_{Pion} their mobility.

The second function requires that the electronic conductivity of the medium is much lower than its ionic conductivity, generally more than three decades.

Consequently, the electrolytes will be differentiated by their ionic carrier, by their phase (liquid, solid, polymer), and by the temperature rate in which they present acceptable conductivities for the cell applications. Thcy furnish the generic designation of the different cells.

Four main types of electrolytes must be considered:

– electrolytes with acid or basic liquid solutions, which are more or less diluted. The ionic carrier is either H_3O^+, or OH^-; the temperature range: 20–200°C; the conductivities higher than 0.1 $\Omega^{-1}.cm^{-1}$. The fuel cells using this type of electrolyte are alkaline cells (KOH) and phosphoric acid cells (H_3PO_4);

– melted salts. The fuel cell application is illustrated by the melted carbonates (Li_2CO_3, Na_2Co_3, K_2CO_3). In this case, the charge carrier is CO_3^{2-}. The conductivity is higher than 1 $\Omega^{-1}.cm^{-1}$. The temperature range depends on the fusion temperature of the salt, here 600°C;

– ion-conducting polymers. Represented by the perfluorosulfonate polymers of the Polymer Exchange Membrane Fuel Cells (PEMFC). In this case, the charge carrier is H_3O^+, and the classical use temperature range is 0–100°C. The conductivity is about 0.1 $\Omega^{-1}.cm^{-1;}$

– ion-conducting solids. Represented by solid solutions of oxides (for example,e Zircon-doped Yttrium) used in the SOFC. The charge carrier is the oxygen blank. The temperature range is 800–1,000°C. The conductivity is around 0.1 $\Omega^{-1}.cm^{-1}$.

5.2.3.2. *Fuel cell electrodes: process description*

The elementary mechanism defining the electrochemical process is the charge (electron) transfer[4] through an interface between an electronic conductor and an ionic conductor.

This charge transfer which corresponds to the electric current transit through the interface has the specificity of being related to chemical transformation; the latter is called an electrochemical reaction. In fact, the ionic conductor, or electrolyte, is a medium that has no or few free electrons. Electron transfer requires that a substance

4. A substance transfer is defined by the transit of this substance from one phase to another (or from one medium to another) through their interface. Adsorption and solubilization are transfer mechanisms.

located near the interface, and which may or may not belong to one of the conducting media, captures the electrons released by the electronic medium and fixes them, so turning them into compatible ions in the ionic conductor medium.

This process is represented in Figure 5.5 in the case of a plane electrode immersed into an electrolyte solution. This process is generally accompanied by different elementary mechanisms of electro-active species transport or transfer: diffusion of the species from the supply channel to the reaction point in the electrode, surface adsorption, etc.

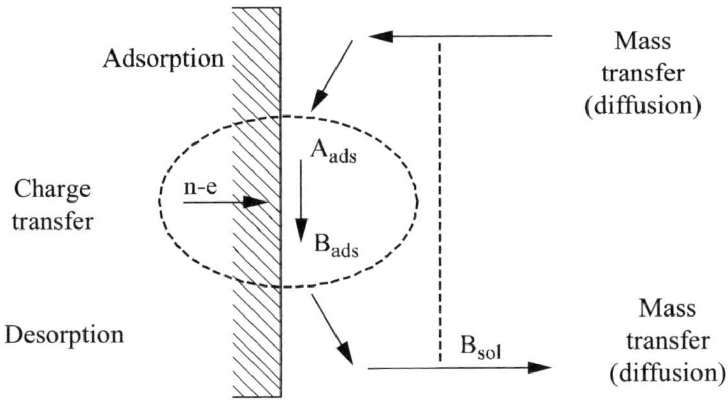

Figure 5.5. *Representation of processes on one electrode*

In theory, this charge transfer requires three conditions:

1) The triple contact: on the reaction location, which means at the same point, it it necessary to have the three phases or media: the electronic phase, the ionic phase, and the gaseous phase (gas electrode). This is the interface concept (point, line) of triple contact (Figure 5.6).

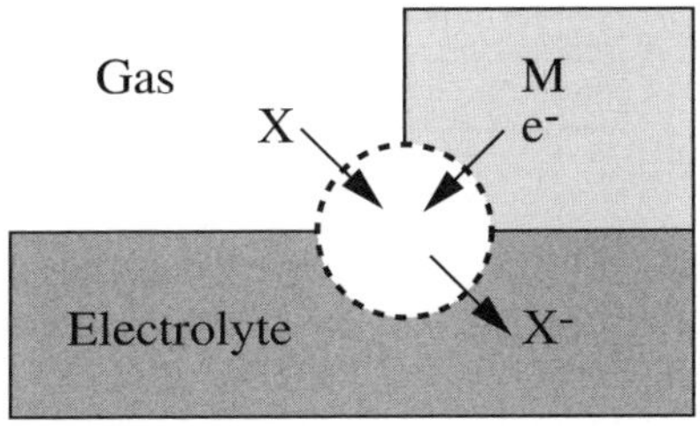

Figure 5.6. *The triple contact*

This figure is a theoretical view that would, in reality, give bad kinetics. Fortunately, the kinetics are usually improved in the electrodes where at least two of the species involved in the process are present in the same phase. The choice of the carrier phase, the realization, and, above all, the multiplication of the triple contact points, lines, or surfaces will strongly depend on the electrolyte nature (liquid, solid, polymer).

In the case of the liquid electrolyte, as shown in Figure 5.5, the electronic phase is immersed in the solution. The gaseous phase is generally introduced by bubbling in the electrolyte solution. It is dissolved and diffused inside the solution to reach the electronic interface on which the charge transfer reaction will occur.

It is also true for the electrodes associated with a polymer electrolyte. The latter is a thin layer laid on a metallic phase. The gaseous species located outside adsorbs itself onto a polymer surface, and then diffuses to the metal, where the reaction is present.

In the case of solid electrolyte, the two conducting phases are solid. It is impossible to dissolve and diffuse the electro-active gaseous species in the volume of this phase. It is therefore difficult to generate a large triple-contact surface. The concept consists of making and using a composite and porous mix conductor (Figure 5.7).

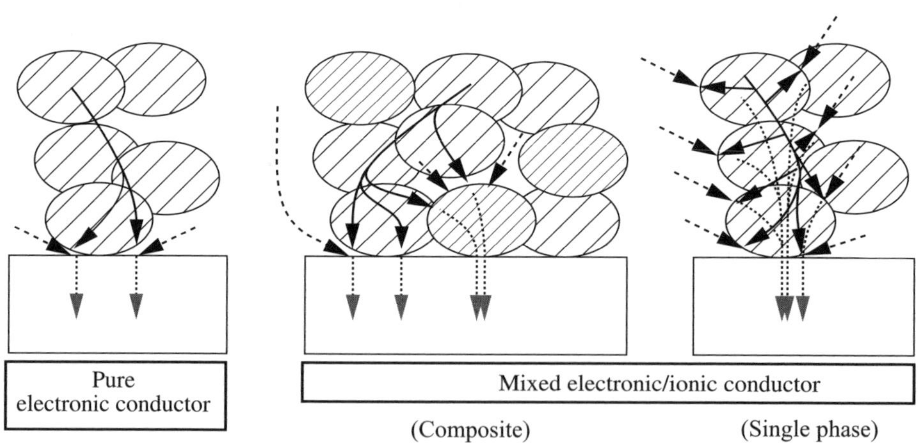

Figure 5.7. *The different realization modes of the triple contact in the electrodes of SOFC (solid lines: electronic conduction path; dotted lines: ions path; dashed lines: gas path)*

2) A local electrochemical activity sufficient to have rapid and less energy dissipative kinetics. This local electrochemical activity strongly depends on the temperature and on the electronic phase of the material. At low temperature, the weakness of the electrochemical activity requires the use of catalyst which will be an activator of the reaction. Unfortunately, at low temperature (T < 200°C) these catalysts for the interesting reactions in the area of the fuel cell (oxygen reduction) are precious and expensive materials, the most relevant being platinum. At higher temperatures, nickel replaces it.

The relationship between the reaction speed[5] (i.e. the local delivered current) and the resulting local overvoltage is the following:

$$i = i_O \left\{ \exp(\alpha \frac{nF}{RT}(\eta) - \exp\left[-(1-\alpha)\frac{nF}{RT}(\eta) \right] \right\} \quad [5.23]$$

The constant i_0 and the coefficient α are reaction coefficients. α lies generally between 0.3 and 1, and i_0 strongly depends on the presence or the absence of a catalyst. This formulation, at high overvoltage values, is written as:

$$\eta_C = \frac{RT}{(1-\alpha)nF}\ln i_O - \frac{RT}{(1-\alpha)nF}\ln(|i|) = a + b\ln(|i|) \quad [5.24]$$

3) The development of a triple contact area in the electrode volume. This supposes the realization of a volume electrode ensuring an internal connectivity to each phase, thus limiting at the minimum the transport dissipation of the moving charged and uncharged chemical components. The electrodes are generally made of a porous medium, sometimes with bi- or tri-constituents (for example the composite electrodes of SOFC).

The electrode may also be made of a porous medium with many layers of different functions as, for example, in the case of the PEMFC in which the electrode (active layer) is laid on a porous substrate made by felt or by a carbon tissue used as an electronic conductor and as a gas distributor at the electrode. The active part of the electrode is made of a skeleton of platinum carbon and of electrolyte.

5. Here limited to the charge transfer step.

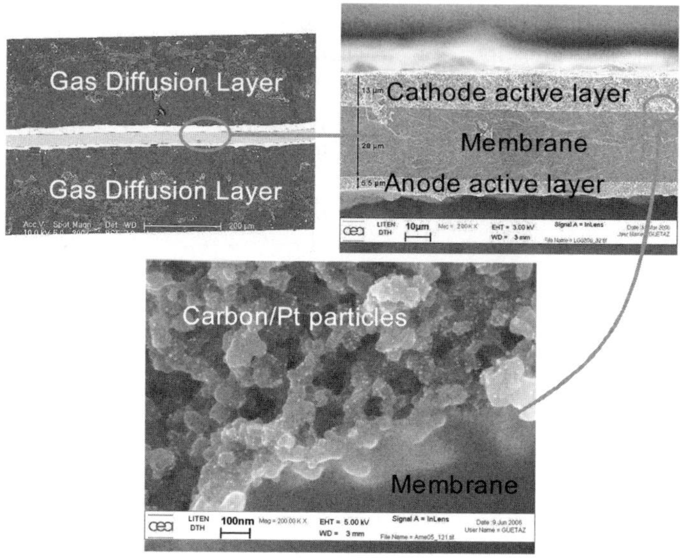

Figure 5.8. *SEM images of an electrode-electrolyte assembly for a PEMFC*

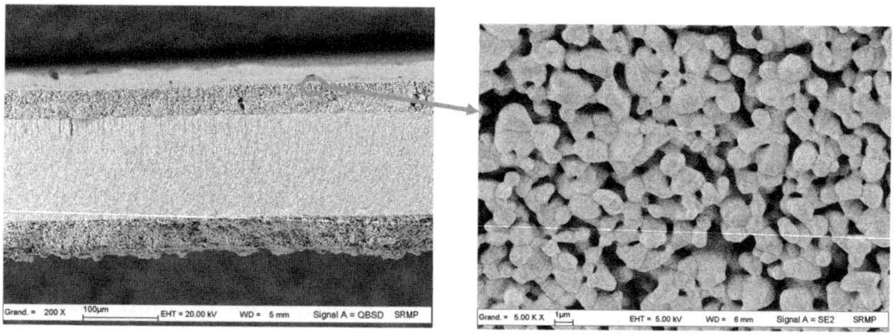

Figure 5.9a. *SEM images of an electrode-electrolyte assembly for a SOFC*

NOTE: The complexity of a reaction process is added to this structure complication. The global reaction at the electrode is, in fact, a succession elementary steps, which are more or less numerous according to the reaction and to the electrode. As an example, the reduction reaction of the oxygen on a SOFC cathode, illustrated in Figure 5.10 in the case of a mixed electric conductor electrode, may show the steps or elementary mechanisms shown below in Table 5.2 and classified in species.

Surface reactions	**Surface and gas phase diffusion mechanisms**	**Electrolyte transfers**
Adsorption of the oxygen molecule	Surface diffusion/migration of the oxygen ions	Adsorbed oxygen ions incorporation in the triple point
Dissociate adsorption of the oxygen molecule	Surface diffusion of the oxygen atoms	Oxygen ions transfer from the cathode phase to the electrolyte phase
Dissociation of the adsorbed oxygen molecule	Phase diffusion/migration of the oxygen ions	Adsorbed oxygen atoms incorporation in the triple point
Transfer of the oxygen ion from the surface to the phase		
Ionization of the oxygen atom		

Table 5.2. *Basic steps of the reaction mechanism*

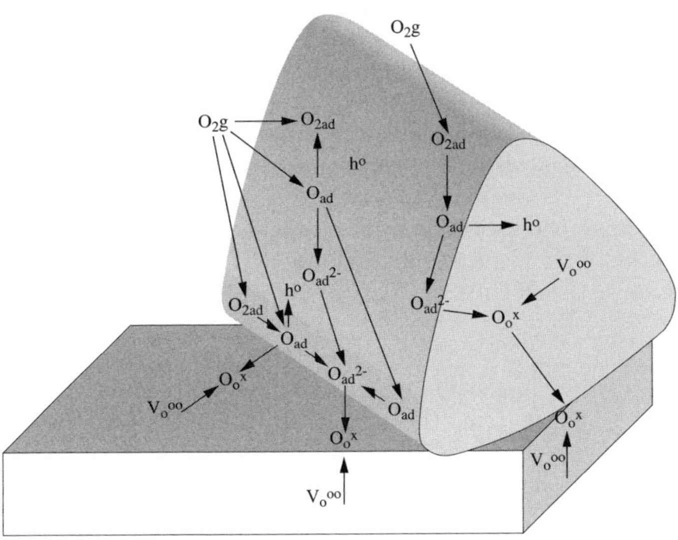

Figure 5.9b. *Porous volume electrode sections of PEMFC and SOFC*

5.2.3.3. *Fuel cell architecture and interconnections*

At best, the voltage generated by a fuel cell is in the order of 1 V. As such, it does not meet the needs of electrotechnics, which require high voltages. An electric series connection of several cells is necessary to reach the required voltage.

The stack architecture must combine this electric series connection with the two other functions which are the gas supply of the anodes and cathodes and the gas tightness between anodes and cathodes. The three typical configurations are presented in Figure 5.10.

The first consists of connecting the electrodes by the small side, or by a cable fixed at a given point of each electrode. This is the tubular architecture configuration that is operated, in particular, in Westinghouse SOFC technology.

The elementary cell is made from a porous cylinder on which are successively deposited the cathode, then a dense layer of electrolyte, and finally a porous layer of anode. This last layer does not completely cover the surface of the cylinder in order to create an inactive surface. An electrical access to the cathode is created there and connected to the anode of the next cell.

The oxidizer (air) flows inside the tubes; the fuel outside. The anodes and the cathodes of the different tubes are electrically connected by means of nickel felt.

The advantage of this architecture is to separate the gastightness and electrical connection functions, which enables us to treat them with better-adapted materials and under more adequate conditions (temperature in particular).

The second configuration consists of ensuring these functions (gastightness and electrical connections) using the same component.

The electronic current issued from a cathode is transferred to the anode of the next cell in the main plane of the electrode by means of an electronic conductive plate which also ensures the tightness between the reactive gases located on both sides.

The plate connecting the plus terminal of a cell to the minus terminal of the same cell is called a bipolar plate. The architecture is called planar (in the sense of the electric connection). Its major advantage is to minimize the electronic conduction path and thus the electric resistance between two cells.

This configuration is that most used in fuel cell stacks. Its major drawback is the difficulty of achieving, on the same component, the three functions needed to run the stack.

The used materials are then subject to corrosion and electronic conduction constraints. It can become difficult to make them compatible (e.g. in case of the high temperatures).

In these two configurations, the gas supply of the stack is created in parallel.

The third configuration combines the electric series connection with the hydraulic series connection of the cells. It operates on the SOFCs and PEMFCs.

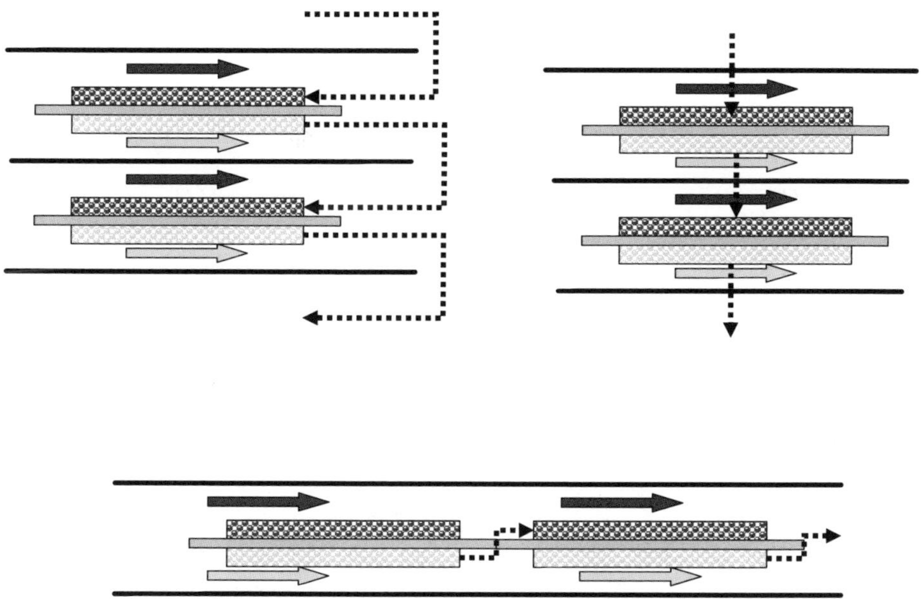

Figure 5.10. *Architecture configuration of stacks (black continuous line: the gas tightness function; black broken line: the electric connection function; arrows: reactive gas flows for the anode and the cathode respectively)*

5.2.4. *Main cell types*

The cells differ in the choice of electrolyte and the reactions they produce. The main visible consequences of these choices are the operation temperature levels and the fuel and oxidizer they can accept.

The following table shows, in a non-exhaustive manner, the expected performances and major problems of the standard cells.

As there is today a strong interest in the PEMFC and the SOFC, a separate section is dedicated to each of these cells.

<table>
<tr><th colspan="5">PEMFC (Proton Exchange Membrane Fuel Cell)</th></tr>
<tr><td>Combustible</td><td>Oxidizer</td><td>Temperature</td><td>System efficiency</td><td>Power density</td></tr>
<tr><td>H_2
(CO<50 ppm)</td><td>O_2
Air</td><td>0–100°C</td><td>30–45%</td><td>1.2 W/cm² (O_2)
600 mW/cm² (Air)</td></tr>
<tr><td colspan="3">Anode reaction</td><td colspan="2">Cathode reaction</td></tr>
<tr><td colspan="3">$H_2 \rightarrow 2H^+ + 2e^-$</td><td colspan="2">$\frac{1}{2}O_2 + 2H^+ + 2e^- \rightarrow H_2O$</td></tr>
<tr><td colspan="3">Electrolyte</td><td colspan="2">Catalysts</td></tr>
<tr><td colspan="3" rowspan="3">Perfluorosulfonate (Nafion®) membranes
T<100°C</td><td colspan="2">Pt</td></tr>
<tr><td colspan="2">Interconnector or bipolar plates</td></tr>
<tr><td colspan="2">Graphite, metal</td></tr>
<tr><td colspan="3">Advantages</td><td colspan="2">Drawbacks</td></tr>
<tr><td colspan="3">High surface power
Large dynamics
Instantaneous starting</td><td colspan="2">Cost of the electrolyte, of Platinum
Thermal and hydraulic management
Lifetime</td></tr>
<tr><td colspan="3">State of the technology</td><td>Actors</td><td>Realizations</td></tr>
<tr><td colspan="3">Pre-industrial demonstrators for automotive and stationary applications.</td><td>Ballard, Daimler, GM, Nissan, etc.</td><td>Cell and system: up to 250 kW</td></tr>
<tr><td colspan="3">Applications</td><td colspan="2">Automobile and small stationary applications</td></tr>
</table>

Table 5.3. *PEMFC characteristics*

<table>
<tr><td colspan="5">DMFC (Direct Methanol Fuel Cell)</td></tr>
<tr><td>Combustible</td><td>Oxidizer</td><td>Temperature</td><td>System efficiency</td><td>Power density</td></tr>
<tr><td>Methanol</td><td>Air</td><td>10–120°C</td><td>20–35%</td><td>300 m W/cm²</td></tr>
<tr><td colspan="3">Anode reaction</td><td colspan="2">Cathode reaction</td></tr>
<tr><td colspan="3">$CH_3OH + H_2O \rightarrow CO_2 + 6H^+ + 6e^-$</td><td colspan="2">$\frac{1}{2}O_2 + 2H^+ + 2e^- \rightarrow H_2O$</td></tr>
<tr><td colspan="3">Electrolyte</td><td colspan="2">Catalysts</td></tr>
<tr><td colspan="3" rowspan="3">Perfluorosulfonated (Nafion®) membranes
T<100°C</td><td colspan="2">Pt</td></tr>
<tr><td colspan="2">Interconnector or bipolar plate</td></tr>
<tr><td colspan="2">Graphite, metal</td></tr>
<tr><td colspan="3">Advantages</td><td colspan="2">Drawbacks</td></tr>
<tr><td colspan="3">Use of a liquid fuel
Low temperature
Dynamics, starting</td><td colspan="2">Requires a layer to mix methanol and water
Permeability of the present electrolytes to methanol
Low efficiency</td></tr>
<tr><td colspan="3">State of the technology</td><td>Actors</td><td>Realizations</td></tr>
<tr><td colspan="3">Prototype demonstrators (small vehicles)
Application to portable fuel cells and micro-cells (low power, compactness)</td><td>Ballard, IRD, Siemens,
Motorola, Samsung</td><td>3 kW systems
micro-Fuel cells</td></tr>
<tr><td colspan="3">Applications</td><td colspan="2">Laptop and automobiles</td></tr>
</table>

Table 5.4. *DMFC characteristics*

<table>
<tr><td colspan="5">PAFC (Phosphoric Acid Fuel Cell)</td></tr>
<tr><td>Combustible</td><td>Oxidizer</td><td>Temperature</td><td>System efficiency</td><td>Power density</td></tr>
<tr><td>H_2
(CO < qq%)</td><td>Air</td><td>150–220°C</td><td>40%</td><td>200 mW/cm²</td></tr>
<tr><td colspan="3">Anode reaction</td><td colspan="2">Cathode reaction</td></tr>
<tr><td colspan="3">$H_2 \rightarrow 2H^+ + 2e^-$</td><td colspan="2">$\frac{1}{2}O_2 + 2H^+ + 2e^- \rightarrow H_2O$</td></tr>
<tr><td colspan="3">Electrolyte</td><td colspan="2">Catalysts</td></tr>
<tr><td colspan="3" rowspan="3">Concentrated acid solution
Phosphoric acid (H_3P0_4) 100%
Containment matrix: silicon carbide + PTFE</td><td colspan="2">Pt (0.1 mg/cm² at the anode,
0.5 mg/cm² at the cathode).</td></tr>
<tr><td colspan="2">Interconnector or bipolar plate</td></tr>
<tr><td colspan="2">Graphite</td></tr>
<tr><td colspan="3">Advantages</td><td colspan="2">Drawbacks</td></tr>
<tr><td colspan="3">Less sensitivity to CO than PEMFC (avoids shift layers at the reforming level)
Easy co-generation (200°C)</td><td colspan="2">Cost of platinum
Bipolar plates in graphite to avoid corrosion by H_3PO_4
High cathode activation losses.</td></tr>
<tr><td colspan="3">State of the technology</td><td>Actors</td><td>Realizations</td></tr>
<tr><td colspan="3">Generation set with external reforming for gas: marketed at more than 200 units (P > 50 kWe to 250 kWe).</td><td>UTC (IFC)</td><td>Plant up to 11 MW (Japan)</td></tr>
<tr><td colspan="3">Applications</td><td colspan="2">Stationary applications</td></tr>
</table>

Table 5.5. *PAFC charateristics*

<table>
<tr><td colspan="5">MCFC (Molten Carbonate Fuel Cell)</td></tr>
<tr><td>Combustible</td><td>Oxidizer</td><td>Temperature</td><td>System efficiency</td><td>Power density</td></tr>
<tr><td>H2, CO, CH_4</td><td>Air</td><td>600–700°C</td><td>40–45%</td><td>200 mW/cm²</td></tr>
<tr><td colspan="3">Anode reaction</td><td colspan="2">Cathode reaction</td></tr>
<tr><td colspan="3">$H_2 + CO_3^{2-} \rightarrow H_2O + CO_2 + 2e^-$</td><td colspan="2">$\frac{1}{2}O_2 + CO_2 + 2e^- \rightarrow CO_3^{2-}$</td></tr>
<tr><td colspan="3">Electrolyte</td><td colspan="2">Catalysts</td></tr>
<tr><td colspan="3" rowspan="3">Melted salts: carbonates mixture (Li_2CO_3, Na_2Co_3, K_2CO_3)
Fixed in the Li Al_2O_3 matrix</td><td colspan="2">Ni, NiO</td></tr>
<tr><td colspan="2">Interconnector or bipolar plates</td></tr>
<tr><td colspan="2">Ni and steels for high temperature</td></tr>
<tr><td colspan="3">Advantages</td><td colspan="2">Drawbacks</td></tr>
<tr><td colspan="3">Inexpensive catalysts
Internal reforming possible for some hydrocarbon (natural gas, synthesis gas (carbon gasification))</td><td colspan="2">Interconnected materials must resist to The electrolyte at 700°C (cost)
Contact resistance → low surface power density
Lifetime
Necessity of a CO_2 source at the cathode (recuperation at the anode)</td></tr>
<tr><td colspan="3">State of the technology</td><td>Actors</td><td>Realizations</td></tr>
<tr><td colspan="3">Generator set with external reforming for gas: pre-industrial demonstrator</td><td>Siemens, ERC, Ansaldo, Mitsubishi, BCN</td><td>Demonstrator plant up to 1 MW (150 kW units)</td></tr>
<tr><td colspan="3">Applications</td><td colspan="2">Massive generation of electric energy, carbon product chain</td></tr>
</table>

Table 5.6. *MCFC charteristics*

SOFC (Solid Oxide Fuel Cell)				
Combustible	Oxidizer	Temperature	System Efficiency	Power Density
H_2, CO, CH_4	Air	700–1,000°C	50–65%	400 mW/cm^2
Anode reaction			Cathode reaction	
$H_2 + O^{2-} \rightarrow H_2O + 2e^-$			$\frac{1}{2}O_2 + 2e^- \rightarrow O^{2-}$	
Electrolyte			Catalysts	
Solid oxide solutions ZrO_2 doped Y_2O_3 (8–10%).			Cermets Ni/ZrO_2, strontium doped lanthanum manganite.	
			Interconnector or bipolar plate	
			Lanthanum chromites, high-temperature steels, nickel alloys	
Advantages			Drawbacks	
Cheap catalysts Internal reforming possible for some hydrocarbon (natural gas, synthesis gas (carbon gasification))			Interconnector materials to resist at 800°C (cost) Thermo-mechanical compatibility of the electrode-electrolyte assembly (ceramics) Gas tightness (glass gasket, etc.) Contact resistance Lifetime Start-up	
State of the technology			Actors	Realizations
Demonstators of cell and stack prototypes			Westinghouse, Siemens, Mitsubishi, etc.	Power cells 100 kW
Applications		Massive generation of electric energy, chain of carbon and natural gas products, auxiliary power unit (APU)		

Table 5.7. *SOFC characteristics*

5.2.4.1. *Specificities of the SOFC*

The SOFC uses an oxygen-depleted solution of solid oxides as electrolyte. It operates at a temperature range between 700 and 1,000°C, which is the temperature range necessary to obtain an ionic conductivity sufficient for the electrolyte usually used.

Its main advantages, related to its operating temperature, are:

– have only two phases (solid and gas);

– accepts fuels other than pure hydrogen. CO, but also light alcanes (methane), which may be directly vapor-formed in the cell, may be used;

– do not require expensive catalysts;

– generate a high-quality heat (high temperature) for a co-generation use.

Its major drawbacks (lifetime, operation difficulties, cycling, cost, etc.) are due to:

– the use and the assembly of different oxide materials to make a cell;

– the high temperature to which the gaskets and bipolar plate materials are subject;

– the oxidative and reductive atmospheres of the anodes and cathodes.

5.2.4.1.1. Electrolyte characteristics

The two main functions of the electrolyte are to transport the oxygen ions and to separate the reactive gases.

The main constraints, in addition to having a very high ionic conductivity, are to resist very oxidative conditions on one side and very reductive conditions on the other sides, and not to react with the materials of the two electrodes.

The charge carrier which transports the oxygen ion is generally made of oxygen holes present in the oxide solution which makes the electrolyte.

These holes are made by dissolving a metallic oxide in another oxide in which the metal valence is lower. This doping leads to a predominance of oxide ions holes, which unlike developed cationic faults, are very mobile.

The most common electrolyte is a yttria stabilized zirconia with a fluorite structure $(ZrO_2)_{(1-x)}(Y_2O_3)_x$ in which x ranges between 0.03 and 1. The doping in yttrium oxide creates oxygen holes.

$$Y_2O_3 \xrightarrow{ZrO_2} 2Y_{Zr}^{'} + 3O_O^x + V_O^{\bullet\bullet} \text{ (6)} \qquad [5.25]$$

The substitution of *Zr (IV)* by *Y (III)* creates a net negative charge in the structure. The electro-neutrality condition requires the formation of a double positive oxygen hole.

At low dopant concentration, the *YSZ* ionic conductivity increases with the dopant concentration, but presents a maximum around 10%. Moreover, the negative effect of this concentration on the mechanical properties limits its interest. The nature of the dopant has an effect on the conductivity. Table 5.8 gives the conductivity dependency with respect to the temperature.

The structure's electronic conductivity is related to the electronic charge carriers, electrons, and electron holes, which are generated by the oxidative and reductive atmospheres located on both sides. Consequently the conductivity varies with the oxygen pressure, but its value remains negligible (more than three orders of magnitude less than its ionic conductivity) over a very large range of oxygen pressures.

$$\sigma_{e-} = 10^{-8} P_{O_2}^{-\frac{1}{4}} + 3 \cdot 10^{-6} P_{O_2}^{\frac{1}{4}} \ll 10^{-3} \text{ S/cm} \qquad [5.26]$$

Other solutions and oxide structures have been tested. Some of them present higher ionic conductivity, particularly at lower temperatures. Nevertheless, this is generally followed by a lower stability and a growing dependency on the oxygen pressure; or by an increase in the electronic conductivity. Moreover, there can be problems of cost and of temperature in the realization of the material.

The most experienced materials are gadolinia doped ceria and lanthanum gallate which has a Perovskite structure and, more recently studied, apatite-structured phases.

Table 5.8 shows the conductivities at different temperatures of some oxide solutions tested in this field.

---

6. O_O^x represents the oxygen on an oxygen site of the electrolyte network with a zero net charge; $V_O^{\bullet\bullet}$ represents the oxygen hole in the network and its positive net charge(+2); $Y_{Zr}^{'}$ represents *Y* on a *Zr* site with a negative net charge.

	600°C	700°C	800°C	900°C	1,000°C
Zircone solution					
YSZ (8%)	2.82 10^{-3}	8.29 10^{-3}	2.00 10^{-2}	4.13 10^{-2}	1.04 10^{-2}
Ceria solution					
GDC: $Ce_xGd_{1-x}O_y$ (x: 0.8, y:1.8)	2.82 10^{-2}	7.3 10^{-2}	1.58 10^{-1}	3.01 10^{-1}	5.18 10^{-1}
Lanthane solution					
LSGM: $La_xSr_{1-x}Ga_yMg_{1-y}O_3$ (x:0.9, y:0.8)	3.16 10^{-2}	7.69 10^{-2}	1.58 10^{-1}	2.89 10^{-1}	4.79 10^{-1}

Table 5.8. *Conductivities in S.cm-1 at different temperatures of some oxide solutions [WIN 05]*

NOTE: The improvement of the material conductivity being faced with stability problems, the main performances gains have been obtained recently reducing the electrolyte thickness. In the case of solid oxide fuel cells, this is done using a thick porous electrode on which a thin layer of dense electrolyte is deposited, the other porous electrode is then added to the first. Electrolyte layers with a thickness of a few microns are produced. This cell structure is called "electrode supported".

5.2.4.1.2. Electrode characteristics

– Cathode

The cathode is the most penalizing electrode in terms of overvoltage.

At the cathode, the oxygen molecules must be ionized, then integrated into the electrolyte.

The precious metals which present a catalytic activity are today abandoned in favor of solid metallic oxides which present an electronic or mixed conduction.

$$O_{2,g} + 4e^- + 2V_O^{\bullet\bullet} \rightarrow 2O_O^x \quad [5.27]$$

– Electronic conduction oxides

In this case, the oxides are associated with electrolyte (*YSZ*) to make a porous composite electrode (Figure 5.7). The perovskite structure oxides (ABO_3, A and B being metals, A being generally a rare earth and B a transition metal) generally present high electronic conductivities at high temperature. The reference material is *LSM* ($La_{0.8}Sr_{0.2}MnO_3$). It is an electronic conductor and does not react very much with *YSZ*.

The formation at the contact of *YSZ* of an electrically insulating phase ($La_2Zr_2O_7$ and $SrZrO_3$) must be avoided. This last point eliminates some materials more electro-catalytic and better electronic conductors, such as $LaSrCoO_3$, $LaFeO_3$.

Specific surface resistances of 0.04 Ω.cm² at 950°C and 0.23 Ω.cm² at 750°C have been measured.

To limit the interface resistance at the current collector (interconnector or bipolar plate), to increase the number of triple points, and to avoid a reaction between some materials, electrode structures optimized with more than nine material layers with different porosities have been made.

– Mixed conduction oxydes

Here, the metallic oxide produces both conductions (electronic and ionic) enabling a double contact structure and surface. The used materials are often perovskite, which have been made by lack of substitution oxygen on the *A* and *B* sites, thus inducing an ionic conductivity (example: $La_{1-x}Sr_xMn_{1-y}Co_yO_{3+d}$). More recent studies have tested phases A_2MO_{4+d} of structure type K_2NiF_4.

The use of MIEO (mixed ionic electronic oxide) tends to be increasingly common.

NOTE: The recent mixed oxides operate as they are, especially at high overvoltage, due to their low conductivity.

– Anode

It is the place where hydrogen oxidizes. The global reaction is generically written (independently of the fuel):

$$\mathrm{Re}\,d_g + nO_O^X \rightarrow O_xO_n + nV_O^{\bullet\bullet} + 2ne^- \qquad [5.28]$$

The reference electrode is made of a Nickel and *YSZ* composite porous matrix, called "crèmet". The Nickel percentage is about 40% which enables a good percolation of the electronic phase and an expansion coefficient of the cermet close to that of the electrolyte. The surface resistance is about 0.15 Ω.cm² at 850°C.

With pure hydrogen, this cermet gives good electric performances. With hydrocarbon fuel, these performances are decreased by the formation of coke because of the nickel activity. In that case, alternative materials among mixed conductor metallic oxides (ZrO_2- Y_2O_3-TiO_2 solutions) are under investigation.

– Interconnectors

The requirements of these components are challenging. They must exhibit:

– a chemical stability under oxidative conditions (O_2, H_2O) and under reductive conditions at high temperatures (700–1,000°C);

– a chemical stability in presence of electrodes materials;

– a good conductivity;

– a low contact resistance with the electrodes;

– a thermal expansion coefficient compatible with that of the cell.

Two types of materials may potentially be used depending on the temperature.

Above 900°C, ceramic materials are used with Lanthanum chromites as a reference.

Below 900°C, active researches on refractory steels are under progress. The chromium-based alloys present the drawback of a generally high contact resistance. Ferritic steels, which may be used below 800°C, have a moderate price and a high thermal expansion coefficient.

Contact resistances of more than 100 $m\Omega.cm^2$ have been measured.

NOTE: Gas tightness is a difficult problem in these ranges of temperature and atmosphere. The used materials are glasses, mica, and precious metals.

5.2.4.1.3. Performance levels and lifetime of SOFC

– Planar technology

The planar technology used to increase the power density faces many technological difficulties concerning the bipolar plates and the gaskets, along with the realization of large planar cells. This technology has not yet reached the technological maturity, power and lifetime of the tubular technology, which was developed long before it.

Figure 5.11 illustrates the evolution of the performances obtained over the last ten years and the power of the planar technology. These results have been achieved at the Julich (Germany) Research Center, which is one of the leading laboratories in Europe on this technology.

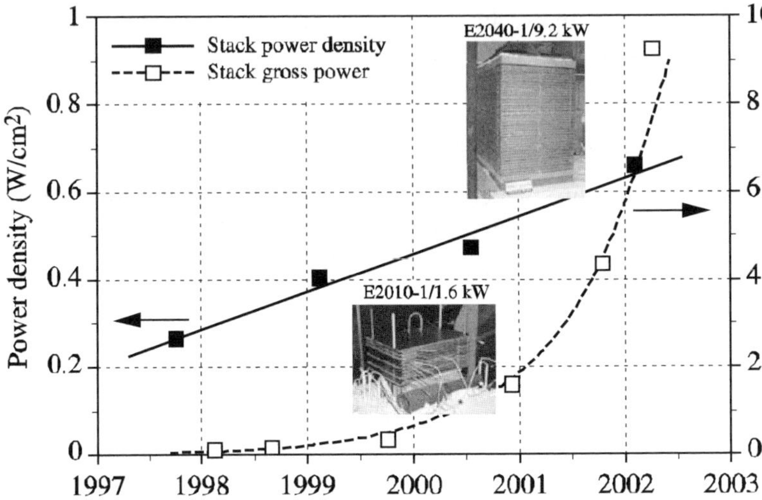

Figure 5.11. *Power and performance evolution of the stacks manufactured by FZJ in the last 10 years*

Performances with supported anodes cells of 0.6 W/cm² at efficiencies (based on LHV (low heating value)) of 59% have been obtained (see Figure 5.12). The lifetimes exceed 3,000 hours.

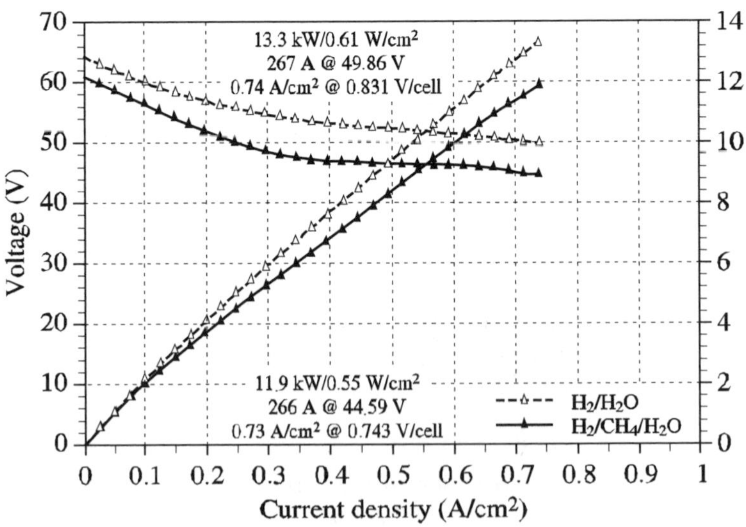

Figure 5.12. *Voltage-current curves of FZJ cells*

– Tubular technology

This technology developed rapidly in the 1980s (Figure 5.13). In principle, it raises fewer material and gasket problems than the planar technology. The problems due to the differential thermal expansions' interconnector-cell are avoided and the number of gaskets is reduced. Furthermore, the gaskets are sent back in the cooled areas. If the volumic power denities are less than those given by the planar technologies (ratio 2 to 3), the power levels of the stacks and the lifetime obtained are much better (factor 10).

Flattened tubular geometries with internal reinforcement have been developed, particularly by Rolls Royce (Figures 5.14 and 5.15). It enables significant increases in the power density and the cost.

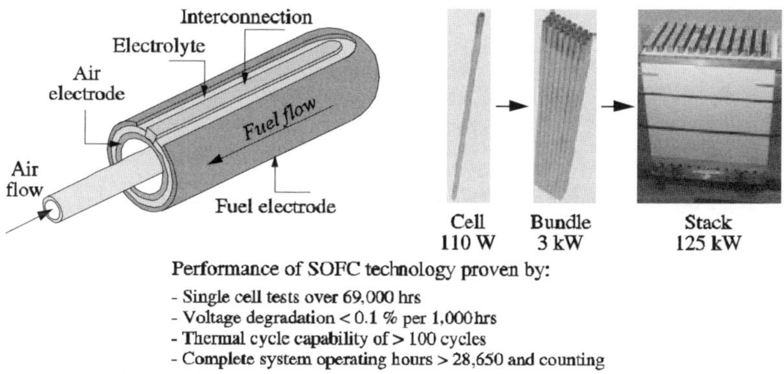

Figure 5.13. *Performances of the tubular technology by Siemens (from Lucerne Fuel Cell Forum 2006/ Operation of the Siemens SOFC generator CHP100/Gianmichele Orsello)*

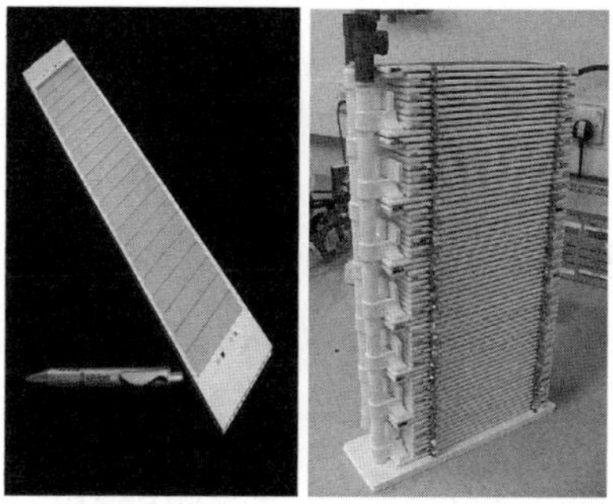

Figure 5.14. *Cell and stack from Rolls Royce (Lucerne Fuel Cell Forum/ Development of a 1 MW SOFC at Rolls Royce/Gerry D. Agnew)*

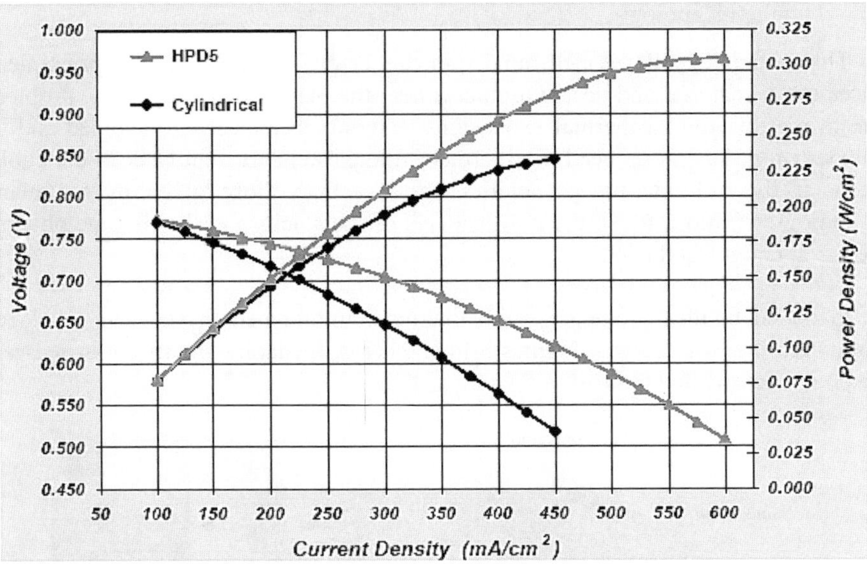

Figure 5.1.5. *Performances obtained on a standard tubular technology, called flattened tube, from Siemens Westinghouse (source: [TEC 04])*

5.2.4.2. *Specificities of the Polymer Exchanger Membrane Fuel Cell (PEMFC)*

The major advantage of the PEMFC is its high power density and its low operating temperature; this is why it is dedicated to transport applications. The high power density is due to the use of a polymer electrolyte with, on one side, a high conductivity and, on the other side, a good aptitude to be transformed into a membrane of a few tens of microns thick.

Moreover, its low temperature operation and its polymer membrane make it easy to manufacture the cells, and the interconnector materials and the gaskets are classical.

The drawbacks are:

– a high activation overvoltage, even with a catalyst like platinum;

– a high cell-cost due to the cost of polymer and platinum;

– the difficulty of using another gas than pure hydrogen, which has to be free of carbon and sulphur monoxide;

– a particular difficulty associated with water management.

Electrolyte specificities

The reference electrolyte of the proton exchange membrane fuel cell is Nafion®, developed by Dupont de Nemours. It is a perfluorosulfonic polymer.

There are many competitor commercial polymers of similar structure. Their main trademarks are Flemion® (Asahi Glass), Aciplex® (Asahi Chemical), Hyflon® (Solvay), Gore (Gore).

The membranes' thickness ranges from 20 microns to 100 microns and even more, according to the final application. The membranes are made of either polymer alone or a porous weave supporting the proton conductor polymer (in the case of Gore). The protons' conduction inside these membranes is due to the presence of a network of SO_3^- anionic sites creating hydrophilic pores in the polymer matrix (Figure 5.16).

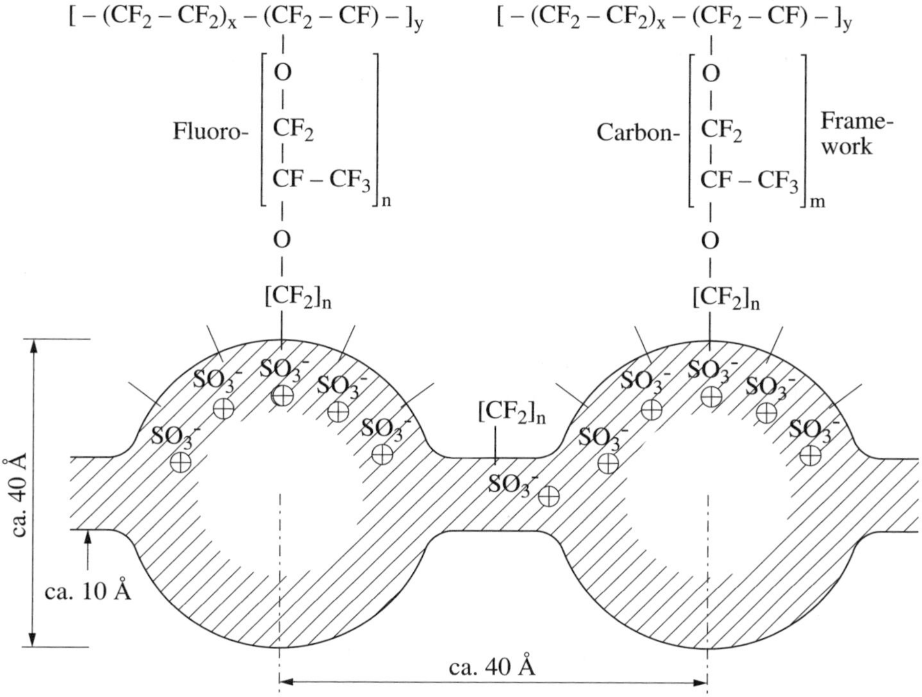

The charge transport mechanism through the Nafion membrane : x = S-f3 ; y ≈ 1 000 ; m = 0-3 ; n = 2-6

Figure 5.16. *Representation of the perfluorated polymer structure of Nafion® type*

Ionic conductivity

The ionic conductivity of this type of electrolyte depends on its water content, which is characterized by the λ parameter. The λ parameter is equal to the number of water molecules per sulfonated site contained in the membrane.

The water content depends on the material and the water activity (a) in the membrane.

$$\lambda = \sum \alpha_i a^i \quad [5.29]$$

This activity depends on the interface between the environment and the membrane. In particular, it depends on the thermodynamic properties of the water included in this environment.

If we consider a plane[7] membrane in equilibrium with a hydrated gaseous phase, the equilibrium condition connects the activity to the partial pressure of the water in the gas and to the local saturated steam pressure, which means that near the membrane:

$$a \approx \frac{P_{H2O}}{Psat(T)} \quad [5.30]$$

The activity is maximum ($a=1$) when the steam partial pressure equals the saturated steam pressure.

NOTE: The saturated steam pressure depends on the temperature only, in an exponential way. In the temperature and pressure case of the PEMFC, the saturated steam pressure can be given by the Antoine formula:

$$P_{sat}(T) = \exp\left(23{,}1961 - \frac{3816{,}44}{T - 46{,}13}\right) \quad [5.31]$$

The relationship between λ and the activity depends on the membrane material. In the case of Nafion® membranes, the relation between λ and the activity is [ZAW 91]:

7. In the electrodes, where the gaseous phase membrane interface in not plane (porous area), the water activity depends on the local capillary pressure, which in turn depends on the local curvature radius (pore radius).

$$\lambda_g = 0.043 + 17.81a - 39.85a^2 + 36a^3 \qquad [5.32]$$

Moreover, the relation between the membrane conductivity and its water content is linear and is written as follows:

$$\sigma = C(33.75\lambda - 21.41)\exp(\frac{-1268}{T}) \quad (\Omega m)^{-1} \qquad [5.33]$$

where C is a coefficient that depends on the membrane thickness. C is equal to 1 for a Nafion® 117 membrane with 175 μm thickness.

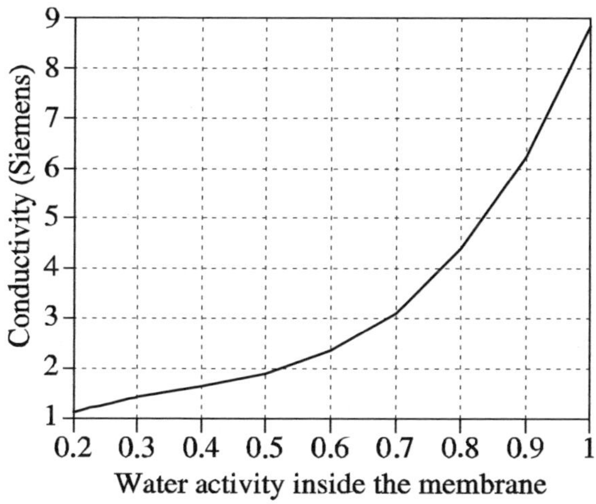

Figure 5.17. *Evolution of the ionic conductivity (S/m) of a Nafion® 117 membrane versus the water activity in the membrane*

The evolution of σ with respect to a is parabolic (Figure 5.17). The conductivity drop is rapid, between 1 and 0.6.

Below 0.2, there is a lack of conductivity, with the number of moles per sulfonated site being less than 2.

NOTE: In the case of a membrane in equilibrium with liquid water, this activity suddenly ranks to a value higher than 1 (Schrœder paradox) and the water content is thus given by:

$$\lambda_l = 9.38 + 0.138T \qquad [5.34]$$

The membrane conductivity follows this increase, as shown by the resistance measurements results obtained on a 43 cm^2 cell in equilibrium with a gaseous atmosphere saturated with steam or with liquid water.

Membrane type	Cell environment	Temperature	$R(m\Omega)$
Nafion® 112	Saturated gas	60°C	3.0
Nafion® 112	Liquid water	60°C	2.4

Table 5.9. *Resistance measurements at the 43 cm^2 cell terminal; $I = 0$; membrane thickness 50 μm*

This dependency of the electrolyte conductivity on its water content, which in turn is related to the water content in the gaseous environment, is one of the water management problems faced in the PEFMC. Note that the temperature is the dominant parameter in this dependency.

Electro-osmosis coefficient

This type of electrolyte is similar to a liquid electrolyte when the water content is important: a water transport is induced by the proton transport through the membrane. The proton is solvated by a certain number of water molecules. Values between 1 and 6 are generally used. These water molecules migrate with the proton from the anode to the cathode of the cell and may lead to two feared effects:

– drying of the anode and consequently a local reduction of the ionic conductivity of the electrolyte. The consequence is an increase of the electrode global overvoltage;

– additional supply of water to the water produced at the cathode. The consequence is the premature flooding of the cathode.

This effect is counterbalanced by the water diffusion through the membrane, which is due to the water concentration gradient in the membrane between cathode and anode.

The electro-osmosis effect can be observed with strong membrane (thickness >100 microns) and high current densities, conditions for which the diffusion flux cannot counterbalance the electro-osmosis.

Presently, the membrane thickness of the cells being less than about 50 microns, the electro-osmosis effect, except at very high current density, is rarely observed today.

Reactive gases solubility in the electrolyte

The reactive gases' solubility in the electrolyte contained in the electrode is a critical parameter in electrochemical performance of the electrode.

For Nafion®, the reference experimental values [AMP 95] lead to the following equations between the electrolyte concentration and the partial pressure in the gaseous phase:

$$c_{O_2}^* = \frac{P_{O_2}^*}{5.08 \cdot 10^5 \exp \frac{-498}{T}} \qquad [5.35]$$

$$c_{H_2}^* = \frac{P_{H_2}^*}{1.09 \cdot 10^5 \exp \frac{77}{T}} \qquad [5.36]$$

Particularities of the electrodes

The operation at low temperature requires:

– in the triple contact zone, the presence of a performing catalyst: platinum;

– the evacuation of liquid water generated outside the electrode because of the flooding risk of the electrode. In the standard operating conditions, most of the generated water may be liquid because of the low operating temperature. If it is not eliminated from the electrodes, it can fill the pores, preventing the reactive gases from reaching the triple point and consequently reducing the reaction rate. This elimination is generally made by means of a hydrophobic agent in the electrode and of a corresponding local capillary pressure.

The electrode is generally made from a felt or a carbon fabric backing; on one face of this backing, a thin porous layer is deposited by different methods (pulverisation, enduction, etc.). This layer is usually a mixture of a platinized carbon, Nafion®, and generally of a hydrophobic agent (PTFE).

The ideal reaction scheme considered in modeling is to assume that the electronic conducting media is made of carbon and platinum particles which cover the surface and that the ionic conducting media is Nafion® deposited on platinum as a very thin film (less than one micron). The whole represents a porous solid skeleton in contact with the reactive gas. This gas is supposed to dissolve in the Nafion® film and to diffuse towards platinum, which is the triple contact location. The generated water is evacuated in the opposite direction by Nafion®. Then it is transported in the porous area of the active zone, and from the backing to the bipolar plate distribution channel, either by the gaseous phase, if it is not saturated, or under the effect of the capillary pressure generated by the liquid water in this porous media.

The gas solubility in the electrolytes and their diffusion coefficient are thus important parameters.

To take into account the resulting two-phase flow is one of the major problems in modeling the PEMFC electrodes.

The thickness of the backing is generally from 100 to 300 microns. Their porosity is about 60%. The active zone has typically a thickness of about 20 microns.

The platinum loading for both electrodes is typically about 0.2 to 0.3 mg/cm^2.

5.3. The system aspect

A fuel cell stack is a device that converts chemical energy into electric energy. The cell is only the place of this conversion, and many functions must be ensured by auxiliary equipments in order to obtain electric power from a given fuel source, and which is available to and compatible with the user needs. Finally, the fuel cell stack is only one element of the system containing all the auxiliary subsystems necessary for its operation.

The most common definition of a fuel cell system is applicable to each type of generator set: it consists of an autonomous system which delivers electric energy according to the demand of a user network, and from a given fuel source.

The diversity of the possible applications (stationary, automobile, portable) and of possible fuels (hydrogen, natural gas, methanol, gas oil, etc.) explains why this type of converter is becoming a matter of interest. It also requires the early integration of the constraints coming from the needs of the dedicated application in the design of the stack and of the associated system. To reach this goal, a functional analysis of the system is necessary; a functional analysis enables us, on one hand, to define which type of cell is the best adapted and, on the other hand, to steer the design of the stack and of the system.

5.3.1. *Functional analysis of a fuel cell system*

The expression of the requirements of the targeted application enables the determination of all the interactions between the system and the outside environment, whether they are physical (pressure, temperature, vibrations, etc.) or functional (starting time, control mode, etc.). The differences from one application to another may be very important: the environment constraints are very different depending on whether the fuel cell is integrated in an automobile vehicle, in an Ariane rocket or in a submarine!

Generally speaking, the functional analysis usually starts with an investigation of the needs of the global system (a vehicle, for example), then goes on to the analysis of subsystems (drivetrain, powertrain, etc.), up to the analysis of the fuel cell system, which is itself divided into subsystems.

As an example, a generic system and its main "interactors"[8] is presented in Figure 5.18. The fuel cell system contains five subsystems. Their composition very much depends on the characteristics of their interaction with the external environment. Thus, the type of fuel will have a strong impact on the conditioning subsystem of the input-output anode (with or without a reforming, for example). For the same reason, the comburant tank may be the external air in many applications, with significant consequences on the cathode input-output conditioning subsystem.

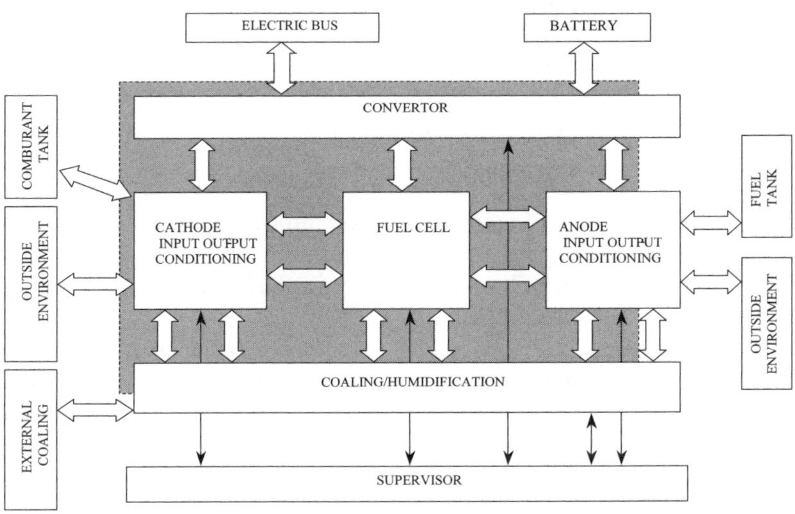

Figure 5.18. *Functional decomposition of the fuel cell as a generator*

8. External environment elements in interaction with the considered system.

5.3.2. *The major components of the generator*

5.3.2.1. *The cathode input-output conditioning*

In the cathode input-output area the air is compressed (or at least pulsed towards the cell if under atmospheric pressure), and humidified if necessary. It may possibly be oxygen enriched (selective membrane [JAM 94] or oxygen tank under pressure for a short periods).

In the cathode output, the air is charged with water and sometimes with a liquid water flow. In most cases, the recovery of part of the water is necessary to continuously satisfy the water needs for the humidification of the gases or of the membrane. Only a low operating temperature enables this point to be avoided.

The use of an expander at the cathode output enables the recovery of a non-negligible part of the compression energy (up to 50%). According to the variation of the cell performances with the operating conditions, the generator set can exhibit an increased efficiency, a reduced mass (the addition of equipment is compensated by the size reduction of the unit), and a reduced cost. But this increases the system complexity and the noise.

5.3.2.2. *The anode input and output conditioning*

The fuel used in a PEMFC is hydrogen. It can be stored or generated directly from a hydrocarbon in the system. Actually, compared to gasoline storage, hydrogen storage is penalizing in terms of performance, cost, and global system volume. In the case of reformed liquid fuels, the technology of onboard reformers has had many improvements during recent years and has led to performing and low-volume systems [MIT 06].

The development of a hydrogen distribution infrastructure, which is necessary if hydrogen is stored onboard a vehicle, is a constraint which has, over many years, stimulated the development of liquid fuel reforming [GIL 97].

Nowadays, the safety problems related to methanol use have led to a reduction of research efforts in this direction. The fuel nature influences the architecture of the generator at the anode input-output conditioning. Two cases must be considered: reformed fuel and molecular hydrogen.

5.3.2.3. *Onboard molecular hydrogen*

This is the most commonly used fuel in the fuel cell vehicle prototypes, and most of the stationary system demonstrators use it also, whatever its storage mode. If compressed storage (bottles of 700 bars and above are increasingly common) is the most commonly used, other alternatives are possible: liquid, or hybrids adsorbed in solid or liquid.

Generally speaking, a pure hydrogen supply may be considered for the optimal use of onboard energy quantity. Actually, the hydrogen concentration at the anode in the cell is thus limited only by the steam presence and by the nitrogen permeation from the cathode compartment. Even if a purging system is necessary, its optimization would enable the reaction of a minimum of 95% of onboard hydrogen, [PIC 98] to be compared with the efficiencies of hydrocarbon reformers (see next section).

Two operating modes are possible: the recirculation mode (with pump or ejector), where the gas mixture at the cell output is reinjected at the input, and the dead-end mode which involves periodic purges of the anode to eliminate water and possible pollutants.

Finally, humidifiers and condensers are sometimes present in the systems, but today the tendency is to eliminate these components, especially when improving the MEA tolerance to dry gases.

5.3.2.4. *Hydrocarbon reforming*

The generation of gaseous hydrogen requires a reactor that converts the hydrocarbon fuel into hydrogen.

The main processes are the reforming of methanol, of natural gas, or gasoline, or, more seldom, ammoniac [GIB 98] and hydrazine dissociation. Hydrocarbon reforming is the most commonly used process. It operates at temperatures from 300 up to more than 1,000°C depending on the hydrocarbon.

Some additional reactions, such as the selective oxidation of carbon monoxide usually requires a second reactor.

Many methods for reforming hydrocarbon exist:

– steam reforming (SR), which is endothermal and leads to the best efficiencies;

– partial oxidation (POX), which is exothermal and is well adapted to heavy hydrocarbons;

– a combination of both (auto-thermal reforming (ATR)), which facilitates the thermal management of the whole set.

This fuel transformation leads to a gas mixture which must supply the anode of the fuel cell. However, the oxidation of the hydrocarbon will generate excess CO_2, but also a non-negligible quantity of CO, which is capable of poisoning the catalyst. Indeed, even the platinium-ruthenium alloys, more tolerant to CO (up to 100 ppm) than pure platinium, are not enough to avoid a mixture treatment.

This cleaning is often done by a CO preferential oxidation (PROX) on a catalyst and in the presence of oxygen. Another option consists of extracting hydrogen from the mixture through a membrane [BOR 96].

	Steam Reforming (~300°C)				Partial Oxydation (~700°C)			
	Theory		Practice		Theory		Practice	
	Input	Output	Input	Output	Input	Output	Input	Output
H_2O	50%	-	60%	12.5%	-	-	-	11%
CH_3OH	50%	-	40%	-	29.5%	-	29.5%	-
H_2	-	75%	-	65%	-	41%	-	30%
O_2	-	-	-	-	14.5%	-	14.5%	-
N_2	-	-	-	-	56%	38.5%	56%	38.5%
CO_2	-	25%	-	21%	-	20.5%	-	9.5%
CO	-	-	-	1.5%	-	-	-	11%

Table 5.10. *Molar composition of gas mixtures at the output of a methanol reformer*

Whatever its composition, the mixture supplying the anode excludes the operation in recirculation mode or dead-end mode. Actually, the presence of unburned species requires the permanent elimination, at the anode output, of a poor hydrogen solution. The latter may then be burnt in order to use the maximum of the energy available in the fuel in the heating of the reformer when the reaction is endothermal.

In the case of methanol, as the steam-reforming temperature is exceptionally low (about 300°C instead of 800°C or more for the other hydrocarbons), this technique is generally used [NON 97], [KRA 97]. However, it may be coupled to the partial oxidation [EDW 98].

In the case of gasoline or diesel, the method used is partial oxidation. Such a reactor may require many different stages, with many catalysts to make up to five different reactions. These reactions aim, at the same time, to increase the hydrogen concentration in the gas mixture and to reduce the pollutant rates.

5.3.2.5. *Impact on the architecture of the generator*

The complexity of the system is greatly increased when a fuel reformer is used (for an example of methanol steam-reforming, see Figure 5.19). Compared to a system using pure hydrogen as onboard fuel, a generator set containing a hydrocarbon reformer presents two major differences:

– the thermal architecture of the system includes one component, the temperature of which is far higher. It often consists not only of one but of many components at different temperatures;

– the response time of a reformer penalizes the response time of the whole system. Its high temperature is a major drawback on this particular point.

Moreover, during transient power demands, the efficiency of the reaction drops and the quality of the gas obtained from the selective treatment units is reduced (for example, CO preferential oxidation).

A drop of performance and of global efficiency caused by CO poisoning of the electrodes will follow [JOS 00].

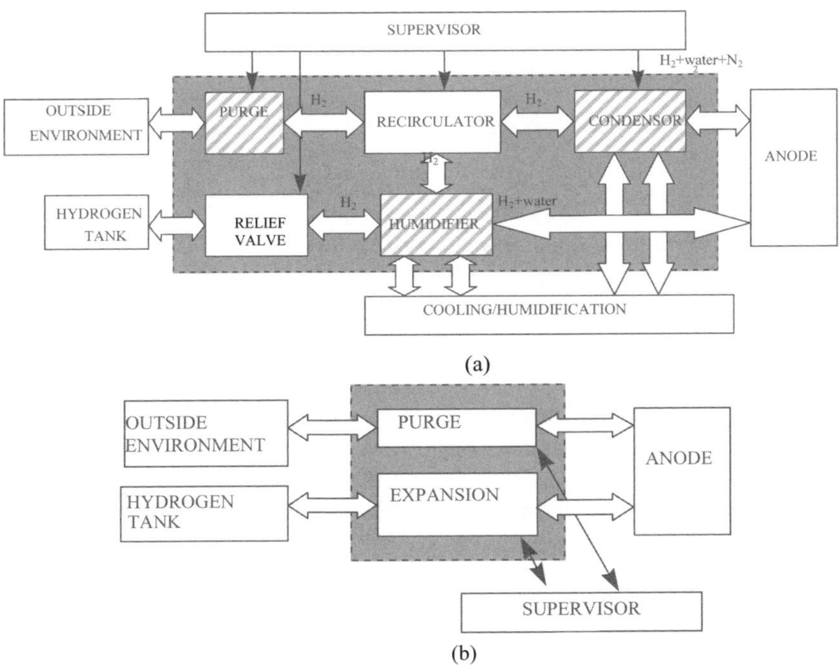

Figure 5.19. *"Anode conditioning" system decompositions: a) recirculation; b) dead-end*

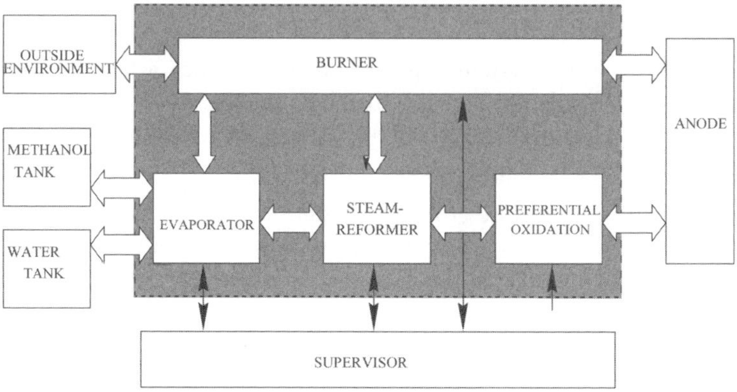

(c)

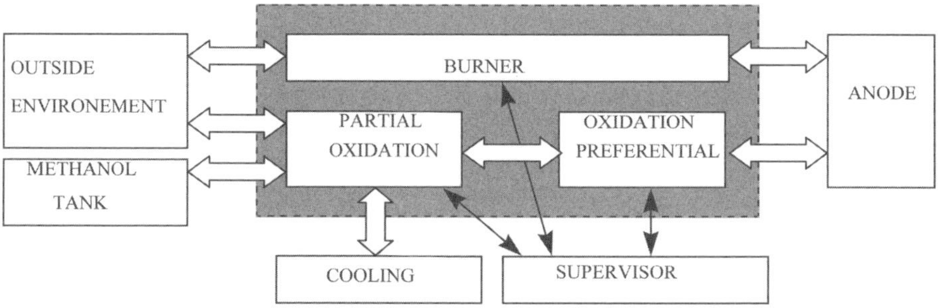

(d)

Figure 5.19. *"Anode conditioning" system decompositions: c) methanol steam-reforming; d) methane reforming*

5.3.2.6. *Electric converters and hybridization*

The electric converter connected to the stack terminals enables the use of electric energy at the output of the generator set. Actually, not only the cell voltage varies with the load, but the stack design does not always meet the requirements of the electric powertrain in terms of nominal voltage.

Many converters may be necessary in the generator set because the output voltage of the main converter may not be convenient for the supply of the auxiliaries (compressors, circulation pumps) or for the energy management strategy in the case of a hybridization of the stack with power batteries.

For many applications, including automotive, the generator must be able to respond very quickly to the power variations. The response time of the cell is theoretically very short, but its auxiliaries (compressors, etc.) are often much slower. The problem is more acute if the vehicle uses a reformed fuel (see section 5.3.2.4).

The hybridization ratio of the generator may be more or less significant. Two options must be considered:

– assisting the fuel cell stack with a battery or a supercapacitor;

– permanent recharging of a battery ensuring the power supply to the powertrain (also called range extender), in the case of a vehicle.

In the first case, the fuel cell is designed to supply the power needs over a long period (for a vehicle: stable maximum speed on highway).

It is coupled to the battery in every case of high power demand for a short time (acceleration, steep slopes of limited length).

This technique enables the improvement of the low power efficiency, generally reduced due to high energy consumption of the auxiliaries, especially the compressor (Figure 5.20). The problem of efficiency drop at low power, which is mainly faced in urban driving cycles in the case of the vehicle, is then less important.

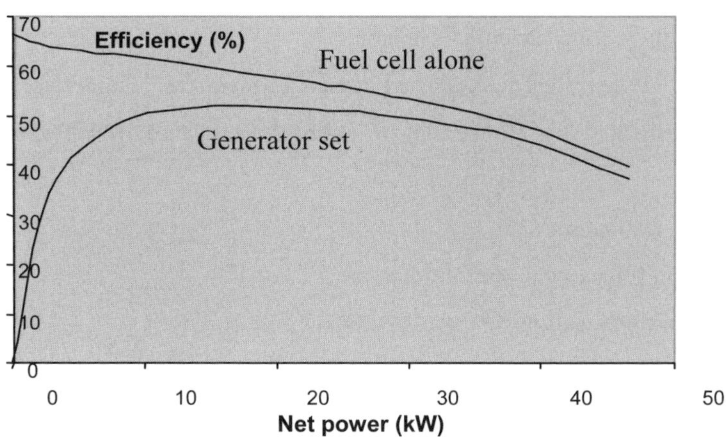

Figure 5.20. *Parasitic load impact example on the efficiency*

The range extender technique avoids the generator operating in a transient state, since it is designed, in this case, to generate a constant power. It is therefore designed in such a manner that its efficiency is optimum. However, the energy density of such a system is, most of the time, lower than that of a simple generator, which can raise volume problems.

The advantages and disadvantages of both options are outlined in the following table.

	With hybridization	**Without hybridization**
Advantages	Better efficiency at low power Breaking energy recovery	Reduced volume Simplified energy management
Drawbacks	Volume, significant weight and costs, mainly because of batteries	Poor efficiency at low power Longer response time and starting time, especially with a reformer

Table 5.11. *Advantages and disadvantages of a fuel cell system with and without hybridization*

5.3.2.7. *Cooling and humidification*

In a fuel cell generator, there are many heat sources:

– thermal losses from the electrochemical reactions;

– air heating during the compression;

– heating of the different electric equipment (converters, pump motors, etc.);

– heat generated by exothermal reactions in the case of reforming (partial oxidation).

There are also heat sinks:

– heat necessary to gas humidification;

– endothermal reactions in the case of steam reforming.

The cooling circuit is thus complex and is different according to the operating mode of the generator. It can be divided into many circuits, with or without the same fluid, and having different temperature levels.

The cooling liquid used in the cell must have a very low electric conductivity ($R > 50$ kΩ), in order to avoid any electric short-circuit risk between the cells. The water used for the reactive humidification generally comes from the water recovered during the reaction in order to avoid any filling need of an external water tank.

5.3.3. *The key points of a system operation*

5.3.3.1. *Air compression*

The operating pressure has an impact on the efficiency of the electrochemical conversion (Figure 5.21), the reactants humidification, and the energy consumption of the compressor.

The choice of a low-pressure operation enables the reduction of the complexity of the system by elimination of the output expander, although it may increase the size of the humidification/ condensation set.

In fact, the percentage of steam in the air increases when the pressure decreases, which raises problems in relation to the size of the humidifier and the drying out of cell membranes, which leads to a continuous drop in the performances.

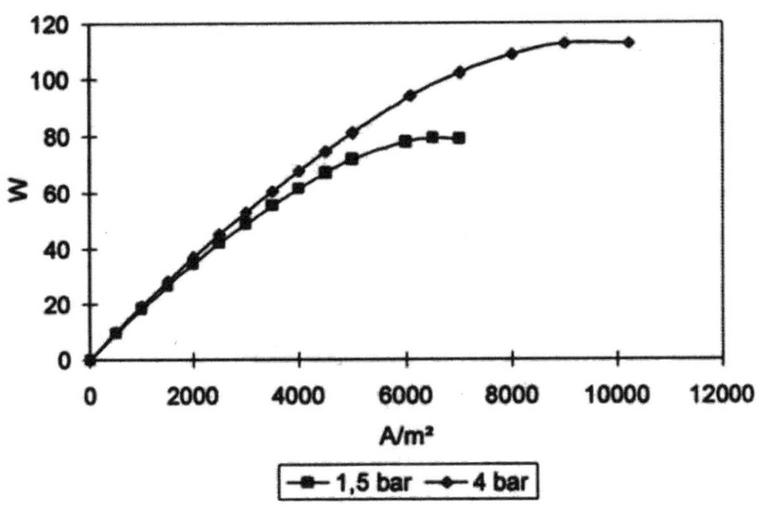

Figure 5.21. *Effect of the operation pressure on cell power [BET 98]*

A high-pressure operation requires an efficient recovery of the mechanical energy of the gases at the output of the cell; otherwise a poor global efficiency of the system, especially at low power, is obtained. However, it enables easier thermal and water management and higher-power densities for the cell. This is possible due to an operation in high current density ranges, because the effects on the electrochemical conversion are not visible at low current density (shown in Figure 5.21 from 2 to 3,000 A/m²).

The air compressor, which absorbs energy, also has an impact on the efficiency of the motor generator set. The theoretical work needed for an adiabatic compression of the air quantity corresponding to the production of one mole of water may be written as:

$$W_{comp} = \frac{1}{2}\frac{St_{air}}{x_{O_2}}\frac{\gamma}{\gamma-1}RT_{input\ comp}\left[\left(\frac{P_{output\ comp}}{P_{input\ comp}}\right)^{\frac{\gamma-1}{\gamma}}-1\right] \qquad [5.37]$$

The mechanical energy, theoretically recoverable with the adiabatic expansion, is written in the same manner:

$$W_{recov.} = -\frac{1}{2}\left(\frac{St_{air}}{x_{O_2}}-1\right)\frac{\gamma}{\gamma-1}R \text{ turbine input}\left[\left(\frac{P_{turbine\_output}}{P_{turbine\_input}}\right)^{\frac{\gamma-1}{\gamma}}-1\right] \qquad [5.38]$$

In practice, the compression and expansion of air are have an efficiency between 50 and 80%. The example of a compressor-expander presented in Figure 5.22 has a full load efficiency of 80% for compression and 70% for expansion.

Calculations starting with these values, which also appear in Figure 5.22, show the advantage of an expansion energy recovery at the cathode. In this figure, the work recovered by the expander is subtracted from the total compression work to give the actual work supplied.

This work is expressed as a function of the LHV of hydrogen, denoted LHV. This enables us to directly see what will be the compression unit impact on the global efficiency of the generator. For example, at an operating pressure of 3 bars and a stoichiometric ratio of 2, the necessary work at compression represents an efficiency drop of 9% if the expansion energy is not recovered, against only 5.7% if it is recovered.

The expansion energy calculation does not take into account the load losses between the compressor output and the turbine input. The error margin, in Figure

5.22, is low since at these operational pressures, losses represent only a fraction of the relative pressure.

For lower pressures, the usefulness of such a system becomes more questionable: because of the system the efficiency gain is low compared to the volume increase.

As a comparison, a 1% gain of LHV is equivalent to a voltage gain of 12.5 mV for a basic cell.

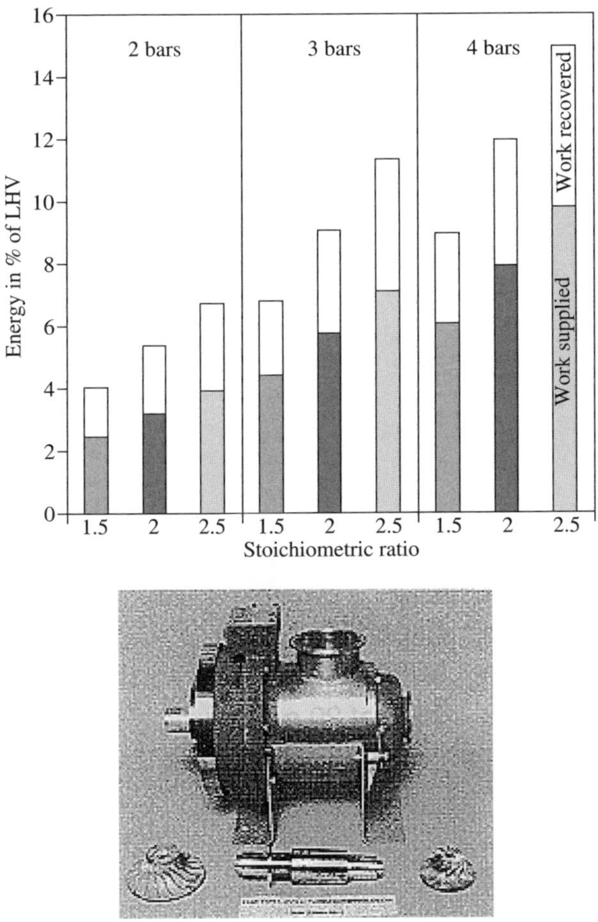

Figure 5.22. *Mechanical energy recovery as a function of the absolute pressure and of the stoichiometry; example of compressor-expander coupling [HEN 98]*

5.3.3.2. *Problem of water management from cell to system*

At the core of the cell, the problem of water management in the cell of a PEMFC is expressed by the following paradox:

– the water produced at the cathode must be eliminated to prevent flooding of the cathode (reaction interruption);

– enough water must be saved in the membrane and electrode electrolyte; otherwise its conductivity drops and local ohmic overvoltages appear.

This problem is illustrated in Figure 5.23 where modeling has enabled us to understand the current limits observed in with two cells in which the only difference is electrolyte thickness (7 and 2 milli-inch). In the case of thick electrolyte (Nafion® 117: 175 microns), the anode dries out until stopping the electrochemical half-reaction. The λ value at the anode drops because of the electro-osmosis effect due to the flow of electrical charges. The water diffusion cannot balance this effect. With the thin electrolyte (Nafion® 112: 50 microns), the diffusion completely counterbalances the electro-osmosis. The water content at the anode is constant and it increases is greater. However, at high current density, the water generated at the cathode is no longer eliminated. The electrode is flooded.

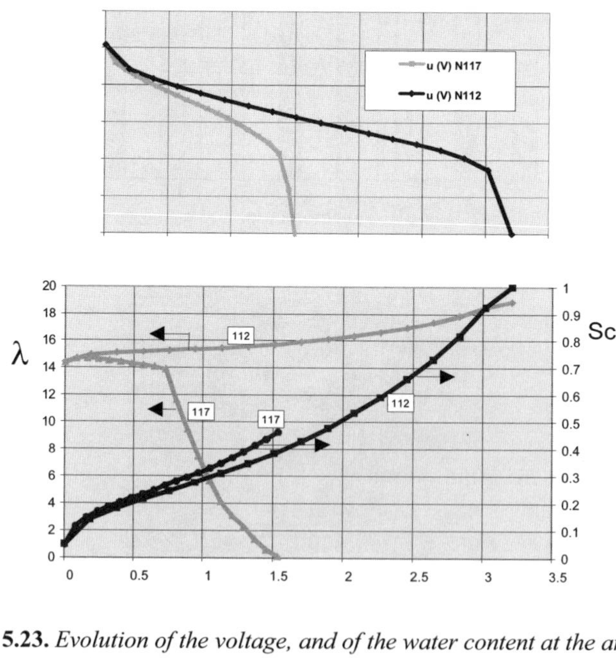

Figure 5.23. *Evolution of the voltage, and of the water content at the anode with the current for a Nafion® 117 and Nafion® 112. Modeling results experimentally validated*

To save enough electric conductivity of the electrolyte, it is necessary to ensure at the electrolyte interface, which means in the gaseous medium, minimum water activity.

In the past, in a conservative manner, the gases entering the cell at the anode and at the cathode were systematically humidified at saturation, which was a paradox in most cases; the water quantity produced by the reaction was enough to humidify the gases crossing the cell. On the anode side, this enabled the counterbalancing of the drying effect of the electro-osmosis. On the cathode side, this prevented the drying out of the cell input area due to the normally high gas flow. The consequence was that it was difficult to achieve high current densities because of the cathode flooding.

The arrival of membranes with a thickness of less than a few tens of microns and also the arrival of electrodes with capillary pressure gradient have simplified this management.

On most applications today, the hydration rate of the gases at the cathode inputs is less than 50%. The operating conditions (temperature, pressure, and current ranges) are adapted to avoid the drying out.

Remark on low temperature operation (<60°C)

At these temperatures, the steam pressure is low and the amount of water produced by the reaction is generally higher than the pressure needed for the water saturation of the gases entering the cell. A performant operation is then possible without humidification.

The problem of the reactants' humidification is in terms of water and thermal energy quantities requested. Technically, gas humidification may be made either outside the cell by permeable membrane exchangers (bubbling may be considered only for laboratory cells), or inside the cell by direct water injection or via permeable membranes (association of the membrane humidifier and the cell in the same stack).

The energy necessary for humidification may come either from the cooling liquid of the cell, or from the output gas flow of the cell at the cathode or at the anode. Thus, heat and mass exchangers have been adapted to fit the fuel cell application; they are called gas-to-gas humidifiers.

To recover the necessary water for humidification, two solutions exist: either a standard condenser cooled by the cooling circuit of the cell is used, or a gas-to-gas humidifier is integrated in the system. This last option is presently the most promising since it has many advantages. This equipment is generally compact, it does not need connection with the cooling fluid circuit and it does not require liquid water in the system (interesting for freezing situations). The only drawbacks are: the associated cells must accept a partly steam-saturated atmosphere (relative humidity above 80% at the cell input is difficult to reach) and, for some equipment, a connection to the electric network is necessary, using a consumption of a few tens of watts.

Two types of gas-to-gas humidifiers exist:

– enthalpy wheels: porous cylinders that retain water by adsorption. The cylinder rotates on its axle, and each sector is thus successively crossed by the humid gas and by the dry gas. Steam is adsorbed with the humid gas, then is desorbed with the dried gas (Figure 5.24);

– membrane exchangers: a water-permeable membrane separates the dry and the humid gases. The water is adsorbed on the humid side, crosses the membrane by diffusion, and is desorbed on the dry side. The currently available devices use Nafion® tubes.

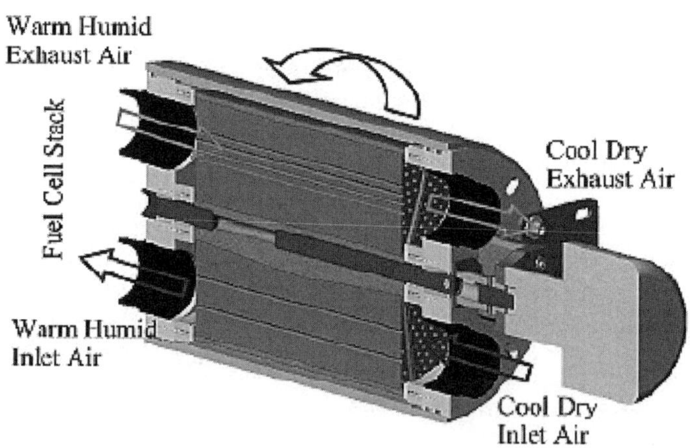

Figure 5.24. *Schematic of an enthalpy wheel (Emprise Company)*

5.3.4. *Present performance levels of PEMFC systems*

Nowadays, many fuel cell system prototypes have been developed worldwide. Some systems have been commercialized for low-power stationary applications by the Canadian firm Ballard.

In terms of pure performance, the evaluation criteria of the systems are linked to their power density (either kW/L or kW/kg), associated with an energy efficiency. In fact, the same system could present a high power density and low efficiency or a low power density and high efficiency, depending on the operating point on the polarization curve. At the moment, it is often difficult to obtain simultaneously associated values of power density and efficiency. Moreover, the power density is often given for the cell and not for the system.

In the automotive field, the performances are better because of the importance of the volume criteria for this application. Some examples of the performances achieved by these cells (and not systems) are given in Table 5.12.

	Power (kW)	**kW/L**	**kW/kg**
CEA/ /PSA	80	1.5	1.0
GM	102	1.75	1.25
Honda	50	1.4	1.0
Ballard	85	1.1	0.9

Table 5.12. *Performances of some automobile-specific PEMFC prototypes*

From a lifetime point of view, improvements have to be made to reach the objectives given for both automobile (5,000 hours) and stationary (many tens of thousands hours) applications. The lifetime evaluation very much depends on the test protocol used, which makes it difficult to compare figures from different sources. The operation type (stabilized, cycling, including or excluding stop periods) is a crucial point.

For the automotive application, the present lifetime, on an operation mode including cycling and stop periods, is in the order of 1,000 to 2,000 hours. The lifetime is higher in stationary applications under a stabilized operation. Some manufacturers claim more than 15,000 hours, with losses of 6 μV/hour. The lifetime tests are conducted with thin membranes (20 to 25 μm), which are the only ones that enable the performance goals to be reached.

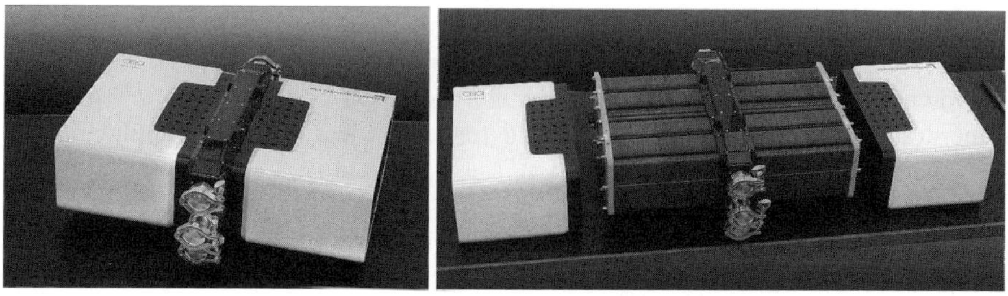

Figure 5.25. *Genepac 80 kW stack by CEA/PSA*

5.4. Energy conversion efficiency

Efficiency is defined as the ratio between the product of the action to be assessed and a reference unit to be defined.

For the fuel cell, as a converter of chemical energy into electric energy, the product is generally the delivered electric power. The reference unit may be:

– a value representing the total potential power entering the cell, which means the power emitted by the total combustion of the entering fuel. It is therefore equal to the product of the combustion enthalpy by the flow of the fuel input;

– a value representing the chemical potential power input. It is equal to the total power emitted by the combustion of the entering fuel less the uncompensated power, which means the entropic loss due to the transformation of the fuel and the comburant in a third body. It is then equal to the product of the free enthalpy of the combustion by the fuel flow input.

In both cases, the value of these enthalpies depends on the thermodynamic conditions in which the combustion has happened. Then the reference choice corresponds to the choice of these reference thermodynamic conditions. In the first case, the efficiency is based on the total energy conversion and is written as:

$$\eta_{total} = \frac{U_{stack} I_{stack}}{N_{comb,e}\left(-\Delta H_{comb}\left(T_{ref.}, p_{ref.}\right)\right)} \quad [5.39]$$

with U_{stack} and I_{stack} taking into account only the usable electric energy; the comb indicator designates the fuel.

In the second case, it is the electrochemical conversion efficiency. It is written in the following form and essentially translates the quality of the electrochemical design, through the gap with an ideal reactor.

$$\eta_{conversion} = \frac{U_{stock} I_{stock}}{N_{comb,e}\left(-\Delta G_{comb}\left(T_{ref.}, p_{ref.}\right)\right)} \qquad [5.40]$$

5.4.1. *Energy efficiency equation*

5.4.1.1. *General equation*

For a standard energy chain, the fuel and the oxidizer are available or stored under the atmospheric pressure and at the ambient temperature in the phase where they are stable under these conditions. The reaction product is ejected into the atmosphere at atmospheric pressure.

We can then generally consider the standard conditions as reference conditions for the evaluation of the combustion enthalpy and the atmosphere temperature (25°C, for example) in which the water is finally discharged. If this temperature is low, the major part of the generated water is liquid and the combustion enthalpy corresponds to the high heating value (HHV).

NOTE: This is only an arbitrary choice of reference power. It is not necessarily the most significant. Its relevance to, for example, the case of the efficiency of an onboard energy chain in which the fuel is stored in a specific phase or condition can be considered.

In another convention, used for historical reasons on internal combustion engines, instead of using the HHV, the LHV is used considering that at the atmosphere and at the ambient temperature, the water produced is not liquid but steam.

Depending on the fuel, the difference between both efficiencies may or may not be significant: 2% for gas, 15% for hydrogen.

The efficiency of the energy conversion in the reference cell, with respect to the HHV of the fuel, is written as:

$$\eta_{energy} = \frac{U_{stack} I_{stack}}{N_{comb,e} \Delta H_{HHV}} \quad [5.41]$$

The ratio between the input fuel flow and the fuel consumed for the useful current generation I may be written as St_{comb}:

$$St_{comb} = \frac{N_{comb,e}}{n_{cell}} \frac{2F}{I_{stack}} \quad [5.42]$$

We obtain:

$$\eta_{energy} = \frac{1}{n_{cell}} \frac{1}{St_{comb}} \frac{U_{stack}}{\left(\frac{\Delta H_{HHV}}{2F}\right)} = \frac{1}{St_{comb}} \frac{U_{cell}}{U_H^{HHV}} \quad [5.43]$$

where U_H^{HHV} [9] is a "symbolic" voltage corresponding to the total conversion of the combustion energy into electric energy which cannot be achieved (total contradiction with the second principle of thermodynamics).

The conversion efficiency is thus directly proportional to the cell voltage. It depends on the supplied current by the cell and decreases with it because of the energy dissipation due to the irreversible phenomena caused by the current supply.

5.4.1.2. *Taking the system into account*

The efficiency value which will be used in practice is that of the complete system, since only the system makes the energy conversion in an autonomous manner.

However, important differences may appear between the efficiency calculation of the cell subsystem and the one of the generator. For example, the value of $N_{H2,e}$ may lead to a cell efficiency that is artificially low in the case of an anode conditioning which integrates a recirculation: the input flow, in fact, integrates hydrogen which is not issued from the tank, but is recycled in the stack output.

It is therefore necessary to consider the entire system and to integrate all the losses in the auxiliary equipment.

9. This symbolic voltage is often called thermoneutral voltage.

The reference power injected into the system will consider for $N_{H2,e}$ the input flow in the anode input-output conditioning coming directly from the tank. This may take into account the hydrogen losses in the anode-circuit purges.

This reference must be compared to the power really available at the output of the system: it thus integrates the electric converter efficiency and the sum of the electric power consumed by the auxiliaries (compressor, pumps, supervisor, etc.). Finally, the equation is the following:

$$\eta_{system} = \frac{\eta_{convert.}\left(U_{stack}I_{stack} - \Sigma U_{aux}I_{aux}\right)}{N_{comb,e}\Delta H_{HHV}} \qquad [5.44]$$

5.4.2. *Some efficiency values*

5.4.2.1. *Zero current energy efficiency*

The maximum value of the conversion efficiency is obtained at zero current. It is equal to:

$$\eta_{(I=0)} = \frac{1}{St_{comb}} \frac{U_{rev}(T.P)}{U_H^{HHV}} \qquad [5.45]$$

It grows with pressure and decreases with temperature; this last parameter is the most significant of the applications. It, of course, varies with the fuel used.

The values of this efficiency in these standard conditions are shown in Table 5.13 for the main fuels used. The high values obtained are only of theoretical interest; the current supply by the cell quickly degrades this value.

The comparison with internal combustion engines, which transform chemical energy into mechanical energy, must be cautiously made because different final energies are compared. If we assume that the electrical–mechanical conversion may be made with a 100% efficiency, the zero current efficiency obtained with a low temperature cell corresponds to the Carnot efficiency (zero supply) of an internal combustion engine operating with temperature sources higher than 1,500°C.

Data on standard fuels

The following tables give the combustion enthalpies values and voltages at zero current for some standard fuels.

Combustible/Reaction	ne^-	A^0_{298} $(= -\Delta G^0_{298})$ (kJ/mole)	ΔH^0_{298} (kJ/mole)	U^0_{rev298} (V)	$\rho^0_{298}(0)$ (%)	U^0_H (V)
$H_2 + \frac{1}{2}O_2 \rightarrow H_2O(l)$	2	237	-285	1.23	83	1.48
$H_2 + \frac{1}{2}O_2 \rightarrow H_2O(g)$	2	229	-242	1.18	94	1.25
$CO + \frac{1}{2}O_2 \rightarrow CO_2$	2	257	-283	1.07	91	1.47
$CH_3OH(l) + \frac{3}{2}O_2 \rightarrow CO_2 + 2H_2O(l)$	6	704	-727	1.21	97	1.26
$CH_4 + 2O_2 \rightarrow CO_2 + 2H_2O(l)$	8	818	- 891	1.06	92	1.15

Table 5.13. *Thermodynamic function values for some standard reactions in fuel cells (standard value per mole of generated product)*

Combustible/Reaction	E^0_{rev298} (V)	$\rho^0_{298}(0)$ (%)	$E^0_{rev1273}$ (V)	$\rho^0_{1273}(0)$ (%)
$H_2 + \frac{1}{2}O_2 \rightarrow H_2O(g)$	1.18	94	1.00	78
$CO + \frac{1}{2}O_2 \rightarrow CO_2$	1.07	91	1.01	69
$CH_4 + 2O_2 \rightarrow CO_2 + 2H_2O(g)$	1.04	100	1.04	100

Table 5.14. *Variations with the temperature of original e.m.f. in some standard reactions*

5.4.2.2. *Energy efficiency under operation*

The current supply leads to a quick efficiency drop because internal dissipations of energy occur due to chemical irreversibility (Figure 5.4).

In the case of a low-temperature cell, the efficiency drops by about 20% when a current appears because of, in particular, the importance at low temperature of the activation overvoltage.

If for compactness and cost constraints, the current density delivered by the cell becomes important, about 10% is lost, again because of the ohmic overvoltage (Figure 5.23), even with very thin electrolytes.

Taking into account the major contribution to these losses by the ohmic and activation overvoltages, respectively located in the electrolyte and the electrodes, the main sources of improvements are *a priori*:

– in the increase of the operating temperature because of the beneficial effect on the electrochemical kinetic and the conductions;

– in the search for high-conduction electrolytes;

– in the reduction of the electrolyte thicknesses;

– in the search for catalysts to reduce the activation overvoltage.

Major improvements have been made in the last 10 years, in particular for the PEMFC and SOFC. As an example, the present PEMFC, in competitive volumes of the order of the kW/liter, can reach gross conversion efficiencies of more than 55%.

From a practical point of view, the experimental determination of voltage/current characteristic (polarization curve) of a basic cell in conditions closed to the stack operation is the unavoidable step in the assessment of the conversion efficiency, the cell number, and the cell size of the future stack.

At the same time, the efficiency value dimensions of the energy chain for the overall performance, the cost, and volume must be noted.

	kJ	**kWh**
1 mole of hydrogen (HHV)	<u>285.5</u>	0.0793
1 mole of hydrogen (LHV)	<u>241.6</u>	0.0671
1 kg of hydrogen (HHV)	$14.27\ 10^4$	<u>39.65</u>
1 kg of hydrogen (LHV)	$12.08\ 10^4$	<u>33.56</u>
<u>Electric energy supplied by a stack with a 0.75 V/cell</u>		
1 mole of burnt hydrogen (no losses)	144.7	0.0402
1 kg of burnt hydrogen (no losses)	$7.24\ 10^4$	<u>20.10</u>

Table 5.15. *Some values of hydrogen energy quantities*

5.5. Main applications

5.5.1. *History*

The first application of electrochemistry is dated long before the 19th century. Between 250 BC and 230 AD Persians used electrochemical reactors. The reactors consisted of cooked-earth vessels (Figure 5.26) in which a couple of Fe/Cu electrodes where inserted and separated by an acid solution (maybe vinegar) which acted as an electrolyte. This device was dedicated to medical or metallurgical applications (galvanic deposit).

Figure 5.26. *The Baghdad "cell" discovered in 1938*

During the 19th century, many scientists investigated electrochemical cells and proposed many devices. The most well known are Alessandro Volta, John Frédéric Daniell, Johann Christian Poggendorf (1796–1877), Robert Wilhelm Bunsen and Sir William Grove, who was generally considered the father of the fuel cell. Indeed, this Welsh judge and physicist was one of the first to publish the schematic of an experimental device capable of making the inverse electrolysis of water (*Philosophical Magazine and Journal of Science*, 1839).

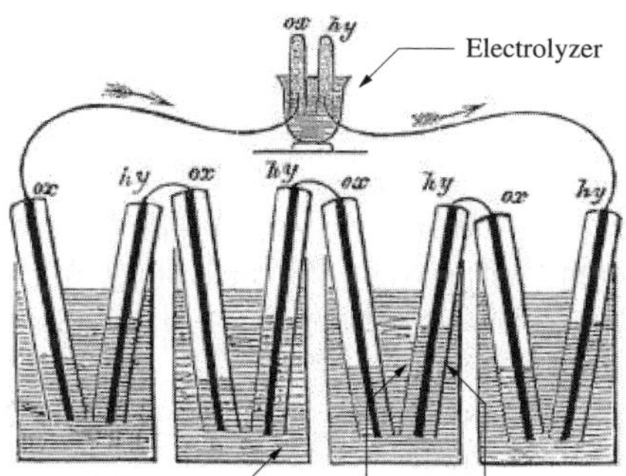

Sir William Grove
(1811-1896)

Figure 5.27. *The Grove cell (4 cells in series); and picture of Sir William Grove*

These works are at the origin of the development of PAFC fuel cells; the first 5 kW demonstrator appeared in 1965 (Figure 5.28).

Figure 5.28. *The Allis-Chalmers 5 kW PAFC, 1965*
(source: Smithsonian Institute, USA)

This type of cell was used for the first time in 2000, in France, for a stationary application. Today there are more than 300 PAFC installations worldwide.

The gas electrochemical generator described and developed by Sir William Grove was called a fuel cell half a century later (that is, after it was developed by Grove) by the English chemist and industrialist Ludwig Mond and his assistant

Charles Langer who, in 1889, tried to manufacture a fuel cell capable of operating with air and gas extracted from coal. That was the first true industrial application of the fuel cell.

We had to wait for the 20th century and the 1950s before the fuel cell reaches performance levels that make commercial application realistic. These prototypes we developed mainly in military and space programs, which today are important developers of the AFC, PEMFC, and DMFC.

In 1932, Francis Thomas Bacon developed an AFC prototype which used, for the first time, porous electrodes and potash as electrolytes. In 1959, he used a 5 kW prototype to operate a soldering machine. His works interested Pratt & Witney who supplied NASA (National Aeronautics and Space Administration) with AFCs for its Apollo program.

Figure 5.29. *The AFC developed by F. T. Bacon in 1959*

It was also in the 1930s in Switzerland that Emil Baur and H. Preis experimented with high-temperature fuel cells equipped with a solid oxide electrolyte using materials from zirconium, yttrium, cerium, lanthanum, and tungsten oxides. But it was only after 1962 that Westinghouse researchers succeed in operating this kind of cell, called SOFC.

In the 1960s, Thomas Grubb and Leonard Neidrach finalized details for a PEMFC type cell with General Electric for use in the Gemini space program. But the membrane used was not stable enough and this solution was abandoned to take advantage of the AFC technologies. It was not until 1987 that Ballard Power System used a sulfonite polymere membrane and thus developed prototypes that performed well enough for automobile manufacturers to be interested in the technology.

Since the middle of the 1990s, the development of better-performing demonstrators accelerated and the fuel cell, in particular the Polymer Exchange Membrane type, now covers a power range from W to MW.

5.5.2. *Special applications*

These applications, mainly military and space applications are at the beginning of the AFC, PEMFC and even DMFC type cell developments.

Since 1990, vehicle prototypes for niche markets also appear; for example, golf carts, and power auxiliaries for camping cars or urban transport-system vehicles. In most cases, atmospheric and acoustic pollution reduction motivates their development.

5.5.3. *Military applications*

With the development of submarines at the end of World War I, the military was permanently searching for compact electric generators able to operate silently in an isolated environment (Figures 5.30 and 5.31).

Fuel cells presented this potential and military agencies financed cell development for attack submarines, as well as observation submarines.

The growing needs in communication and observation on the operation field also led the infantry to be interested in PEMFC and DMFC technology to supply the air conditioning of the NBC combinations, and the communication needs (walkie-talkies, computers, see Figure 5.32) of the infantry personnel of the future.

Figure 5.30. *German navy fuel cell submarine U32, delivered in October 2005*

Figure 5.31. *Launching of the U212 fuel cell in 2004; and internal view of the Italian navy submarine*

Figure 5.32. *Walkie-talkie battery recharging unit*

5.5.4. *Space applications*

No manned flight could have been taken place without the development of the fuel cells operating with hydrogen and oxygen to supply both electricity and water for the astronauts. Fuel cells are so necessary that even if many incidents or accidents during space missions have been caused by a fuel cell or the failure of its auxiliaries (the Apollo 13 flight incident was due to the explosion of an O_2 tank supplying the fuel cells), this equipment is nevertheless still used.

Although AFC technology has for a long time replaced PEMFC technology its compactness, lightness, and strength should now impose this technology thanks to the progresses made in the material field (catalyst, membrane, etc.). Therefore, NASA is currently preparing a PEM fuel cell for use in its space shuttle program.

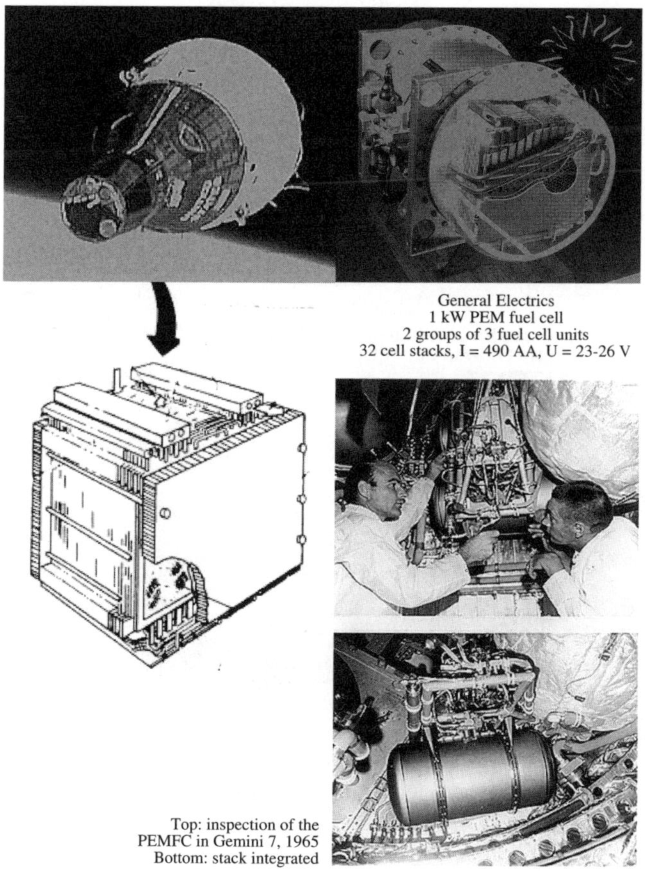

Figure 5.33. *1962–1966, the Gemini program*

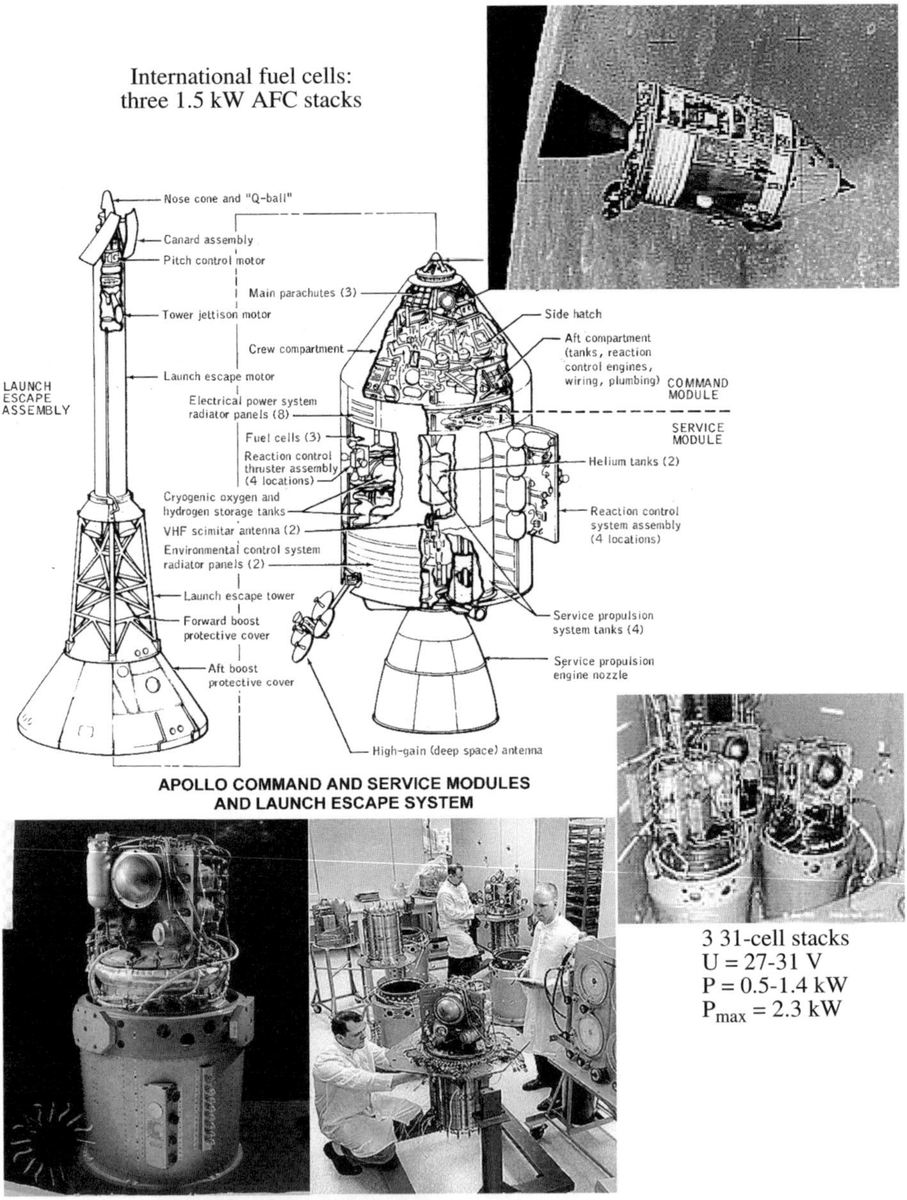

Assembly of an AFC stack by
Pratt & Whitney technicians for the Apollo modules

Figure 5.34. *1966–1972, the Apollo program*

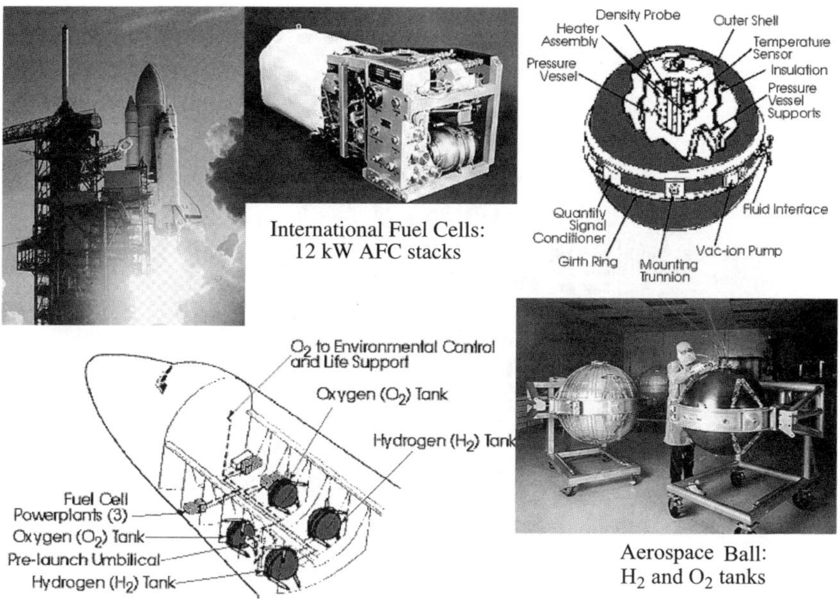

Figure 5.35. *1981–200x, the Space Shuttle*

5.5.5. *Special engines*

There are many examples, and with the nomad applications they are at the beginning of the development of tens start-ups. Some examples are: motor-homes, forklift trucks, observation balloons, small utility vehicles.

Figure 5.36. *DMFC as APU on Hymer motor home (2005)*

Figure 5.37. *Siemens forklift truck powered by a PEMFC*

Figure 5.38. *NASA, Glenn Research Center, scientific observation balloon powered by a 200 W PEMFC*

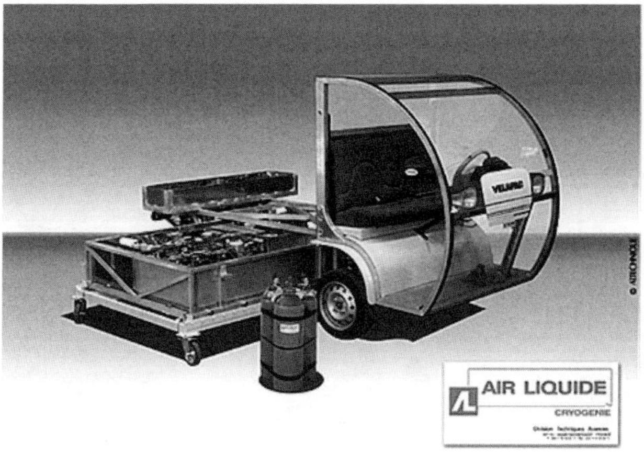

Figure 5.39. *Velapac prototype of a utility vehicle for use in a post office*

5.5.6. *Stationary applications*

Energy deregulation makes the implementation of decentralized electric power generation stations easier; the fuel cell has a potential for development in this market.

Some post offices, banks, computer centers, and many factories already profit greatly from this very safe energy supply which has no interruptions and is decentralized.The energy is supplied by the fuel cells, or they are used as auxiliary generators.

Figure 5.40. *200 electric kW and 220 thermic kW PAFC operating with natural gas. First cell installed by EdF-GdF at Chelles (Seine et Marne – France) in March 2000*

Figure 5.41. *International fuel cells. Installation of five PC25 units of 200 kW at the Regional USPS Mail Sorting Center at Anchorage, Alaska*

Figure 5.42. *Motor generator set fuel cell of 3.2 kW developed by the CEA*

Figure 5.43. *2 kW motor generator set developed by Axane (source: AIR LIQUIDE)*

Figure 5.44. *Hélion stationary generator for 20 kW auxiliary supply (2005) (source: CEA)*

Figure 5.45. *MCFC developed by MTU: 250 kWe + 180 kWth*

5.5.7. *Transport applications*

5.5.7.1. *Fuel cell bus*

The captive fleet market is a very favorable environment for the development of the fuel cell. In fact, it is an industrial environment that respects the security necessary for hydrogen use and for which the reduction of polluting emissions is important.

The United States, Canada, and the large European nations (except for France) – soon to be joined by China and Korea (Figure 5.48) have equipped themselves, through the Clean Urban Transport for Europe (CUTE) project, with experimental fleets of buses. It concerns an experiment return necessary for the deployment of this kind of bus which presents more than the accumulated 500,000 km performed, 100 tons of burnt H_2 and 2.5 billion people transported in 30 PEMFC Ballard Daimler Chrysler buses distributed in 10 European cities (Figures 5.46 and 5.48).

Figure 5.46. *CUTE program buses*

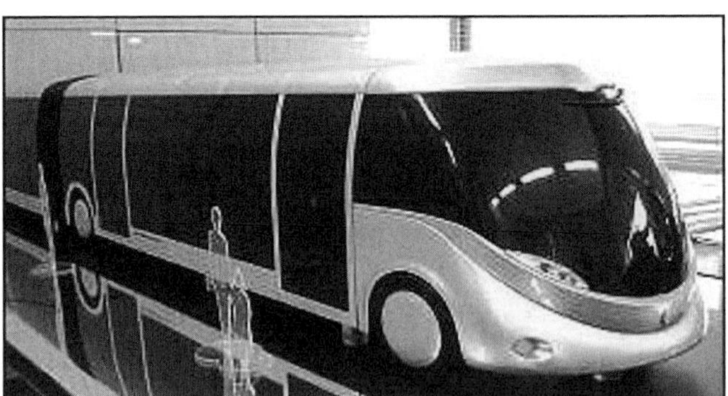

Figure 5.47. *Korean fuel cell bi-mode public transport project*

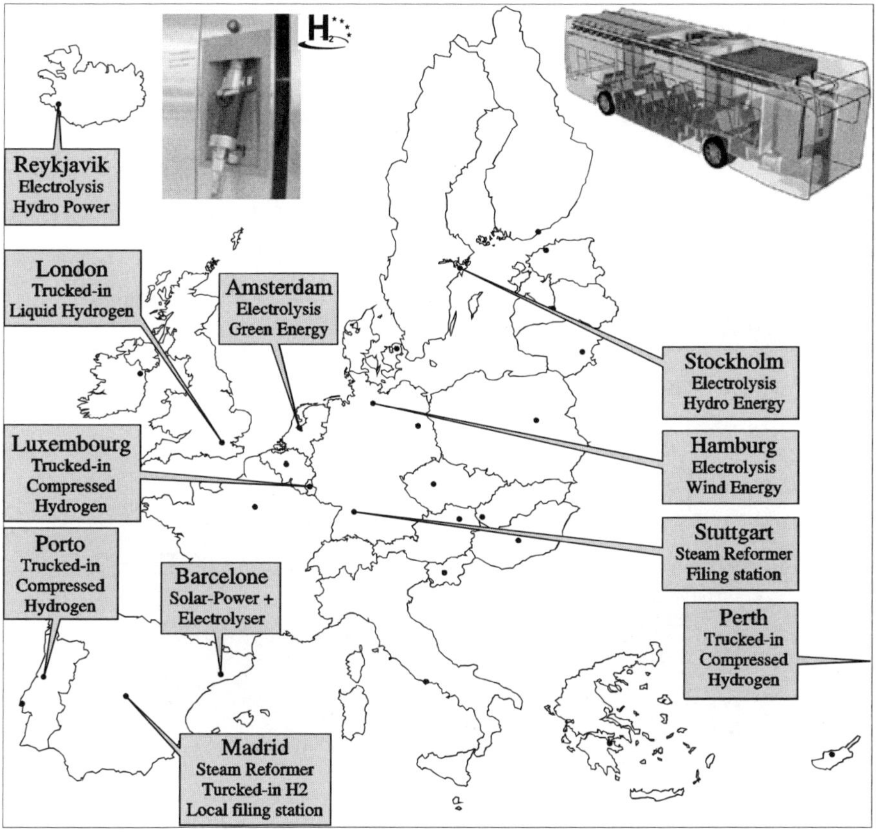

Figure 5.48. *The CUTE European project: Clean Urban Transport for Europe*

5.5.7.2. *Vehicles for individuals*

Since 1994, Daimler Chrysler has developed more compact and performing models of electric vehicles which are powered by a Ballard PEMFC. Up to now, 60 vehicles have been delivered worldwide (Figure 5.49).

This manufacturer is now joined by most of its competitors who all have at their disposal one or more demonstrators (Figures 5.51 and 5.52). Although at the beginning the cell was a Ballard design, the automobile manufacturers now develop their own cells. There are more than 50 models of specific vehicles in the world.

We must also outline the Michelin process (Figure 5.53) which has completely modified the design of a vehicle to be adapted to the specific operation of the cell:

the fuel storage and the electric architecture (rear mirror replaced by a camera, active electric suspension, H_2 tank integrated in the structure).

All the manufacturers agree that the present process of integrating fuel cell into an existing platform has no future and that the vehicle must be reinvented.

The question of whether public opinion will accept this technology is under evaluation through individual renting experiences (Figure 5.49).

Figure 5.49. *Honda FCX rented to individuals*

Figure 5.50. *First garage specializing in fuel cell cars*

Figure 5.51. *HydroGen 3 (GM & Opel), 10,000 km through Europe (14 countries, averaged 500 km per day)*

Figure 5.52. *PSA Group urban utilities vehicles*

Figure 5.53. *Michelin-PSI "HyLight"*

5.5.7.3. *Two wheels*

Ther application of the fuel cells to two wheels is constantly growing, especially in Asia, where this market is important compared to the vehicles market.

Figure 5.54. *Masterflex Bike 2005*

Figure 5.55. *Honda 2005*

5.5.8. *Nomad applications*

Since 2000 there has been a fuel cell miniaturization race in order to compete with the battery's autonomy and life time. This miniaturization concerns the PEM or DMFC type cells which may then be used to operate any device, from the portable computer to the cellular phone; the hydrogen may be located in a charger not bigger than a ball-point pen.

For the PEM, the major difficulty remains the hydrogen storage. For the DMFC, the converter itself and especially its membrane limits its performance.

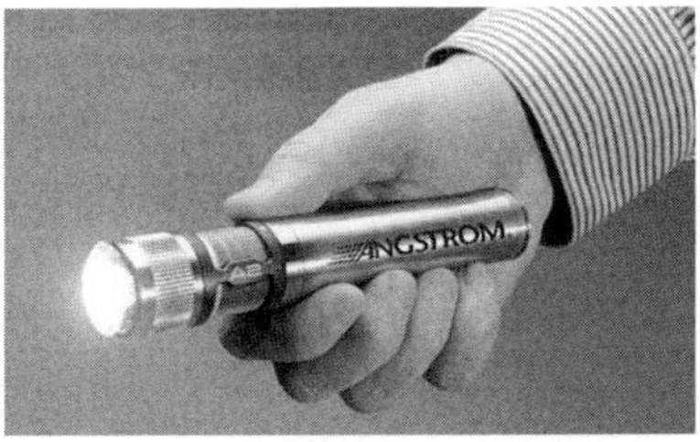

a) Fuel and LED lamp (2006)

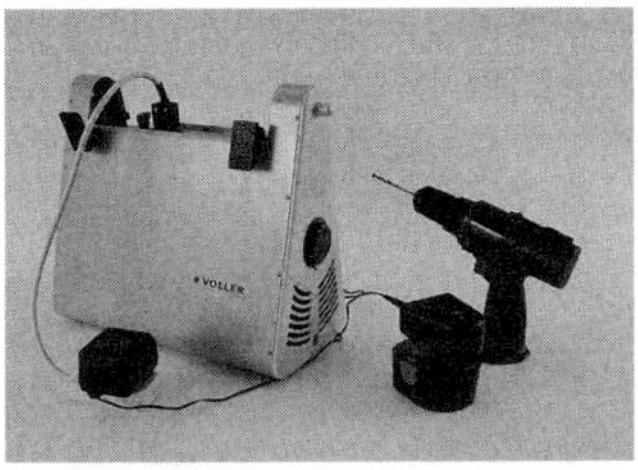

b) 100 W PEMFC portable generator from the British company Voller Energy Group, supplied in hydrogen

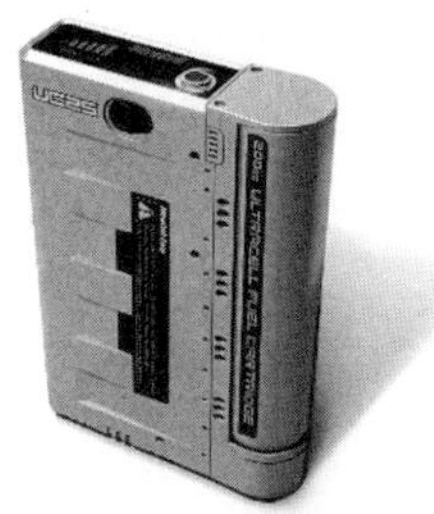

c) DMFC charger XX25 –20W over 72 h (2006)

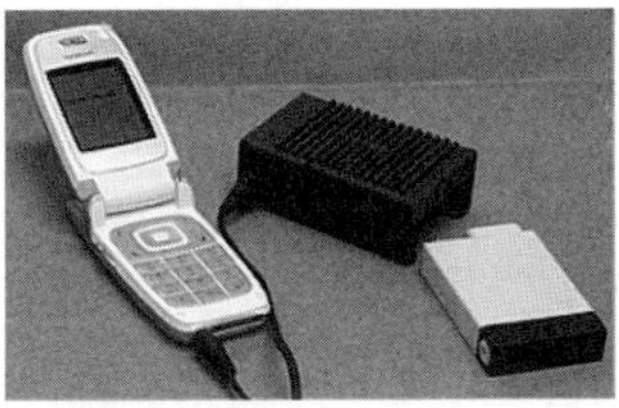

d) Disposable H_2 cartridge with sodium borohydride for miniatures fuel cells developed at the CEA

e) Fuel cell camera from Sinclair Broadcast Group supplied by a Jadoo Power System fuel cell

f) AVC Corp. Computer with a DMFC Antig Tech. (Taiwan) – 45 W for 8 h of autonomy

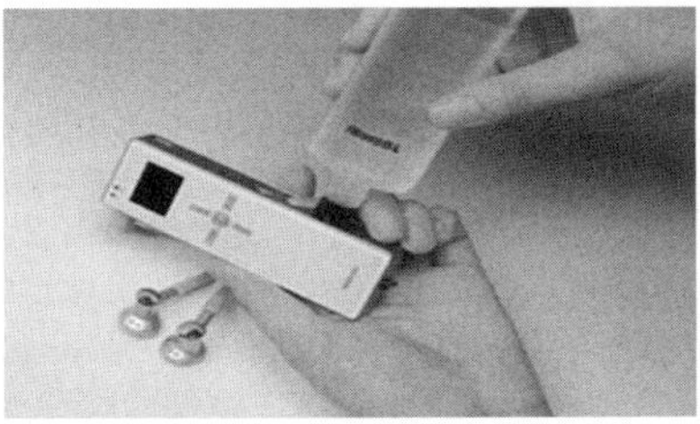

g) Toshiba cell for MP3 reader

Figure 5.56. *Low power mass applications*

5.5.9. *The brakes and motors of fuel cell development*

Beyond the chemical, physical, electrical or technological major limitations, the deployment of the fuel cells remains under conditions of the development of an hydrogen and alternative fuels infrastructure (biogas, synthesis gas, etc.).

The United States are not, the only country dreaming of a *hydrogen highway*, far from it. Other countries, including Canada, China, Germany, Iceland, Japan, Norway, are strongly interested in the potential of hydrogen. Brit Skjelbred, the State secretary of the Norway energy and oil department announced during a meeting in Washington D.C in 2003, a Norwegian hydrogen highway, 500 kilometres long, combined with the elimination of registration and annual taxes costs for vehicles operating with hydrogen. Germany opened its first public station (in Berlin) in November 2004 and China is built in Beijing the infrastructures for the supply of buses operating with the fuel cells during the Olympic Games in 2008.

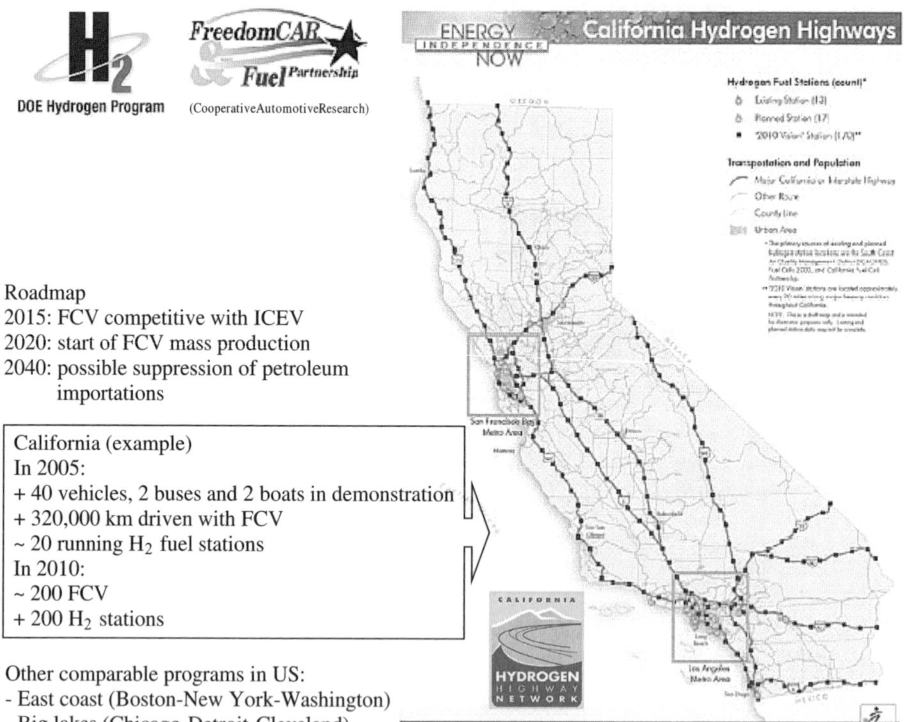

Figure 5.57. *Hydrogen highway projects in the United States*

In parallel, Iceland has the objective to become the first society with hydrogen as an economic basis, and to completely eliminate fossil fuels from now to the middle of the century. Shell Hydrogen opened the first hydrogen supply station in the Reykjavik capital in April 2003. Reykjavik was also the first of the 10 European cities to implement a hydrogen program in the framework of the European program entitled *30 buses for a clean transport system*.

Today, white clouds of steam rather than toxic pollutants are visible behind the buses of some cities. The political authorities hope for the future conversion of fishing fleet and aircraft of the country into hydrogen also. They take benefit of an important popular support: according to a study, the initiative in favor of hydrogen is approved by 92% of the Iceland people.

Generally speaking, in island countries it must be also outlined that the energy policies are interested in a sustainable development of energies available on their territories in order to guarantee the electric supply quality and to preserve their environment (Reunion Island, Corsica).

The electricity suppliers cannot presently integrate more than 20% of intermittent renewable energies in their electric generation park; they should renew and increase them to face the demographic growing and the standard life level. They identified solution to solve this difficulty is the implementation of electric energy storage plants at different power and time ranges (daily, weekly, monthly, season, annual storage).

The fuel cell associated with a hydrogen storage tank and an electrolyzer has the potential to answer this regulation function of the electric network.

Emerging countries like China, India or Brazil, who must increase their electric generation park in order to accompany their present development, will also have to make the choice between fossil or nuclear energy plant, or new technologies of energy.

The experience of the cellular phone has shown in the past that this type of country massively invested in new technologies rather than in more conventional technologies which will be changed in any case sooner or later.

Iceland and emerging countries are for sure the new motors of the fuel cell development which up to now, according to the internet site *Fuel Cell Today*, power around 1,300 companies.

5.6. Bibliography

Books

[AMP 95] J. C. AMPHLETT, "Performance modeling of the Ballard Mark IV solid polymer electrolyte fuel cell", *Journal of Electrochem. Soc.*, vol. 142, n°1 (1995), p. 1-8.

[BAB 05] BABIR F., *Fuel Cells: Theory and Practice*, Elsevier Academic Press, 2005.

[BET 98] M. BETTONI, M. LUCENTINI, V. NASO and L. RUBINI, "Testing of low pressure proton exchange membrane fuel cells", *Proceedings of the 33rd Intersoc. Energy Conversion Engineering Conference IECEC150*, pp. 1–6, Colorado Springs, CO, 1998.

[BIR 05] BIRRAUX and LE DÉAUT, *L'état actuel et les perspectives techniques des énergies renouvelables*, Office Parlementaire des Choix Scientifiques et Technologiques, report n°3415, 2001.

[BOR 96] C. E. BORRONI-BIRD, "Fuel cell commercialization issues for light-duty vehicle applications", *Journal of Power Sources*, vol. 61, 33–48, 1996.

[BOU 06] BOUCHER S., *La Révolution de l'Hydrogène: vers une énergie propre et performante?*, Editions Le Félin-Echéances, Paris, 2006.

[CME] CONSEIL MONDIAL DE L'ENERGIE, "L'énergie pour le Monde de Demain–Le Temps de l'action", *Déclaration 2000 du Conseil Mondial de l'Energie, Editions Technip*, Paris, 2000.

[EDW 98] N. EDWARDS, S.R. ELLIS, J.C. FROST, S.E. GOLUNSKI, A.N.J. VAN KEULEN, N.G. LINDEWALD and J.G. REINKINGH, "On-board hydrogen generation for transport applications: the HotSpot™ methanol processor", *Journal of Power Sources*, vol. 71, pp. 123–128, 1998.

[GAL 01] GALLEY R., GATIGNOL C., *Les perspectives offertes par la technologie de la pile à combustible,* Rapport de l'Office Parlementaire d'Evaluation des Choix Scientifiques et Technologiques, Assemblée Nationale n° 3216, Paris, 2001.

[GIB 98] H. F. GIBBARD, "Development of PEM fuel cells for IST and propulsion application", *6th UECT Electrochemical Symposium*, Ulm, June 1998.

[HEN 98] G. HENRIKSEN, "Compressor technologies for PEM fuel cells", *Congress SAE Fuel Cells for transportation TOPTEC*, Cambridge, MA, March 1998.

[JAM 94] B. D. JAMES, G. N. BAUM and I. F. KUHN Jr., Technology development goals for automotive fuel cell power systems, Report ANL-94/44, Agreement No. 22822402, Argonne National Laboratory, 1994.

[JOS 00] K. JOST, "Fuel cell concepts and technology", *Automotive Engineering International Magazine*, March, 173–185, 2000.

[KOR 96] KORDESCH K., *Fuel Cells and their Applications*, VCH, Germany, 1996.

[KRA 97] R. KRAUSS, D. ZUR MEGEDE and F. PANIK, "The properties of fuel cell vehicle", *14th International Electric Vehicle Symposium (EVS-14)*, Orlando, FL, 1997

[LAR 00] LARMINIE J., *Fuel Cell Systems Explained*, John Wiley & Sons, New York, 2000.

[MIT 06] W. MITCHELL, B.J. BOWERS, C. GARNIER and F. BOUDJEMAA, "Dynamic behaviour of gasoline fuel cell electric vehicles", *Journal of Power Sources*, vol. 154, n°2, 489–496, 2006.

[NON 97] Y. NONOBE, Y. KIMURA and S. OGINO, "A fuel cell electric vehicle with methanol reformer", *14th International Electric Vehicle Symposium (EVS-14)*, Orlando, FL, 1997.

[PIC 98] D. PICOT, Etude numérique et expérimentale des écoulements dans une pile à combustible adaptable aux applications embarquées, Thesis, l'Ecole de Mines de Paris, 1998

[RIF 02] RIFKIN J., *L'Economie Hydrogène*, La Découverte, Paris, 2002.

[TEC 04] EG & G TECHNICAL SERVICES, INC., *Fuel Cell Handbook*, 7th Edition, US Department of Energy, Office of Fossil Energy, National Energy Technology Laboratory, West Virginia, US, 2004.

[VIE 03] VIELSTICH W., LAMM A., GASTEIGER H., *Handbook of Fuel Cells: Fundamentals, Technology, Applications*, John Wiley & Sons, UK, 2003.

[ZAD 91] ZAWODSINSKY *et al.*, *J. Electrochem. Soc.*, vol. 138, n°08, August 1991.

Websites

[ACH] Online store: www.fuelcellstore.com

[AFH] Association française de l'hydrogène: www.afh2.org

[BDD] Data base: www.fuelcellstandards.com

[CEA] Commissariat à l'énergie atomique: www.cea.fr

[EER] *Energy efficiency and Renewable energy* (US DOE): www.eere.energy.gov

[FUE] www.fuelcelltoday.com

[INF] *The Online Fuel Cell Information Center,* www.fuelcells.org

[IRH] Institut de recherche sur l'hydrogène: www.irh.uqtr.ca

Scientific Journals

[CEP] *Fuel cells: reaching the era of clean and efficient power generation in the 21st century*, Srinivasan, Mosdale, Stevens, Yang, *Annual Reviews Energy Environment*, 1999.

[CFS] *Fuel Cells from Fundamentals to Systems,* published by Wiley-VCH Verlag Gmbh & Co.

[JHE] *International Journal of Hydrogen Energy*, published by Elsevier.

[JPS] *Journal of Power Sources*, published by Elsevier.

[NTE] *Nouvelles Technologies de L'Energie,* CLEFS-CEA, n°50/51, Winter 2004–2005.

Chapter 6

Toward Energy Positive Buildings

6.1. Introduction

Over the last 50 years, human beings have radically changed the ecosystem [MEA 05] in which they live in order to satisfy their needs, in particular their need for housing and transportation. Indeed, construction (buildings and civil engineering) and transportation (car, underground rail system, tramway, train, aircraft, etc.) are certainly the two industries that have had the most impact on our environment – just look around!

Buildings and vehicles are nowadays our "artificial living spaces" in which we spend, on average, more than 90% of our time (homes, offices, shops, factories, schools, transportation, etc.). To bring us safety, comfort and well being, they have become real "energy predators" with more than 70% of the total energy consumption (Table 6.1).

When considering the construction industry, let us remember some important figures. First of all, this industry destroys natural areas; it uses more than 50% of raw materials and produces more than 50% of total waste. Furthermore, buildings currently consume nearly half of the total energy demand mainly. Moreover, these buildings emit 122 million CO_2-equivalent-tons annually, which represents approximately 20% of the national emissions and 2 tons of CO_2 equivalent per year and per capita.

Chapter written by Daniel QUENARD.

After the food challenge (water and food), the energy challenge is the major challenge facing human kind, and building is a key component for the reasons outlined above.

To provide for all – rich countries, emerging countries, and developing countries – an average standard of living similar to that of Western countries, a breakthrough strategy is needed in design, construction, use, maintenance, and deconstruction of buildings.

To face this challenge, development and construction programs for low-energy buildings (LEB) are now being developed worldwide, for example by Minergie in Switzerland, PassivHaus in Germany, and Zero Energy House in the USA and Japan. These programs show that is possible to move from "addicted to fossil energy" buildings toward energy efficient ones, even "energy autonomous" ones, and, in some cases, buildings that produce more energy than they consume [VIS 02, MAU 05, QUE 06].

6.2. Energy and buildings: some key figures in Europe

As indicated in Table 6.1 [CE 05], the 160 million European buildings consume more than 40% of the final energy demand, which is more energy that than is consumed by transportation or industry.

2002	Buildings		Industry		Transport		Final demand [d]	
	Mtep	% [b]	Mtep	% [c]	Mtep	% [b]	Mtep	% [b]
Solid fuel (coal, etc.)	12.2	1.1	38.7	3.6	0.0	0.0	50.9	4.7
Oil	96.8	8.9	46.9	4.3	331.5	**30.6**	475.2	43.9
Gas	155.6	14.4	105.4	9.7	0.4	0.0	261.5	24.2
Electricity [a]	121.3	11.2	91.2	8.4	6.0	0.6	218.5	20.2
Derived heat	22.8	2.1	7.5	0.7	0.0	0.0	30.3	2.8
Renewable	29.0	2.7	16.2	1.5	1.0	0.1	46.2	4.3
Total	437.8	**40.4**	306.0	**28.3**	338.9	**31.3**	1082.6	100.0
[a]: 14% of which in renewable energy				[b]: in % of the final demand				
[c]: tertiary and residential area				[d]: per area				

Table 6.1. *Final energy consumption in Europe in 2002*

In Table 6.1, we notice that oil is the dominant energy source in transportation, whereas the energy mix is more diverse in industry and buildings. For buildings, gas contributes to more than one third of the demand. It is also surprising that electricity, a secondary energy, is compared to primary energies (coal, oil, gas, renewable, etc.). We will return, below, to this confusion between primary and secondary energy.

A more detailed analysis of the various types of energy consumption in buildings shows that heating remains the largest, at more than 70%, despite a reduction in percentage (Figure 6.1). There is also a regular increase in the electricity demand. These figures give some orders of magnitude, but there are differences, sometimes significant, between the various types of buildings, whether they are residential, tertiary, private, public, etc. [ADE 04].

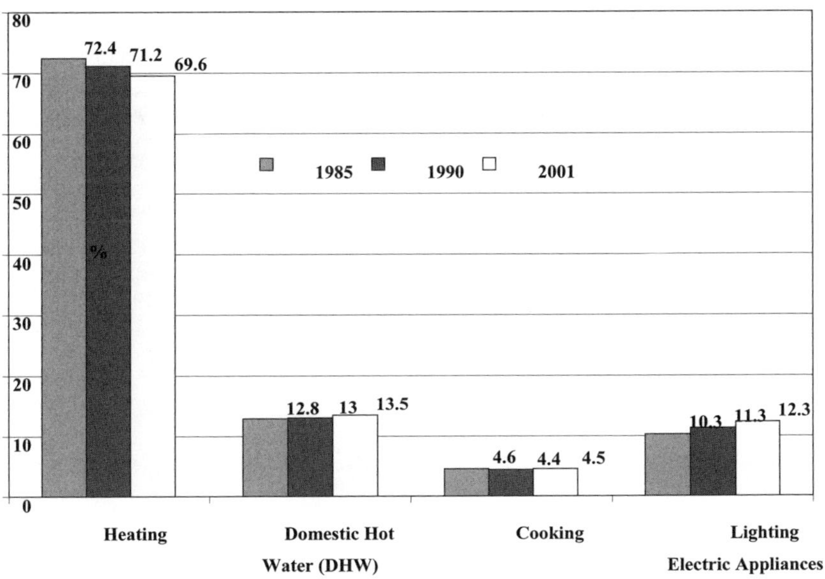

Figure 6.1. *The different types of energy consumption in buildings (15 Member States of Europe) (source: ADEME, 2002)*

6.2.1. *Energy consumption per building type in France*

In France, there are currently 30.6 million homes (including 24.5 million main dwellings, 3.2 million holiday dwellings, and 2 million vacant dwellings) providing a heated surface of approximately 2.1 billion m^2, compared to 814 million heated

m^2, in the tertiary area. The total area of these buildings is about 3 billion m^2 [CAH 03, CAH 05].

Buildings remain the "black sheep" in terms of energy efficiency because, since 2005, the final energy consumption decreased in all sectors of activity (-0.6% for industry, -2.7% for agriculture, -0.8% for transportation), except in the residential-tertiary sector (+0.6%) [MIN 06].

For individual housing (IH), the energy mix is very well balanced between gas, electricity, and the other sources (oil, wood, etc.), while in collective dwellings (CD), gas is dominant (more than 50%).

	Number (millions of dwellings)	Area (millions of m²)	Electricity Consumption (TWh)	Gas Consumption (TWh)	Other Energy Consumption (TWh)	Total (TWh)
IH	14	1442	85.5	99.7	99.9	285
CD	10.5	693	42.1	89.8	35.0	167
Total MD	24.5	2135	127.6	189.5	134.9	452
TB		814	84.4	68.6	55.2	208
Total		2949	212	258.0	190.1	**660**
IH: individual housing – CD: collective dwellings MD: main dwelling – TB: tertiary buildings						

Table 6.2. *Energy consumption of different types of buildings (2002 – normal weather)*

The energy consumption of dwellings (IH+CD) and the tertiary sector is 660 TWh per year, i.e. almost 70 million oil-equivalent-tons per year, or 1.1 oil-equivalent-tons per year and per capita, representing 46% of the total energy used in France (Table 6.2).

The energy consumption of buildings has increased by about 30% over the past 30 years due to the increasing number of buildings (+ 41% increase in dwelling numbers over 30 years), the average dwelling surface, the comfort levels in the buildings, and the emergence of new energy needs. These factors have contributed to a sharp increase in electricity consumption by domestic appliances, lighting, office equipment, computers, air conditioning, ventilation, cooking, drying, etc.

Energy is consumed at a rate of two-thirds by dwellings and one-third by tertiary buildings (offices, shops, educational institutions, health facilities, hotels, etc.). This proportion has remained constant for nearly 20 years.

	Old (before 1975)		Recent (after 1975)		New
	Stock (millions of dwellings)	Consumption (TWh)	Stock (millions of dwellings)	Consumption (TWh)	Stock (millions of dwellings)
IH	9.1	192.6	4.9	92.5	0.22
CD	6.8	123.6	3.7	43.3	0.10
Total	15.9	316.2	8.6	135.8	0.32

Table 6.3. *Distribution of main dwellings by age category and total energy consumption (2002 – normal weather)*

"Old" buildings are those built before 1975, which was the time of the first "oil crisis" and the first French Thermal Regulation. "Recent" buildings are those built since 1975. These old and recent buildings are described overall as "existing" buildings and represent the current stock of buildings. "New" buildings are those built during the current year. They represent the annual flow of construction, an average of about 300,000 dwellings and 12.4 million m^2 of tertiary buildings (1990–2002 average value of new construction, air-conditioned or heated). The rate of renewal of old buildings by new ones is thus less than 1% per year. Moreover, as about 450,000 old dwellings are sold each year, it should be necessary to retrofit all dwellings offered for sale, each year for more than 65 years, if we want to refurbish the whole stock.

6.2.2. *Energy consumption per type of user in the residential sector*

In housing, energy is used, on average, at a rate of 70% for space heating, 12% for electricity (specific uses), and 13% for Domestic Hot Water (DHW), 5% for cooking (Table 6.4). The units of energy consumed by dwellings have been decreasing since 1973.

For the residential sector, the average unit consumption of space heating in the existing stock, per dwelling unit or per m^2, has decreased by about 40% in 25 years whereas the unit average consumption of electricity has increased by 75%.

Overall, in main dwellings, the growth in total consumption of electricity has been +130%, compared to an increase of +49% in the total consumption, considering all kinds of energy.

	Number in millions	Heating Twh	DHW Twh	Cooking Twh	Specific Electricity Twh	Total Consumption Twh
IH	13.94	200.3	28.4	19.1	37.3	285.1
CD	10.68	117	18.8	10.1	21.1	167
TOTAL	**24.62**	**317.3**	**47.2**	**29.2**	**58.4**	**452.1**
%		**70.2**	**10.5**	**6.5**	**12.8**	

Table 6.4. *Distribution of the main dwellings per type of user*

This high electricity penetration is explained, on one hand, by the continuous growth of the specific uses, in particular of household equipment, and on the other hand, by the development of electric space heating and, to a lesser extent, air conditioning. Air conditioning, which concerns about 4% of the dwellings, is included in the specific consumption electricity.

Consequently, for electricity, the specific uses (lighting, appliances, TV sets, computers, etc.) represent the first item of electric consumption (41%), more than space heating (36%), DHW (14%), and cooking (8%). Three specific domestic uses, which are approximately equivalent, account for more than 70% of the specific needs: lighting (11.8 TWh), refrigerators (11.5 TWh), and washing machines (11.4 TWh). The use of computers, TV, multimedia equipment, etc., is growing rapidly and today accounts for more than 16% of the domestic electricity consumption.

6.2.3. *Energy consumption in the tertiary sector*

In 2001, the total energy consumption by the tertiary sector was 214.1 TWh, for 814.6 million m^2 of heated or air-conditioned buildings.

The tertiary sector is divided into eight main areas which cover diversified activities.

Type (1)	Stock	Consumption
	Millions of m^2	TWh
Shops	188.3	49.7
Offices	172.8	51.2
Schools, Educational Institutions	166.4	26.6
Health & Social Sector	93.9	26.6
Sports, Entertainment	61.1	17
COHORE (2)	54.3	22.6
Community Dwellings	53.5	12.3
Transportation (3)	24.3	8.7
Total	814.6	214.1
(1) Excluding army, art and craft, large public research establishments, and public lighting		
(2) COHORE = COffee shops, HOtels, REstaurants		
(3) Railway stations & airports		

Table 6.5. *Tertiary sector: stocks and final energy consumption*

Because of their size, the first two areas, shops and offices account for more than half of the energy consumption. The five areas of greatest consumption account for more than 80% of the total consumption.

6.2.4. *Energy consumption per type of user in the tertiary area*

For the whole tertiary sector, space heating represents the main item of consumption (54% of the total consumption) as in the residential sector, but to a lesser extent. The second main item is the electricity uses (26%).

Most of the electricity consumed in the tertiary sector is used to supply specific usages (lighting, computers, etc.). As far as air conditioning is concerned, its total electricity consumption is estimated at 10.3 TWh/year, which represents about 5% of the total consumption in the tertiary sector. However, the unit consumption, in the tertiary stock remains stable (Table 6.6).

Type	Energy Consumption (TWh)					
	Heating	DHW	Cooking	Specific Electricity	Air Conditioning	Total
Shops	22.2	2.9	1.1	20.5	2.9	49.7
Offices	28.5	1.4	0.8	16.3	4.2	51.2
Educational Institutions	19.9	2.4	1.6	2.3	0.3	26.6
Health & Social Sector	14.9	3.7	1.2	5.3	1	26.6
Sports, Entertainment	9.1	3.8	0.4	2.9	0.8	17
COHORE	9.8	2.6	5.9	3.3	0.9	22.6
Community Dwelling	7	2.1	1.3	1.9	0.1	12.3
Transportation	4	0.5	0.2	3.8	0.1	8.7
Total	**115.4**	**19.4**	**12.5**	**56.3**	**10.3**	**214.7**
%	**54%**	**9%**	**6%**	**26%**	**5%**	**100**

Table 6.6. *Tertiary: final energy consumption per type of user in 2001*

Looking at the tables above, we notice that space heating remains, in Europe, the primary area of consumption which is approximately equivalent to the energy consumed by industry or by transportation.

Therefore, as energy production and consumption are responsible for about 80% of the greenhouse gas emissions, it is clear that the reduction of energy consumption in buildings is the major challenge for a reasonable and intelligent policy for a sustainable development.

6.3. How to move from buildings "addicted to fossil energy" toward "low energy buildings" (LEB) and, further, toward buildings as power plants (BaPP)

The above tables show that, in order to achieve the desired comfort levels, the energy demand of buildings is still mainly thermodynamic (more than 80% in the residential sector and around 70% in the tertiary sector), and concerns low temperature calories for heating and DHW. Indeed, the temperature levels needed for space heating and air conditioning are between 19 and 25°C, and for DHW between 40 and 70°C. These temperatures are relatively low and could be obtained with a limited amount of energy [CFP 02].

For the major part, this energy could be produced by renewable energy technologies (solar, wood, biomass, wind, etc.) and in some cases by fossil energies (gas, fuel, coal, uranium, etc.) using high-performance equipment (condensation boilers, heat pumps, etc.).

However, in most countries, it is clear that we are today witnessing a tremendous waste of energy, on one hand because of losses through the envelope of buildings (roof, walls, and windows) which is a true "spendthrift" (Figure 6.2), and on the other hand, because of the low efficiency of equipment (heating, ventilation, air conditioning, lighting, etc.). Especially in the existing housing, the old equipment does not offer the constantly improving energy efficiency of modern products.

In Figure 6.2, we also notice the relative importance of ventilation which will still increase even if nothing else is done, while at the same time, the thermal insulation and the air tightness of the envelope are reinforced. In addition, ventilation plays a key role in order to ensure adequate air quality in housing and, therefore, to preserve the health of occupants.

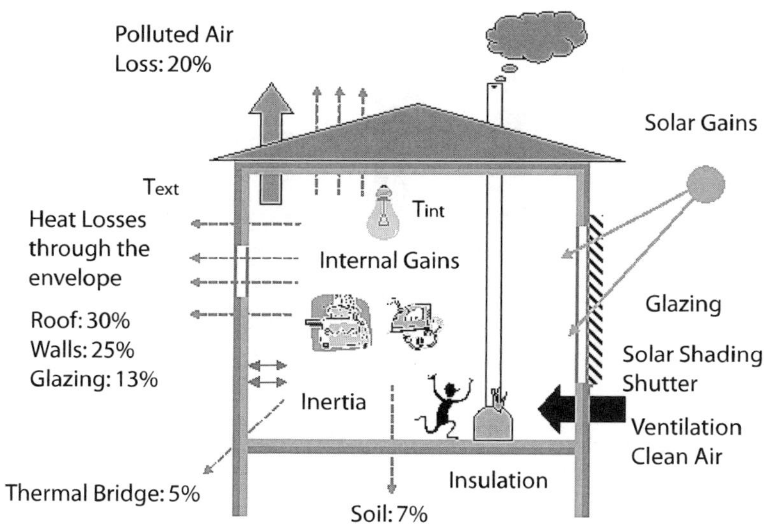

Figure 6.2. *The major heat losses and gains in a traditional building*

First, let us remember that in buildings in general, and in dwellings in particular, space heating (wood, coal, oil, gas, electricity, etc.) was introduced and became usual (e.g. central heating) well before thermal insulation (the first Thermal Regulation came into force in 1975 in France). Instead of protecting buildings from the outside cold by reinforcing insulation, there was a preference for a long time to fight cold by consuming energy for heating. Things started to change after the first oil crisis (1973) and following increase in the cost of energy. The first Thermal Regulation (1975) introduced criteria about insulation in order to limit heating costs. The next regulations will reinforce criteria on thermal insulation in order to limit the heat load. The following regulation strengthened the criteria on thermal regulation and introduced new requirements for heating systems and DHW.

Furthermore, the reduction of energy consumption depends on a key and unavoidable actor: the occupant. In Figure 6.3, the results from an inquiry, carried out in Germany, exhibit the lack of knowledge of occupants about their own energy consumption.

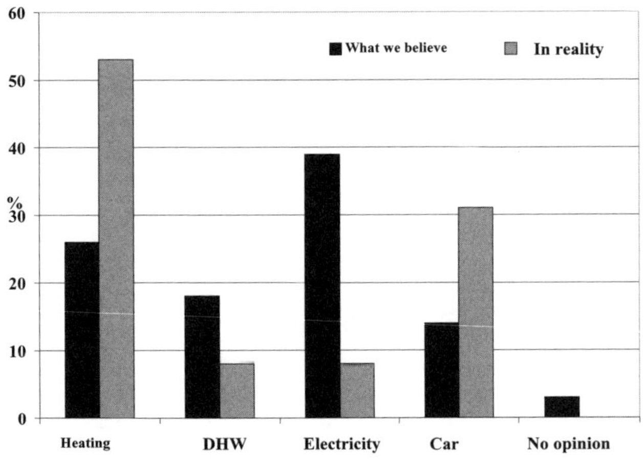

Figure 6.3. *Energy consumption as seen by users (source: German inquiry www.zukunft-haus.de)*

It is surprising to note that occupants underestimate by almost 50% the energy consumption of space heating and car usage and overestimate, by the same proportion, electricity consumption and DHW. Several reasons could explain these discrepancies:

– Two kinds of energy are often mixed up in public opinion: primary energy sources (coal, oil, gas, uranium, etc.) and secondary energy, such as electricity. It may be the same with hydrogen which should not be considered as a new energy source – it is just a vector that is in competition with or complementary to electricity. Indeed, both vectors have similar problems of generation, storage, transportation, etc., but of course they both present their specific losses.

– Each day, all year long, everybody uses electric equipment either at home (refrigerator, washing machine, dishwasher, oven, TV, radio, lights, etc.) or at work (computer, printer, vending machine, etc.). Of course, this electrical environment suggests to us that our electric consumption is important while our interactions with space heating and air-conditioning systems are limited, especially in CD and offices. It could be the same analysis for DHW, which is used many times every day and all year long.

6.3.1. *The approach: from thermal regulation toward energy labeling*

To establish the bases of a simple approach in order to move toward energy-positive buildings – and an approach that involves all stakeholders (customers, bankers, architects, designers, builders, and suppliers) and is easy to understand by everybody – it may be useful to present the energy consumption in buildings in the same way as a bank account, something that everyone knows very well and tries to keep in credit. In Table 6.7, consumptions are grouped in the debit column, whereas all the possible gains are listed in the credit column. The energy storage can be considered as a kind of "cash flow".

	Debit (Consumption – Losses)	**Credit (Production – Gains)**
	Heating	Internal Sources (Occupants, Appliances, etc.)
	Ventilation	Passive & Active Thermal Solar Gains
	Air Conditioning	Natural Lighting
	Domestic Hot Water	Heat Recovery from Air & Water
	Artificial Lighting	Pre-Heating/Pre-Cooling
	Domestic Appliances	Photovoltaic System
	Multimedia	Biomass – Wood
	Other Electrical Equipment	Wind – Hydraulics
	etc.	etc.
BALANCE	**D (kWh/m².an)**	**C (kWh/m².an)**

Table 6.7. *The energy balance of a building*

From Table 6.7, and the previous tables, it is easy to identify the priority actions and the different possible strategies to make low-consumption buildings, passive, zero-energy, or, even better, energy positive.

To reduce the energy deficit illustrated in Figure 6.4, it is necessary to first act on the Heating Ventilation Air Conditioning (HVAC) needs in new buildings and especially in old ones, by improving the energy performance of the envelope (airtightness, insulation, glazing, inertia, ventilation, etc.) and the energy efficiency of HVAC equipment.

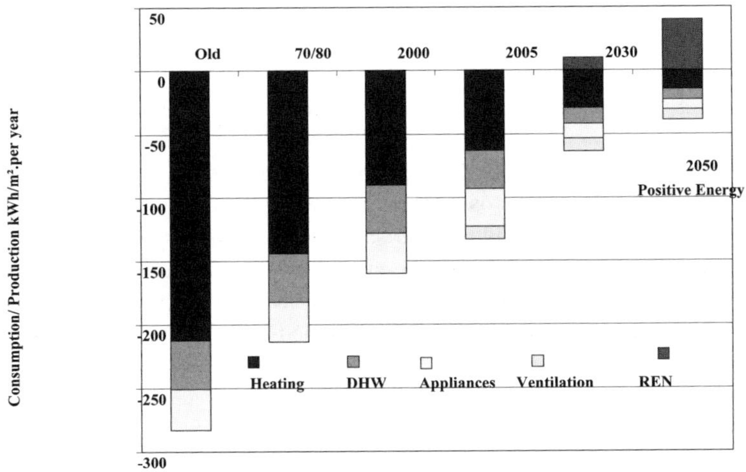

Figure 6.4. *From old buildings towards energy positive buildings, including buildings in accordance with Thermal Regulation and low-consumption buildings*

For DHW, which mainly depends on the lifestyle of the occupants of a building, the reduction will be less important, but the substitution of fossil energy with renewable energies should enable an energy load reduction. For the specific uses of electricity (lighting, household appliances), the reduction will especially depend on the improvement of the efficiency of products (low-energy lamps, class A, A^+, A^{++} equipment, etc.) and on the user behavior.

In Table 6.8, we have listed the strategies used in the Minergie, passivHaus, and zero-energy house programs in order to reduce the debit and to increase the credit. In the case of zero-energy buildings (ZEBs), the search for production/consumption equilibrium starts with a strong reduction of the consumption before an increase of on-site generation (mainly using photovoltaic panels). This process does not necessarily correspond with sustainable development requirements (resources consumption, recycling, etc.) and, for the moment, requires a high investment. However, the feed-in tariffs of PV electricity could change the situation.

Regulation/Label	Actions to reduce (debit)	Actions to increase (credit)
RT 2005 Regulation New Building Label	Maximum consumption – primary energy 80–130 kWh/m² per year using fossil energy 130–250 kWh/m² per year using electricity or heat pump Labels: high performance (- 10%) & very high performance (-20%) Effinergie (-50%) – www.effinergie.org	Internal & passive solar gains (inertia) Solar DHW
Minergie (2006) Label Low-Energy Building New & Retrofitted	Reduction of needs (final energy) HVAC+DHW < 42 kWh/m² per year – new single-occupancy dwelling HVAC+DHW < 80 kWh/m² per year – retrofitted single-occupancy dwelling	Internal & passive solar gains (inertia) Solar DHW Heat recovery from air ventilation Heat pump – wood
PassivHaus Label Passive House New and Retrofitted	Significant reduction of needs Heating needs < 15 kWh/m² per year Final energy HVAC–DHW–lighting–electricity < 42 kWh/m² per year Primary energy HVAC–DHW–Ecl–electricity < 120 kWh/m² per K	Internal & passive solar gains (inertia) Solar DHW – wood Heat recovery from air ventilation Compact systems: heat pump – HVAC+DHW
Zero-Energy Home (ZEH) Label	Reduction similar to Minergie label	C = D over one year Solar PV – heat pump – solar DHW – wood Urban windmills, etc.
Energy-Plus House Label	Reduction similar to PassivHaus label	C > D over one year Solar PV – heat pump – solar DHW – wood Urban windmills, etc.

Table 6.8. *From thermal regulation to energy-positive house*

6.3.2. *Possible actions to reduce consumption*

In order to reduce losses and to increase gains, many actions are possible:

– First, start by minimizing the space heating needs, which the biggest item (Figure 6.1) in Northern and continental countries. This reduction goes through an optimization of the energy performance of the building envelope (roof, wall, floor, and windows) which requires addressing the following key points: improved airtightness, enhanced thermal insulation, use of thermal inertia (as an example, heavy structure), and balancing the ratio losses/gains of windows. The windows play a central role as they fulfill many functions, which are sometimes contradictory: light all the year, insulation and solar gains in winter, solar protection in summer, etc. Thus, these functions must be properly controlled.

– Second, use the correct size of high-energy-efficiency equipment in order to fulfill needs: high-efficiency boiler (condensation), heat pump, low-energy lighting and equipment, energy recovery on air ventilation and water, etc.

– Third, take advantage of the local and renewable energies: thermal solar (mainly for DHW), biomass (wood), photovoltaic panels, windmills and micro-hydraulic plants, etc.

Finally, the multisource and multiuse aspect requires an intelligent Building Management System (BMS) for the buildings in order to help users and managers.

6.3.2.1. *Actions to improve the energy performance of the envelope*

The heat losses Φ^- (W) through the envelope result from the following flows.

1) Air leakage (convection) through the airtightness defaults (cracks, connections, trapdoors, holes due to electrical & water networks, etc.). The heat flux resulting from the air flow through the envelope can be expressed as follows:

$$\Phi_1^- = \rho_a \cdot C_a \cdot q_a \cdot S_d \cdot \Delta T \text{ -w} \qquad [6.1]$$

where ρ_a is the air density, C_a its specific heat, S_d the default surface, and ΔT the temperature difference, with:

$$q_a = K \times (\Delta P)^n \text{ -}\left(m^3/s\right)/m^2\text{, the air flow} \qquad [6.2]$$

K characterizes the airtightness (or the permeability) of the envelope and ΔP the pressure gradient [CAR 06].

2) Conduction and radiation through opaque walls (walls, floors, roofs) and transparent apertures (windows, picture windows), and through thermal bridges (partition walls, balconies, terraces, etc.):

$$\Phi_2^- = (U\ S + \Psi\ L)\Delta T \text{ -W} \qquad [6.3]$$

U ($W/m^2.K$) is the heat transfer coefficient, Ψ (W/m.K) the linear heat transfer coefficient (for thermal bridges), S (m^2) the external surface, L (m) the thermal bridges length, and ΔT the temperature difference K.

The heat transfer coefficient U is expressed as:

$$\frac{1}{U} = \frac{1}{h_i} + \sum_i \frac{e_i}{\lambda_i} + \frac{1}{h_e} - U - W/m^2.K \qquad [6.4.]$$

where h_i and h_e are the thermal exchange coefficients of the external and internal surfaces, e is the material thickness, and λ the thermal conductivity (W/mK).

Test Conditions		Airtightness default 1 mm wide crack	U $W/m^2.K$
Internal temperature External temperature Pressure difference "Equivalent" wind speed	20°C - 10°C 20 Pa 2–3 m/s	Without	0.3
		With	1.44

Table 6.9. *Impact of an airtightness default on the U coefficient*

To illustrate the air permeability effect on the U-value, researchers at the Fraunhofer Institute for Building Physics (FhG – IBP) in Stuttgart have measured heat losses due to a slot 1 mm wide and 1m long. The results in Table 6.9 show that the U-value is then multiplied by a factor of 4.8. In addition, nearly 800 g of moisture may penetrate per day and per m^2 in the roof in such a "thermal tunnel" while only 5 g penetrate where there is a vapor barrier.

The envelope is not only synonymous with heat losses; it can also play the role of thermal solar collector mainly through the glazing.

Glass subject to a radiation 1 transmits an amount τ_e, reflects ρ_e, and absorbs α_e of this energy ($\tau_e + \rho_e + \alpha_e = 1$) (Figure 6.5). Due to the absorption, the temperature of the glass increases; consequently it radiates, in the infra-red range, an amount of energy q_i inside and q_e outside, in a higher wavelength range.

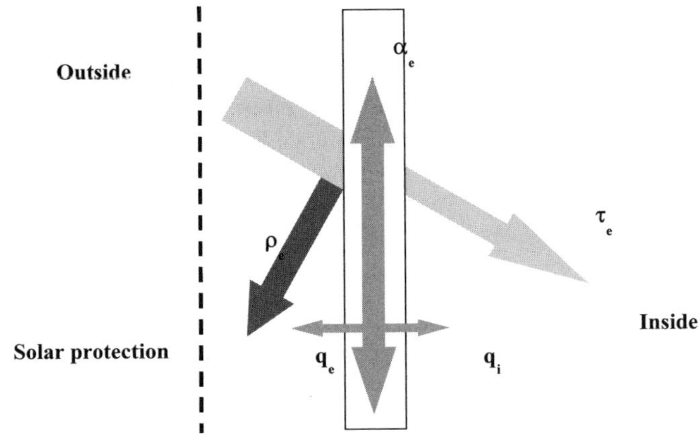

Figure 6.5. *Schematic representation of the thermal transmissions through glass*

The associated heat flux can be expressed as follows:

$$\Phi^+ = g \cdot S_v \cdot I \text{ - W} \quad [6.5]$$

S_v is the glazing surface, I (W/m^2) is the radiative heat flow that arrives on the glazing and g the solar heat gain coefficient (SHGC) defined as:

$$g = \tau_e + q_i \text{ -\%} \quad [6.6]$$

The SHGC factor g of the glazing is the sum of the transmitted part τ_e and the reemitted part toward the inside q_i. In the presence of the solar protection (blinds, shutters, etc.) the g factor, which includes the set "window + solar protection" may be significantly reduced and reach values near 0.1.

The "free" energy entering through the window or resulting from the internal gains could be usefully stored in the partitions (mainly floor and walls). The stored energy will depend on the thermal properties (mass density ρ and specific heat C) of the materials that form the walls; it will also depend of the geometric characteristics (surface S_i and thickness e).

$$Q = \rho \cdot C \cdot S_i \cdot e \cdot \Delta T \quad \text{- J or W.s} \qquad [6.7]$$

Regarding the temperature difference ΔT, it will depend on the energy contribution due to direct radiation and to convection.

The thermal storage potential of a house, characterized by equation [3.7], is called thermal inertia (or thermal mass). The main characteristic of buildings with high thermal inertia is to warm up or to cool down very slowly. Consequently, the internal temperature will be more stable.

At this stage it is worth mentioning the direct solar floor (DSF) which may be considered as an equivalent "active" system to exploit the solar energy and the thermal inertia of a concrete floor as space heating system.

Another aspect of walls with high inertia is the phase shift of the heat wave. For example, when the sun heats the external surface of a wall, the low-inertia buildings follow the external variations (temperatures, heat radiation) without low damping or delay [SID 03]. Indeed, a given time is needed for the heatwave to reach the other internal surface of the wall. This time is called the phase difference.

The order of magnitude of the phase shift may be expressed by the following equation:

$$t \approx e \cdot \sqrt{\frac{\rho \cdot C}{\lambda}} \quad \text{- s} \qquad [6.8]$$

e is the wall thickness (m) and t the phase shift (s).

Thus, walls that have the most important phase shift are thick and heavy walls, made of concrete, rock, adobe, etc. Action to reduce the thermal conductivity λ (insulating material) generally leads to a significant density reduction ρ ($\rho < 200$ kg/m^3).

Recently, phase-changing materials (PCM), generally used in heat (or cold) storage systems, have been adapted to be integrated in the building envelope (ceiling, wall, etc.). Indeed, when the temperature exceeds the melting point of the PCM, the fusion absorbs the overheating, keeping the wall temperature constant. Then, during the solidification, the wall releases the stored energy. Although requiring caution (choice of the fusion/solidification temperature, fire protection, etc.), these materials provide new technical solutions to reinforce the thermal inertia of lightweight structures [AHM 05].

Inertia is particularly interesting for buildings with a continuous use (e.g. housing), but may be more difficult to operate in buildings with discontinuous use and high internal gains, like offices or schools. In this case, the inertia may increase the consumption of energy in winter because, in the morning, the walls must be heated before heating the rooms and it may be not economically feasible when using a high-cost fossil energy. Moreover, in summer, if efficient sunscreens and ventilation are not installed, the risk of overheating, resulting from solar and internal gains simultaneously, becomes high.

Therefore, the solar inputs should be taken into account at the beginning of the building design (architectural masks and/or sunscreens to maximize solar gains in winter and to minimize them in summer, etc.). Furthermore, an intelligent control of the SHGC g of windows and glazing is also necessary, along with the definition of ventilation scenarios, and even more over-ventilation at night. The g factor may indeed vary from a value near zero (closed windows) to near 1 with a single glass without sunscreen. This is why the first solar houses constructed with large glass surfaces, using "non-insulating" glass, were neither comfortable in winter because of the losses, nor in summer because of the solar contributions.

Using the results of the previous equations, it is possible to build a conformity table between the actions to carry out and the parameters to be optimized which appear in the previous equations (equations [6.1] to [6.8]). It will be necessary to modify these parameters in order to reduce the losses, increase the profits in winter, and avoid overheating in summer.

Parameters	**Steps in relation to building design, construction and choice of materials and components**
Geometry S, S_v, S_i, L, e and Ψ	Building compactness to reduce S and L. Optimize S_v and S_i Optimized architectural design and quality implementation to limit and even eliminate thermal bridges (Ψ). A control is necessary (IR webcam).
Boundary conditions ΔT, ΔP, I, h_e	Local weather, sunny periods, dominant winds, building orientation, screens, etc.
Materials, components and envelope properties λ, ρ, X, ε, U, g, K	Choice of material and components adapted to the functions: insulation, inertia, solar contributions and solar protections. Reduced airtightness due to the implementation quality. Airtightness control is sometimes mandatory.

Table 6.10. *Relationship between parameters and steps*

The actions listed in Table 6.10 require, on one hand, a global architectural design which integrates the building into its local ecosystem (sun, dominant winds, etc.) and, on the other hand, a controlled high-quality construction, using high-performance components and materials.

The simplified equations presented in the above sections show the main characteristics to take into account in order to reduce energy consumption. However, they do not represent a rigorous calculation method to estimate the energy consumption of buildings. These calculations must be carried out according to the thermal regulations of each European country. These regulations are all based on a series of European Norms (EN) or international standards (ISO (International Organization for Standardization)) which describe in detail the calculation methods to be used. For example, the next standards can be applied: EN12831 for heat losses from the envelope; EN410 and EN673 for glazing; ISO 9972 for airtightness measurements; EN 13370 for walls that are in contact with the ground and the ground inertia consideration; EN13465 for ventilation losses; EN ISO 13786 for building inertia; EN13790 for the consumption calculation. In France, all these standards are available on the AFNOR website (www.afnor.fr).

6.3.2.2. *Choice of components and material for the envelope*

With traditional insulation materials, such as cellular foam or fibrous materials which have reached their thermal performance limit, the only solution to reinforce the insulation is to increase the thickness. This thickness may vary from 15 cm to 20 cm for Minergie-type houses, and reach 40, and even more 50 cm, for passive houses. In new buildings, and sometimes in the roofs of old building, these high thicknesses may be used, but renovation in existing buildings requires research for new and low thermal conductivity materials in order to reduce the thicknesses of the insulating materials.

With air as "insulating" filling gas, the two possible solutions are "containment" to reduce the gas mobility (nitrogen and oxygen) and pressure reduction to limit the number of molecules which transport energy. Due to nanotechnologies, the development of non-porous materials (for containment) and of very low permeability films (to keep low pressure) enables a large-scale consideration of the appearance of components that integrate super-insulating materials, like vacuum-insulation panels (VIP) [IEA 05].

To achieve an efficient thermal insulation, gas with low thermal conduction has to be contained in small cavities, which are more or less closed by a minimal quantity of material. At first approximation, the equivalent thermal conductivity λ_e of an insulating material is the sum of the respective contributions of the gas conduction λ_g and the solid phase λ_s (mineral, vegetal, animal fibers, polymer foam, etc.) and of the radiation λ_r. Given the cavity size, (around a few hundred microns), convection is negligible:

$$\lambda_e = (1-\varepsilon)\cdot\lambda_s + \varepsilon\cdot\lambda_g + \lambda_r \quad - \text{W/m}^2\text{.K} \qquad [6.9]$$

Because of the high porosity of insulating materials (ε > 90%), the gas contribution represents about 60% of the transfers. The "gas" conductivity is written as a first approximation in the following way:

$$\lambda_g = \cdot\frac{\lambda_{g0}}{1+C\cdot\dfrac{T}{\delta\cdot P}} \quad - \text{W/m}^2\text{.K} \qquad [6.10]$$

λ_{g0} is the gas thermal conductivity under normal conditions (atmospheric pressure P_a and $\delta >> l_m$). δ is the characteristic size of the cavities, T the temperature, and C a constant (around 3 for air). l_m is the mean free path of the embedded gases, about 70 nm for air.

To reduce λ_g in the building conditions, three solutions are available:

– To choose gases with low heat conduction, such as gases known as "heavy" (CFC, HFC, HFA, etc.) or "rare" (argon, krypton, xenon) gases. Some heavy gases are progressively forbidden and are no longer a solution.

– To reduce pressure P (partial vacuum) to limit the number of molecules that transfer energy.

– To use high porosity materials (ε > 90%) with cavities of sizes less than the mean free path l_m to reduce thc gas mobility.

As far as the radiation is concerned, it mainly depends on the mass density ρ and on the extinction coefficient E:

$$\lambda_r = \left(\frac{16}{3}\cdot\frac{\sigma T^3}{E.\rho}\right) \quad - \text{W/m}^2\text{.K} \qquad [6.11]$$

where σ is the Boltzman constant, E the extinction coefficient.

To reduce λ_r, the mass density must then be increased without reducing the porosity or we have to introduce "opacifiers" (dark or reflecting films, aluminum, or carbon powders, etc.) in the materials to increase the E coefficient.

Figure 6.6 shows the performances of the main insulating materials available on the market today or under development.

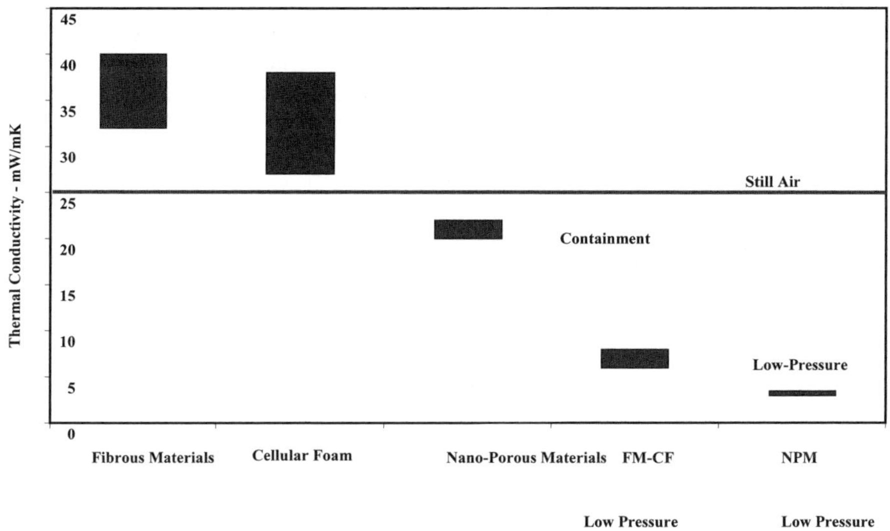

FM: fibrous material; CF: cellular foam; NPM: cellular foam

Figure 6.6. *Different types of insulation materials*

Up to now, insulation has been the major tool used to reduce thermal losses from buildings in winter. Today, the search for comfort all year long requires a new approach to the heat transfer through building walls, in particular taking more account of inertia to manage the passive inputs (internal and solar gains). But an adapted control of solar gains on-site requires a global vision and a systematic approach to take better advantage in winter without having the drawback of overheating in summer. Moreover, the use of convection to homogenize the surface temperatures of walls, inspired by the "hypocaust" heating systems of some Roman houses, could also improve comfort levels.

This holistic (global, systematic) approach of building-envelope science is well illustrated by window and other glazing facades, the functions of which have continuously evolved since their appearance in buildings. Originally installed to ensure watertightness and airtightness, but keeping transparency for natural lighting and vision, windows also play an important role in natural ventilation. With the different thermal regulations published since 1975, windows and glazing play an important role in insulation and the U-value has been reduced by a factor 6 in 30 years (from 6 to 1).

To switch from single glazing which, from a thermal point of view, can be regarded as a thermal "hole" (U ~ 5.6), the usual solutions, already used for opaque insulation, have been implemented. At first, the use of an insulating layer between two panes separated by 12 to 16 mm to reduce the convection has enabled a gain of a factor of 2 (Figures 6.7 and 6.8 and Table 6.11).

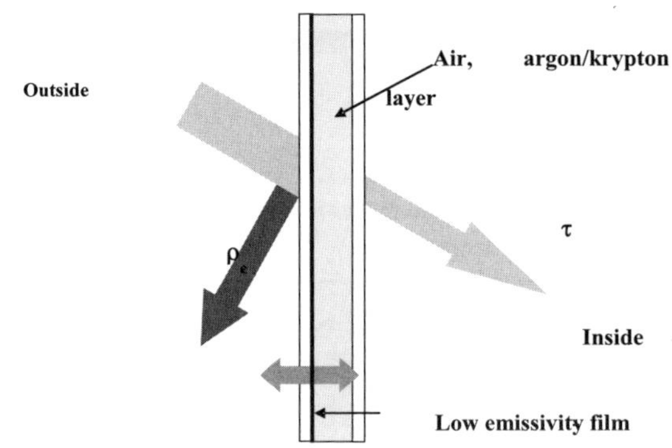

Figure 6.7. *Schematic representation of heat transfers through glass*

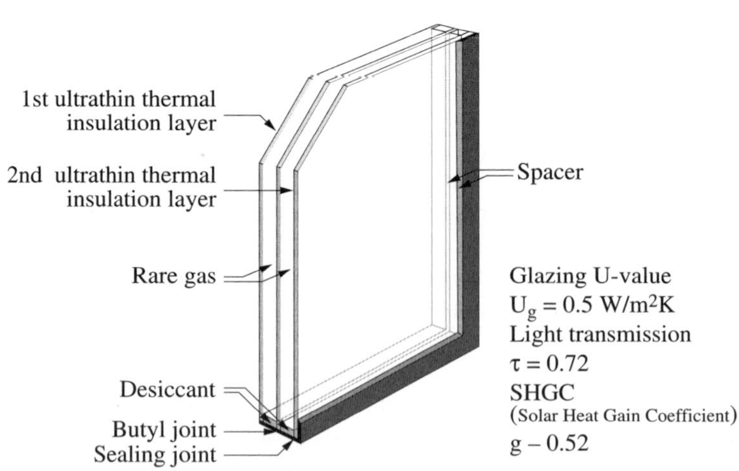

Figure 6.8. *Triple glass (source: www.interpane.de)*

Then, a low emission layer which limits infra-red radiation losses from inside to outside and the replacement of air by a rare gas (argon), which is heavier' and thus less mobile, has again enabled the reduction of the U coefficient by a factor of 2 to finally reach a U-value close to one. This value is still above the U-value considered for opaque walls (close to 0.1), but new and very efficient glazing is emerging, such s triple-glazing or vacuum-glazing with two low-emission layers (Figure 6.8 and Table 6.11).

Table 6.11 presents the key developments in the U-value and gives the order magnitude of the glass surface temperature for severe conditions. These values are very important because the wall temperature is an important parameter for the occupant comfort (cold wall effect due to radiation).

	U W/m².K	Estimated internal surface temperature External Temp: -10°C Internal Temp: +20°C	Solar factor (FS) – g
Simple glass	~ 5.6	~ - 1/-2	~ 0.92
Double glazing, clear glass + air	2.9 / 3.1	~ 8/9	~ 0.8
Triple glazing, clear glass + air	2.1 / 2.3	~ 12	~ 0.7
Double glazing + low emissivity film	~ 1.8	14/15	~ 0.62
Double glazing + low emissivity film + argon	1.1 / 1.3	15	0.6 / 0.62
Double glazing + low-emission film + argon + solar protection	1.1 / 1.3	15	0.1 / 0.2
Triple glazing + two low-emission layers	0.4 / 0.8	17/18	0.48 / 0.65

Table 6.11. *Evolution of U and g coefficients of glass, internal surface temperature*

However, the performance of a window does not depend only on the glazing. The U_w-value of a window (U_w- glass + frame) also depends on the design and on the materials used to make the frame (aluminum, wood, PVC, thermal-bridge breaker, etc.). The coefficient may be lower than one (products are claimed to be

0.8, even 0.7) with the use of glass and frames that are very efficient, but also heavy and voluminous (Figure 6.9).

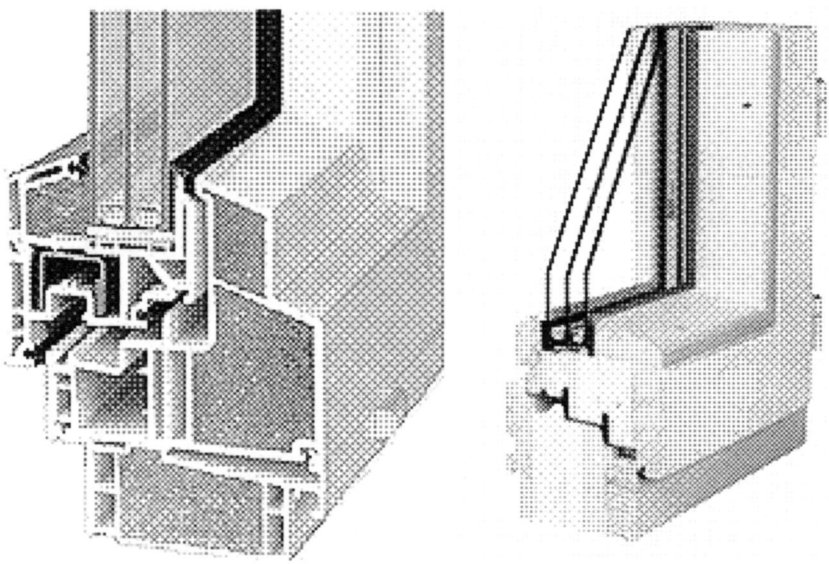

Figure 6.9. *Example of highly efficient windows: PVC and wood frame total thermal transmission coefficient (glass and frame) Uw = 0.7 - 0.8 (source: www.fenetredomex.com and http://www.passivhaus.lu)*

In order to balance its low performance in terms of insulation when compared with opaque walls, the window has some advantages. First, a window can be considered as the first thermal solar collector integrated in the wall. Indeed, the solar inputs are interesting during periods when heating is needed, but they are detrimental during the summer period. Thus, glazing must be associated with architectural solar protection (screens, especially on the east and west walls, etc.) or integrated in the window frame (shutters, Venetian blinds, sunscreens, etc.).

Equations [3.3] and [3.5], enable us to evaluate the loss/gain ratio of a window. Thus, by modifying the characteristics of glazing and the associated solar protections, it is possible to estimate the energy balance of windows and glazed facades.

As an example, Table 6.12 [SID 00] shows the energy balance of a window during the season when heat is needed; it shows different types of glass for different orientations in the city of Chambéry (plain or valley area) and in the city of Bourg-St-Maurice (high-altitude area). The windows used in the calculation are wood-frame windows with a glass/window ratio of 0.7, without a screen. The calculation takes into account losses and solar inputs due to the glass part of the window.

	South		**SE and SW**		**East and West**		**North**
U W/m².K	*1*	*2*	*1*	*2*	*1*	*2*	*1*
Simple glazing **4.95**	- 59	- 48	- 100	- 102	- 172	- 194	- 183
Double glazing (DG) **2.95**	+ 29	+ 49	- 6	+ 4	- 66	- 73	- 84
DG + shutters **2.25**	+ 75	+ 101	+ 40	+ 57	- 20	- 21	- 39
DG low-emission **1.8**	+ 85	+ 111	+ 53	+ 71	- 1	0	- 39
DG low-emission + shutters **1.50**	+ 104	+ 134	+ 73	+ 93	+ 18	+ 23	- 20

1: Plain/Valley: Chambéry (heating period: October 1 to May 10)

2: Mountain: Bourg-St-Maurice (heating period: September 20 to May 25)

Table 6.12. *Energy balance (kWh/m²) of 1 m² of a wood-frame window (installed on the inside edge over a heating season without a screen and a glass/window ratio of 0.7)*

Table 6.12 illustrates the impact of the glazing evolution which makes the window (including the solar protection) the first energy-positive component of the building envelope. For the investigated examples, only the north facade requires a particular treatment with the use of triple-glazing.

Furthermore, we should not forget that the window plays a major role in the quality of the interior spaces, in particular for acoustic insulation and natural lighting, etc.

Finally, functions that the window must fulfill are sometimes in opposition to each other (to see without being seen, light without glare, heat without overheating, etc.). Therefore, these challenges require the development of new types of glass, such as "chromogenic" glasses (electro-chromic, suspended-particles, etc.) or selective glass.

6.3.2.3. *Actions on the equipment*

Despite all efforts made at the building level (envelope and structure), additional equipment is needed to ensure the comfort of the occupants all year long, in particular to guarantee, at the same time, air quality and hydrothermal and lighting comfort during the dark and cold period (winter) and the warm and sunny period (summer).

For the other services which do not depend on the building design, such as DHW and all the electric domestic and multimedia equipment, it is advised, if not required, to use high-performance equipment (high-performance boilers or heat pumps, low-consumption motors, A, A+, A++ class equipment, low-consumption lamps, no warning equipment, etc.). Indeed, the energy that is consumed in Europe by the equipment with warning has been estimated at 53 TWh/year, which represents the electric energy generation of about eight high-power stations.

6.3.2.3.1. Heating-ventilation-air conditioning (HVAC) and domestic hot water (DHW)

With the reduction of heat loss through the walls, the weight of the ventilation becomes important. Indeed, the ventilation that is necessary for the durability of building and the health of occupants generates thermal losses which nowadays represent about 20% of the heating consumption, but tomorrow the ventilation could become dominant.

To meet the development of passive and low-energy house requirements, different systems have been developed, such as double-flow ventilation systems with energy recovery and local or decentralized double-flux systems. These latter are devoted to the ventilation of a single room, as opposed to centralized ventilation systems devoted to the ventilation of many rooms.

The double-flow systems with energy recovery used in the low-energy houses have an efficiency factor higher than 75%. Two major types of heat exchangers are used for this type of system:

– static exchangers with an efficiency between 50%–70% for the cross-flow, 70%–80% for the cross counterflow and 85%–99% for the counter-flow;

– rotating exchangers, which have a heat recovery efficiency in the order of 70%–80%.

Fan engines must also have good performances and DC motors are increasingly used because, on one hand, they enable the reduction of electricity consumption and, on the other hand, they have a longer life-span than standard AC motors. Moreover, DC motors may be more easily installed and maintained. Easier speed regulation is also possible according to needs and to the lifestyle of the occupants: reduced ventilation for periods of vacancy, high ventilation for occupied periods or for the evacuation needs. Indeed, the use of ventilation when needed also enables energy savings.

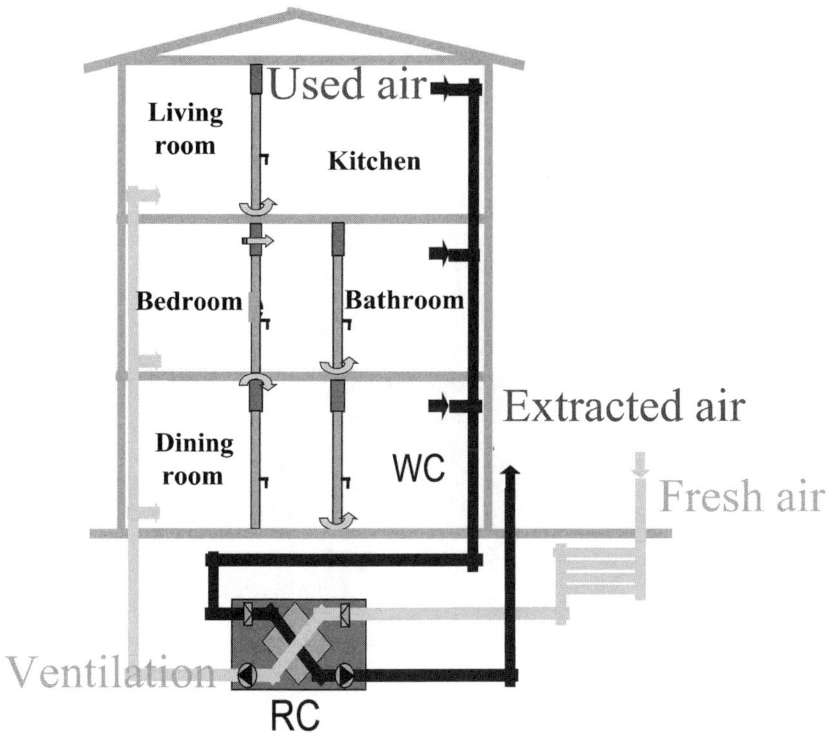

Figure 6.10. *Double-flux ventilation system with energy recovery (source: DIAE-Energy cantonal service, Christian Freudiger, Minergie-Geneva)*

In the case of passive houses, the ventilation system is sometimes also used for heating the air of the house, either by electric resistances or by coupling with a hot water radiator that fully integrated to the device or simply located in front of the ventilation system.

Figure 6.11. *Compact system: HVAC-DHW (source: www.paul-lueftung.net)*

Another way to save energy is to do several things at the same time. This is why, in order to meet the new requirements of low-energy houses, multifunction systems, which called “compact”, are emerging. They gather in a single product the HVAC and DHW functions (Figure 6.11). These products are generally made of:

– a static heat-recovery system: extracted air/fresh air;

– a heat pump on extracted air for heating a DHW tank.

The heat vector is generally air, and a DHW tank may be added to the blown-air heating system. As far as the DHW is concerned, it is generally an electric system and its power is variable according to the product. Sometimes some manufacturers also suggest coupling this compact system with other elements:

– solar collectors which contribute to the storage heating;

– earth-heat exchanger (Canadian wells) to preheat fresh air.

In addition, the night-time ventilation or the coupling of the earth-heat exchanger with the ventilation are also technical solutions which may lead to summer comfort levels which are adapted to the occupants' expectations, but without generating excessive energy consumption.

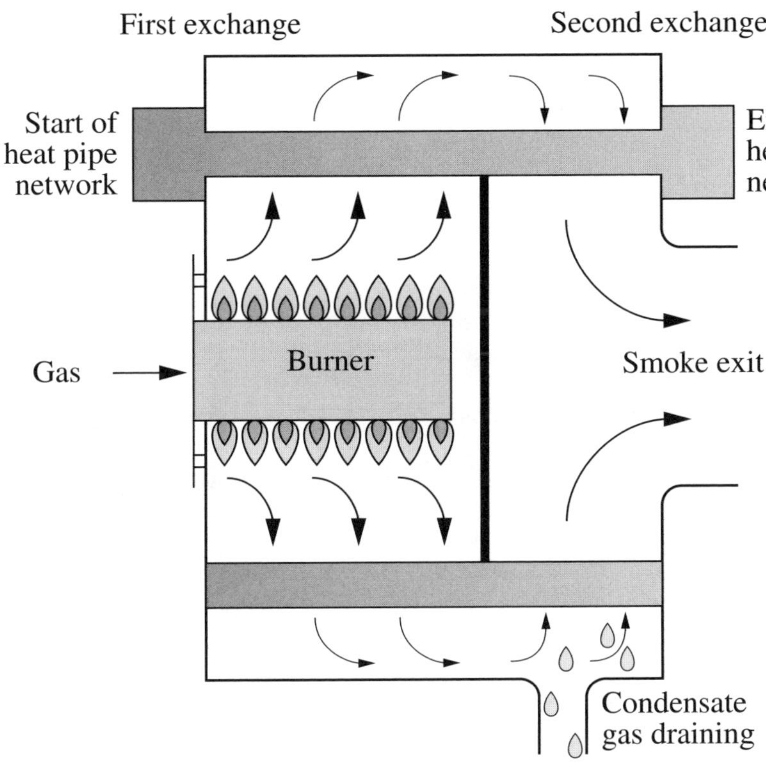

Figure 6.12. *Principle of the operation of a condensation gas boiler (source: www.maison-bioclimatique.fr)*

More "traditional" equipment is also used for heating and for DHW, but always with the goal of reducing energy consumption (high efficiency). For gas (and oil) boilers, nowadays condensing boilers are generally installed. With conventional boilers, a significant amount of heat (called latent heat) is evacuated with the smoke. It implies that the combustion products are at a very high temperature (about 150°C); however, the use of a condensing-type boiler enables the recovery of a very large amount of this latent heat. This energy recovery leads to a significant reduction of the gas temperature to values of about 65°C while at the same time limiting the pollutant gas emissions.

The condensing boiler is so called because, to generate heat, it uses not only the low calorific power (LCP) of a fuel, but also its high calorific power (HCP). For all energy consumption calculations, the European and French standard uses LCP as a reference; therefore efficiencies higher than 100 are obtained because of the latent heat restitution.

As far as the heat pump is concerned (HP), it enables the extraction of heat from air, ground, or water using an electric compressor (Figure 6.13).

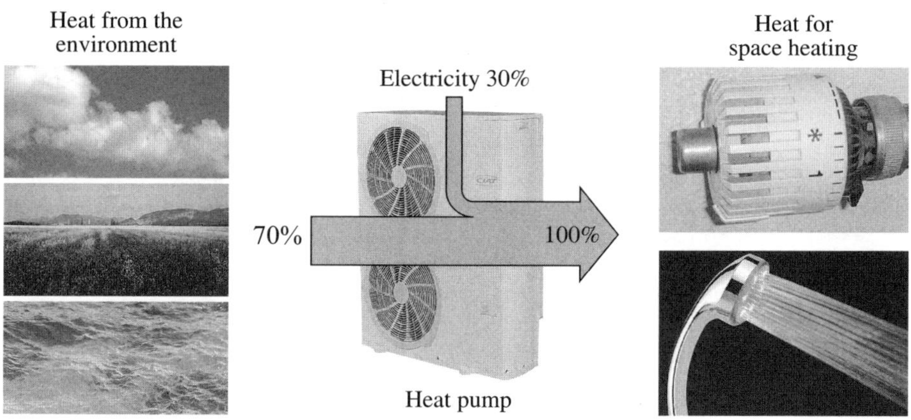

Figure 6.13. *The principle of the heat pump – HP (source: www.pac.ch)*

The heat is transported to the evaporator and transmitted to the fluid. The state of this fluid changes and is transformed into vapor which is then compressed by the compressor, thereby increasing the temperature. At the condenser level, the overheated vapor transmits its heat to the heating system. The fluid becomes liquid again and the valve reduces its pressure. The fluid temperature drops sharply, making the fluid ready to absorb new heat from the environment, and the cycle starts again.

An HP is characterized by its energy-efficiency ratio (EER) which is the ratio between the useful energy supplied and the electric energy consumed by the compressor (Figure 6.14). The EER is generally about 3, or even larger for the most efficient machines. Similar to the case of a window in a building envelope, a heat pump may be considered to be the first qualified as an energy-positive equipment.

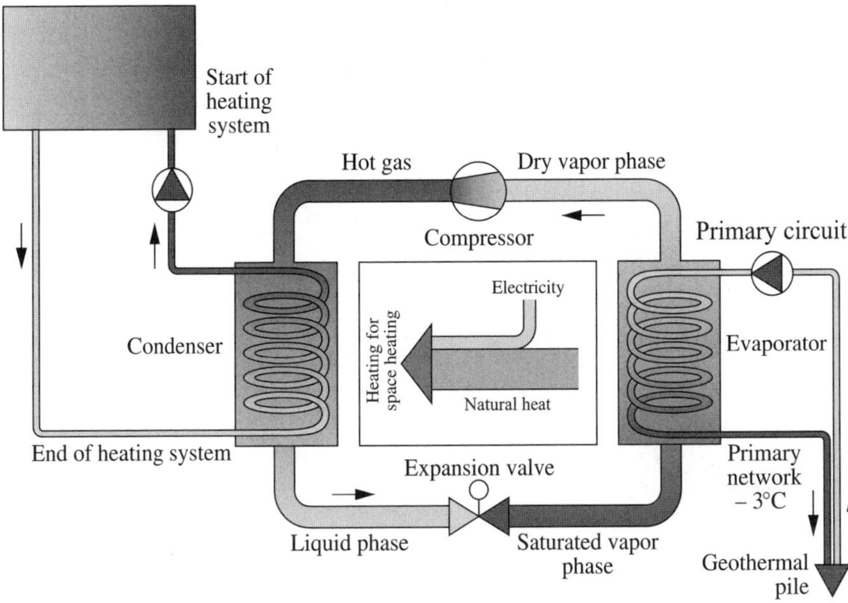

Figure 6.14. *Principle of the operation of an HP (source: www.elco.ch)*

The two weak points of an HP are the refrigerating gas of the compressor and the use of electricity, which is today mainly generated from fossil energy. This is why the general use of the HP as a universal system for heating and DHW could lead to an increase in electricity demand. Nevertheless, there are some interesting developments, such as the use of CO_2 (Japanese Ecocute project and European Sherpa program) as gas and photovoltaic energy to generate electricity (Figure 6.15).

Among future developments is the coupling of a HP with a solar thermal collector, which, according to a recent investigation [TRI 06], could avoid the overheating problems of the solar collector in summer and increase the temperature of the ground when the geothermal stacks are undersized. Finally, other types of HP are under development, such as the adsorption system; but they are still not available on a large scale.

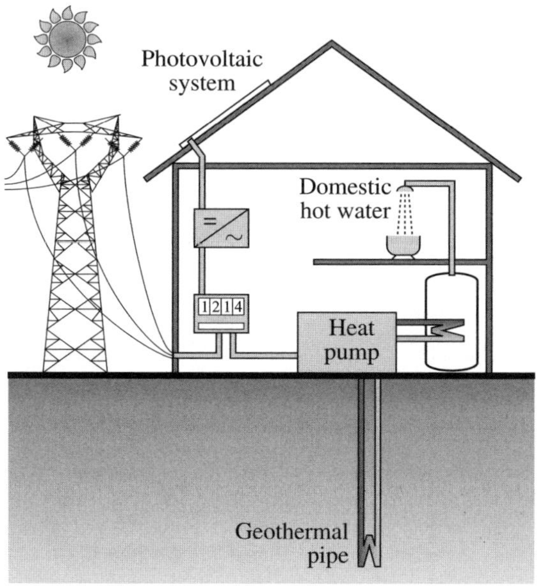

Figure 6.15. *HP+PV coupling (source: www.pac.ch)*

Regarding the use of wood for heating, France is the primary user in Europe, in quantity but not per capita. The use of wood ranks third in the Minergie projects. This resource is of great interest because its combustion generates just a little bit more CO_2 than the forest consumes. Moreover, boilers now have improved performances, such as efficiencies higher than 65% and the automatic mechanisms for wood supply, which make this heating system increasingly attractive for the users.

Many techniques are available:

– Logs are frequently used, but they need considerable handling and the combustion is far from being complete.

– Platelets obtained by crushing branches, trunks, and by-products of the wood industry. Compared to logs, they have the great advantage of being automatically supplied to the boilers.

– Briquettes made from compact sawdust enable higher efficiency than simple logs.

– Wood pellets, which are obtained by compression and compaction, are, due to their high-energy density, seen as a useful combustible for heating systems which operation is completely automatic. They are also very easy to store and to deliver.

Wood pellets are compact cylinders of about 6 mm in diameter and of 5 mm to 30 mm length and contain 100% natural dried-wood residues. Two kilograms of granules represent around 1 liter of oil.

Today, on the scale of a block of flats or a district, a technique is appearing in response to electricity and thermal needs; it is co-generation coupled, in the case of a district, to a local heat network.

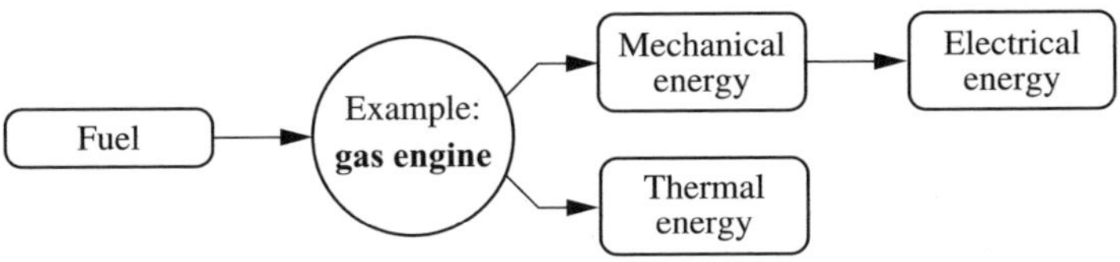

Figure 6.16. *Example of co-generation with a gas engine (source: www.raee.org)*

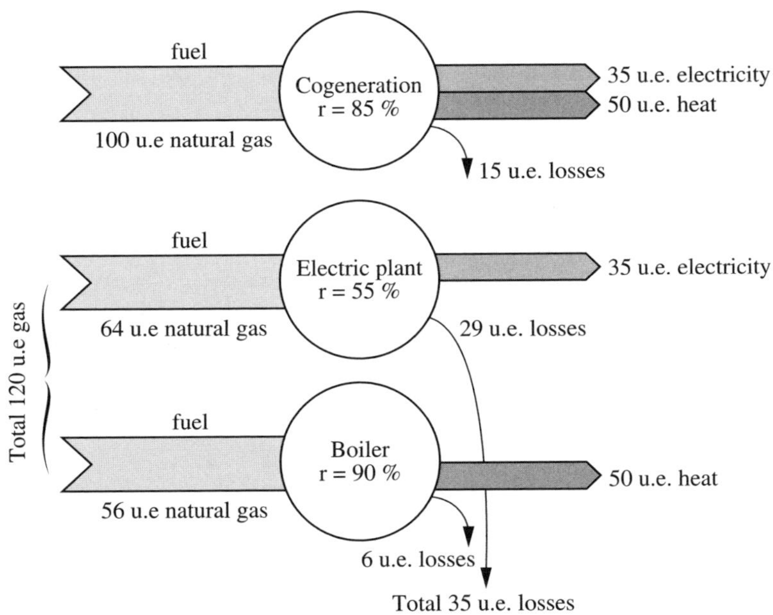

Figure 6.17. *Comparison between a co-generation installation and a separate installation of heat and electricity generation. The heat and electricity generated are identical (source: www.raee.org)*

Co-generation (Figure 6.16) is the simultaneous generation of heat and mechanical energy, usually converted into electricity, from a single source of energy (oil, gas, biomass, waste, hydrogen, etc.). Figure 6.17 compares a co-generation installation with a total efficiency of 85% with two power stations of electric and thermal energy. This set of two power stations is composed of a gas combined cycle with an efficiency of 55% (presently the most efficient means of generating energy) and a gas boiler with an efficiency of 90%. In this scenario, the gain of primary energy is of 17%.

With the majority of the electric power plants currently in operation (coal, fuel, nuclear, etc.), 45% to 65% of the primary energy is lost as heat in the atmosphere (eg, in cooling towers). Co-generation technology, which consists of recovering this heat through thermal exchangers (exhaust gases, cooling water, lubricating oil, etc.), increases the energy efficiency of the installation. Thus, it enables full exploitation of the energy potential of the primary fuel in order to reach an overall efficiency (heat + electricity) of 80% to 90%, compared to 35% to 40% for a conventional power plant.

For low-energy houses, the network efficiency must be analyzed first because of distribution and connection costs.

6.3.2.3.2. Domestic hot water (DHW)

For DHW, which, unlike heating, is a quasi-permanent need, the systems used for heating (oil, gas, wood, HP) generally produce the DHW required by the occupants. But in the case of DHW, the thermal solar energy is a very interesting alternative solution since it can cover between 40% and 70% of the needs, according to the weather and the occupants' lifestyles (Figure 6.18). This is generally complemented by electricity or a gas boiler.

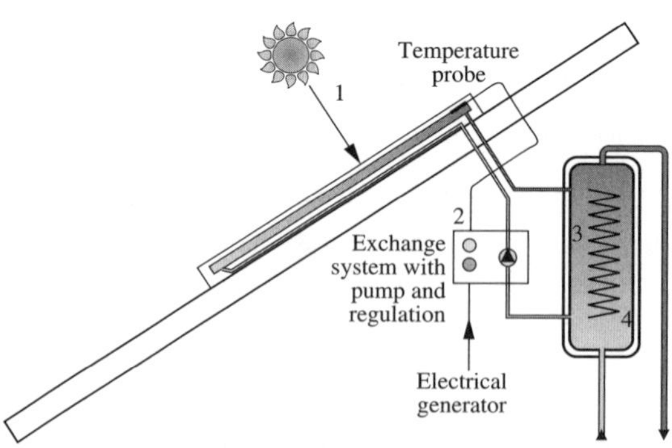

Figure 6.18. *Principle of the operation of a solar hot water system (source: www.outilssolaires.fr)*

The monitoring of 116 installations of ISWH (individual solar water heater) conducted by the Centre Scientifique et Technique du Bâtiment (CSTB) for one year has shown that the facilities were generally oversized and that improvements were possible on the solar loop regulation and on the reduction of heat losses from the tanks [BUS 06]. This study shows that the average ratio "solar production/total energy consumption" is between 60% and 70% as expected, whereas the solar productivity, about 200 kWh/m^2.year, is significantly lower than the usually accepted values (400 kWh/m^2.year). The ISWH consumption at 50°C is around 33 liters per day per capita.

6.3.2.4. *Lighting and other specific electrical equipment*

The energy needs for lighting have doubled in 20 years and represent a little more than 10% of the total electricity consumption. Lighting, on average, is 30% of office energy bills and more 39% of energy bills for buildings related to education. These needs may be reduced by, first, a building design that relies on natural lighting. Moreover, natural lighting has a positive effect on employee productivity and student assiduity [NIC 93]. Second, low-consumption lamps or fluorescent compact lamps should be installed.

Among the innovations that should further reduce electricity consumption, is the LED (light-emitting diode).

There are two main types of LEDs [SID 03]: the inorganic (LED) and the organic (OLED) which is still under development and will probably not be on the market before 2010. Whereas the LED generates point sources like incandescent lamps, the OLED could replace other classic sources like fluorescent tubes.

The conventional inorganic LEDs are P-N (diodes) junctions, doped in order to emit a visible or an ultraviolet radiation when an electric current crosses these junctions. The radiation emitted by a standard LED (red, green) is almost monochromatic (spectral ray). The radiation of a white LED is dichromatic or polychromatic according to the technique used (see Chapter 4 for more information on lighting).

For other electric equipment (appliances and multimedia applications), it is recommended – required, in fact – to use class A, A+, A++ equipment and not to use the standby option.

Furthermore, with regard to on-site production of electricity, the expected reduction of costs coupled with the feed-in tariffs, and not ignoring the direct and indirect financial incentives, should make the installation of photovoltaic collectors more feasible, especially if they are integrated into buildings (roofs, slates or tiles;

industrially designed with or without frames, awnings, window sills, glass walls without rear-protection, window balustrades, balcony or terrace balustrades, boarding or curtain wall). This local production will thus provide a significant part of the electricity needed for lighting and multimedia equipment, and for some means of transportation (bicycles, electric cars, etc.) and considered as new equipment of homes and dwellings.

Two types of installation are possible: the isolated systems (Figure 6.19) which require batteries and the systems connected to the grid (Figure 6.20) which operate according to the feed-in tariffs in force.

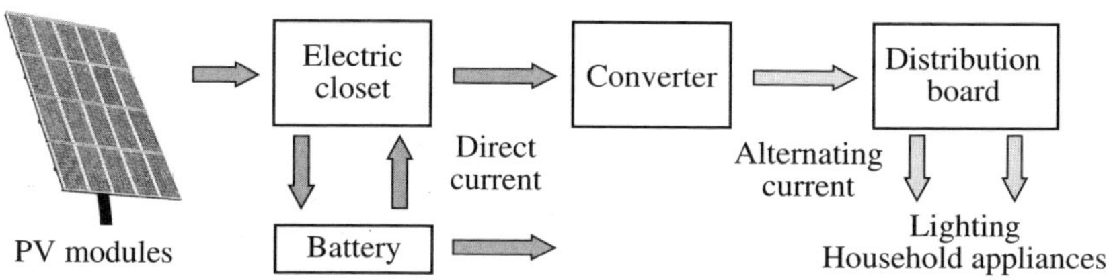

Figure 6.19. *Photovoltaic system for an isolated site (source www.outilssolaires.fr)*

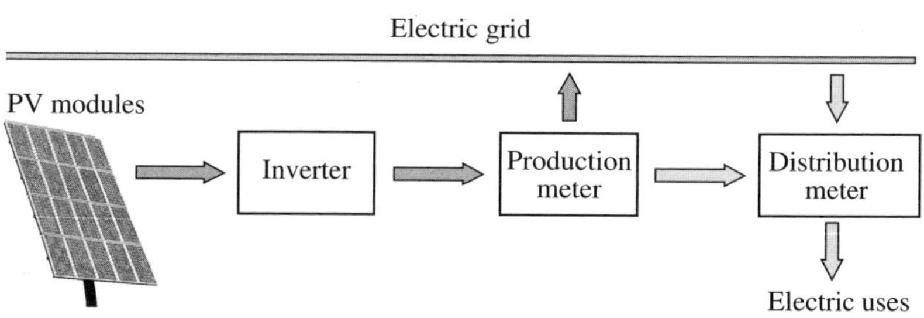

Figure 6.20. *Photovoltaic system connected to the grid (source www.outilssolaires.fr)*

Although it is more difficult to integrate than photovoltaic energy, but complementary in some areas, wind energy has started to be used on buildings (Figure 6.21).

For specific electrical applications, such as lighting and most household equipment (multimedia, electrical appliances, etc.), the occupants, users, and other building managers play a crucial role to reach the expected performances. This role will be much easier than the equipment condition (operating, shutdown, and standby) and their consumption will be known by the users. Tools to help users to be aware of these requirements have to be developed, like the dashboard in a car.

Figure 6.21. *Horizontal shaft wind generator on a flat roof (source www.windwall.nl)*

This dashboard, which is easier to implement for electrical equipment, must also be extended to other components (closed/opened windows, etc.) and to other types of energy and resource consumption: thermodynamic energy, water, etc.

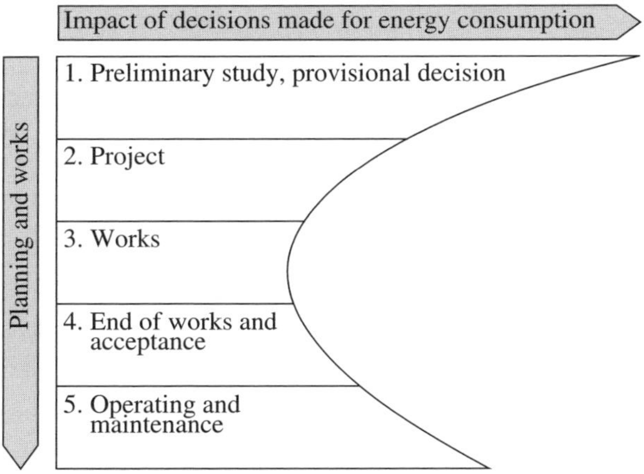

Figure 6.22. *Design – planning – execution – exploitation (source: www.bfe.admin.ch)*

Finally, all the actions described above, whatever their performance, when individually considered do not guarantee a low-energy building. To reach this ambitious goal, a global (holistic, systemic) approach that involves all the stakeholders – architect, engineering office, bankers, insurers, builder, client, etc. –

must be implemented at the beginning of the building process (Figure 6.22). Indeed, it is during the initial stage that the most important decisions are taken and afterwards, during the building process less and less modifications are possible [OFE 06].

6.4. The Minergie trademark

The Minergie® trademark appeared in Switzerland in the late 1990s (1997) as an initiative of a Zurich county, joined by the Bern county, both co-owners of the trademark which is managed by an association. Two levels of energy requirements are proposed: Minergie S and Minergie P. Recently, both labels have been joined by a third one, which includes environmental aspects, Minergie-ECO.

Buildings	**Energy needs for heating and domestic hot water**		
Unit	kWh/m^2.year	MJ/m^2.year	Oil/liters/m^2.year
Buildings built before 1970	120–150	430–540	12–15
Legal limit value SIA 380/1 standard	90	324	9
Legal limit value High Performance SIA 380/1 standard	72	254	7.2
Minergie S limit value New buildings	42	151	4.2
Minergie P limit value New buildings	30	108	3

Table 6.13. *Energy needs for heating and SHW according to the different labeling and regulation levels in Switzerland*

In Table 6.13, it is important to notice that the basic requirements (limit value SIA standard) are closed to those of the RT2005 in France. It is also important to notice that the Minergie label requires a reduction of about 50%.

The energy consumption of the Minergie® houses is calculated according to the Swiss recommendations SIA 180/4 (Society of Engineers and Architects), and the specific characteristics of the different energy carriers are taken into account on the basis of the following three criteria:

– Only the energy delivered on-site (oil, electricity, district heating) is included in the calculation. The energy locally available (geothermic, solar, thermal waste, etc.) is not included.

– The highest value of the electric energy is taken into account by means of a double weighting of the electricity consumption of heating, ventilation, and air conditioning equipment.

– To promote the ecological and economic advantages of wood, the consumption of the heating equipment that uses wood is weighted as only 60%.

The energy consumption index, calculated according to the above described modifications, is called the index of weighted energy expenses (IEE) and takes into account: the energy of room heating, domestic hot water, and ventilation, as well as lighting needs (LIG) (Table 6.14).

Debit – D	**Credit – C**
Losses from the frame: airtightness-insulation	Internal and passive solar contributions
Air change – ventilation	Heat recuperation
DHW	Active solar contributions
$IEE = (D - C)/(\tilde{\eta}/g) + LIG$	

Table 6.14. *Energy balance of a Minergie S house*

The weighting factors η/g are listed in Table 6.15. We notice that wood is favored.

	Useful ratio η	**Weighting g**	**η/g**
Oil	0.85	1.00	0.85
Gas (condensing)	0.92	1.00	0.92
Wood	0.75	0.60	1.25
Direct electricity	1.00	2.00	0.50
HP air-water	2.30	2.00	1.15
HP ground sources	3.10	2.00	1.55

Table 6.15. *Technical and legal weighting factors*

As an example, the indexes of energy expenses of the Minergie® buildings (houses) which are in the "low-consumption building" category must not exceed the following limit values:

– new buildings: IEE < 42 kWh/m^2 (151 MJ/m^2);

– retrofitting (before 1990): IEE < 80 kWh/m^2 (288 MJ/m^2).

Today the average consumption in Switzerland is around 120 kWh/m^2.year compared to 190 kWh/m^2.year in France (2000 Thermal Regulation).

For the other building categories (schools, commercial buildings, restaurants, hospitals, industrial, warehouses, sports facilities, swimming pools, etc.) other IEE are required, both for new buildings (Table 6.16) and updated buildings.

New Building Type	Commercial	Hospital	Industries	Sports facilities
Max final energy – kWh/m^2.year	40	75	20	25

Table 6.16. *IEE for non-residential buildings*

6.4.1. *Examples of solution for a Minergie house*

As the first pillar of the Minergie trademark is the reduction of energy loss, the heat transfer coefficient U of the walls, the windows, and the other envelope components must be less than the listed values of Table 6.17.

To reach these values, the insulation thicknesses vary from 15 m to 20 m according to the type of materials used.

Building elements U in $W/m^2.K$	Building element outside or buried less than 2 m in the ground	Building element separating rooms without heating or buried more than 2 m in the ground
Roof, ceiling	0.2	0.25
Wall	0.2	0.28
Floor	0.2	0.25
Windows	1	1.6
Doors	1.6	2.0

Table 6.17. *Heat Transfer Coefficient U in $W/m^2.K$*

As a comparison, Table 6.18 outlines the reference values given in the French Thermal Regulation 2005 for H1, H2, and H3 areas > 800 m. We note that for roofing, the Minergie/RT2005 values are the same.

Walls	References values U W/m².K
Blind vertical walls	0.36
High floors outside oriented	0.27
Roofs – high floors (other than above)	0.20
Low floors	0.27
Windows and glass doors, window doors – non-residential	2.1
Windows and glass doors, window doors – residential	1.8

Table 6.18. *Thermal transmission coefficient U according to the RT2005*

Regarding thermal bridges, the limit values must be respected, according to the methods described in the SIA 380/1 Swiss standard. Moreover, the ratio of the window surface (ratio between the window surface and the energy reference area (surface of floors in living space, walls included)) must not exceed 30%.

Secondly, in order to fulfill the remaining energy needs (heating, DHW, and electricity), energy-efficient systems are recommended (low-temperature heating, ventilation systems with energy recovery, heat pump, wood, etc.). Although they are not mandatory, renewable energies will be favored (biomass, thermal solar for DHW, etc.), especially due to the weighing factors used during the calculation (Table 6.15). Regarding the generation and distribution of heat, the solutions in Table 6.19 are suggested.

Standard Solution	No. 1	No. 2	No. 3	No. 4
Heat-generation system	Ground heat pump	Wood heating + solar collectors	Automatic wood heating	Air-water heat pump
Output temperature for heat distribution	Maximum 35°C	Maximum 50°C	Maximum 50°C	Maximum 35°C

Table 6.19. *Standard heat-generation system solutions for Minergie® S*

As shown in Figure 6.23, the most frequently used equipment are the high-energy-efficiency systems: heat pump, gas boiler, or renewable energy, for example wood. We notice that district heating is the 4th most frequently used for heating; which is also the position of solar energy for DHW.

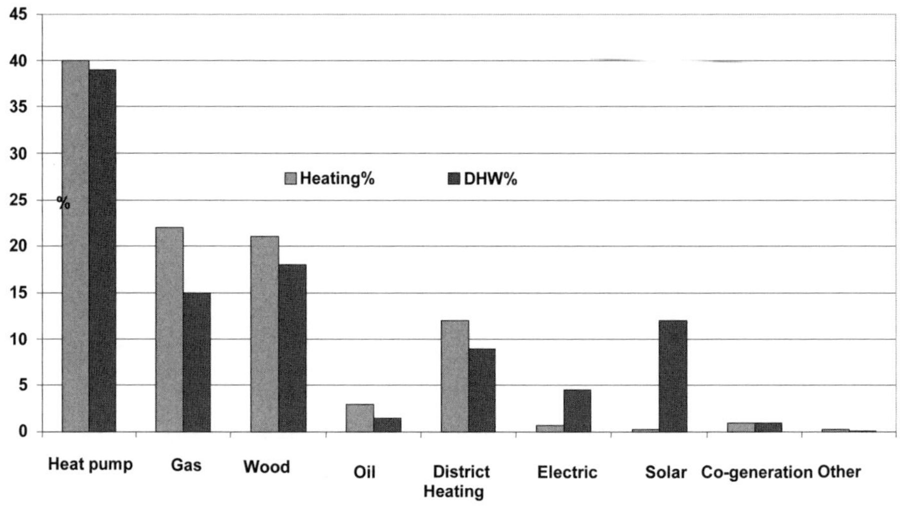

Figure 6.23. *DHW and heating systems used in Minergie houses (source: www.minergie.ch)*

For the ventilation equipment, the following characteristics are recommended.

Double-flow ventilation	Recovery–efficiency: 80%
Air change	0.4 Wh/m^3

Table 6.20. *Characteristics of ventilation systems*

Figure 6.24 outlines all the requirements that must be met in order to obtain the S and P Minergie® label. The P Minergie® label is equivalent to the German PassivHaus label described in section 6.5.

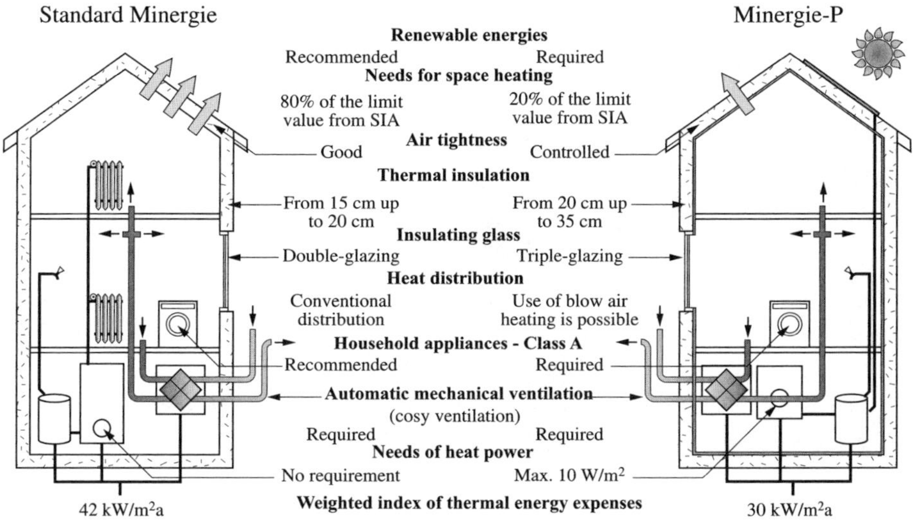

Figure 6.24. *Summary of Minergie S and P requirements*

The satisfaction surveys show that there is a close correlation between a low-energy home and comfortable home. Indeed, a heating system at low temperature, the control of ambient and surface temperatures, together with mechanically controlled ventilation avoid humidity and moisture, air pollution, and external noises (the air exchange taking place through ventilation and not through open windows). According to these surveys, 95% of the occupants declare themselves to be very satisfied.

Moreover, the Minergie buildings have an added value because of their low-energy consumption, their maintenance facility, and their longer lifetime which comes from the quality of the building. The resale value is often above that of traditional buildings (by about 12%) and, if rented, tenants stay longer. The additional investment which must be lower than 10% to gain the Minergie trademark, is in fact in the order of 6% (after more than 5,000 constructions, the additional cost varies between 5 and 8%), and it is rapidly recouped by the generated savings. All these positive aspects are what are often called "co-benefits" by the Minergie label promoters.

Finally, in some counties, public buildings must comply with the Minergie® requirements and the private contractors who undertake to follow the label can gain some of advantages (mainly fiscal incentives) or a larger ground surface (ground occupation coefficient). Several banks also offer preferential loans. At the end of 2006, there were more than 5,000 buildings with the Minergie® label, which represent 10 and 15% of the market according to the counties.

Figure 6.25. *Two examples of Minergie houses*

Both examples in Figure 6.25 show that Minergie houses are much like traditional houses.

6.5. The PassivHaus label (passive house)

The term "passive house" may be considered as a refinement of a low-energy house. Indeed, the original idea about this type of construction was to ensure a comfortable indoor ambiance in winter, as well as in summer, without requiring either a conventional heating system or a cooling system. It combines bioclimatic design, high-performance components, and energy-efficient equipment, high-quality workmanship and the commitment of users [GUE 05].

A passive building is different from a low-energy-building due to the research on a "tunnel effect" on the costs. Indeed, whereas the Minergie label is located at the optimum between the thermal insulation reinforcement to reduce the heating needs and the investment (Figure 6.26), the PassivHaus standard is characterized by an additional investment in "passive" solutions (insulation, windows, airtightness, and internal and solar gains) to reach the breakthrough point ("tunnel effect") which makes the conventional heating system (boiler + emission radiator and distribution system) unnecessary.

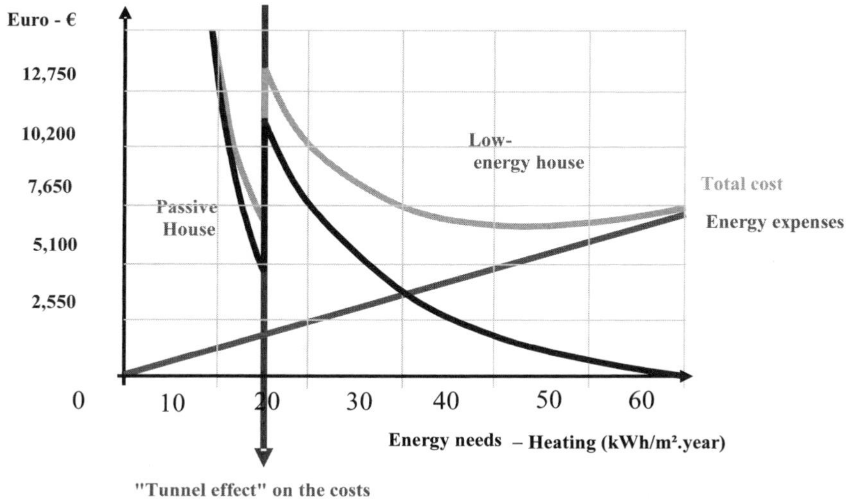

Figure 6.26. *Relationship between investment and energy-saving for a low-energy and passive house of about 100–150 m^2 (source: www.passiv.de)*

This approach, initiated by Professors Feist from Darmstadt and Adamson from Lund, is focused on heating needs and winter comfort. It is well adapted to northern and central European countries, but may be modified and adapted to southern Europe where the summer temperatures last for much longer.

The basic principles of a passive building are very simple: optimizing the essential occupancy needs and minimizing the heat losses before maximizing the heating gains (Figure 6.27).

Indeed, too often the construction concepts developed to optimize the energy demand have required the implementation of overly sophisticated technologies. The interest in the passive house is due to its simplicity. As indicated by its name, the energy saving is mainly achieved through the use of passive strategies with, the reduction of heating losses by a high-performance airtightness (controlled low-air-permeability) and a reinforced thermal insulation of the building envelope. Thus, because the main elements are optimized, a traditional heating system is no longer needed.

Meanwhile, totally sealing the building requires mechanically controlled ventilation in order to guarantee a good air quality and a high-efficiency heat-recovery system to limit the heat losses due to ventilation.

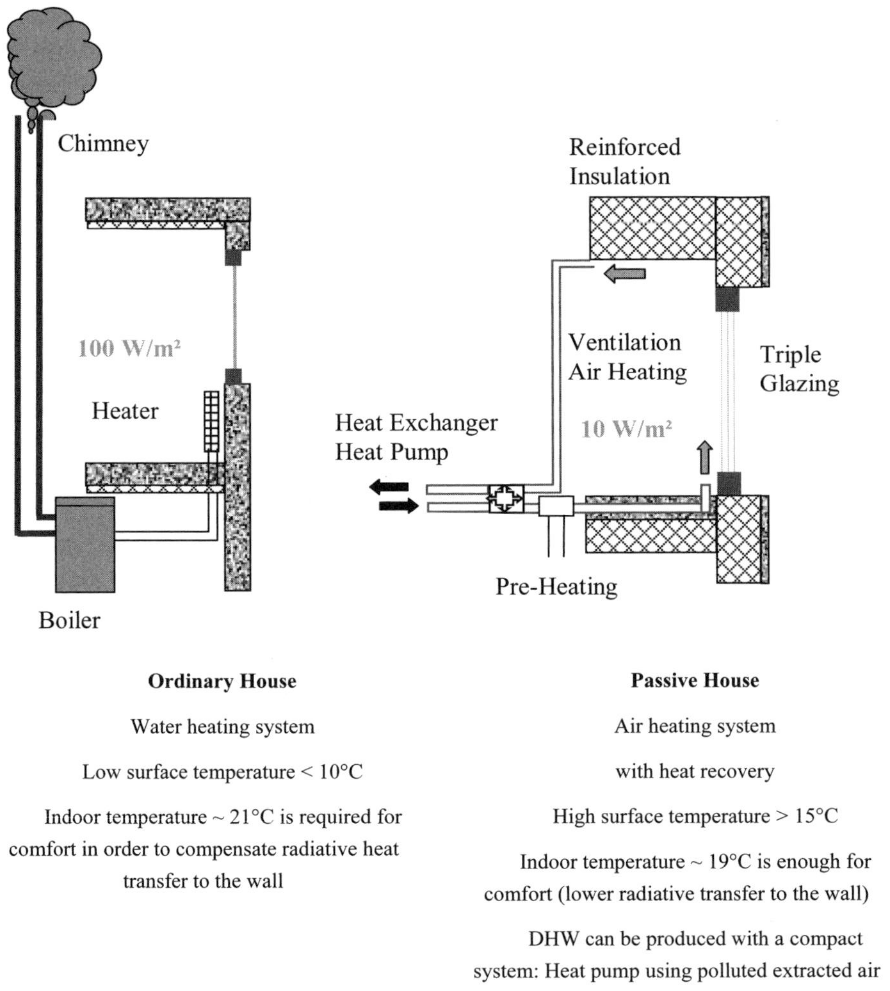

Figure 6.27. *From the conventional house to the passive house*

Finally, due to well-chosen orientation of the well-designed windows, the solar radiation may be used as an external heat input. Efficient windows play the role of passive solar collectors as, in some cases, they bring an energy gain higher than the energy loss. To obtain the benefit it is all the more important that the main windows are properly oriented to the south and not hidden by external elements. Well-designed windows also bring enough light inside and may also reduce the energy demand for lighting.

Moreover, the heat generated internally [LEN 06] by the occupants and household appliances (Table 6.20) contributes to meeting the heat demand. In summer, the internally produced heat must be removed. Nighttime over-ventilation or use of a pre-cooling system, such as an earth heat exchanger of the ventilation air, could be a solution to this problem.

Equipment	**Heat generation during 1 day in kWH**	**Equivalent heated surface m²**
TV on stand-by (15 W) for 20 hours	0.3	2
TV on (75 W) for 4 hours	0.3	2
20 minutes of ironing	0.3	2
30 minutes of oven use	1.5	10
Refrigerator (compressor + heat generated by the condenser)	3	20
PC with cathodic screen for 24 hours (ADSL – 250 W)	6	40
PC with cathodic screen for 8 hours (office – 250 W)	1.75	12
PC with flat screen for 8 hours (office – 125 W)	1	7
Mobile/cell phone for 8 hours (office – 30 W)	0.24	2

Table 6.21. *Examples of heat generated by domestic equipment (source: Fraicheur sans clim', edition Terre Vivante and association AERE)*

Finally, in passive houses energy consumption is generally four times lower than the average consumption in new houses built according to the current regulations or standards.

The previously stated requirements are valid for a passive house built in a northern or central European country where, for winter, comfort is vital. These requirements must be revised to adapt the passive house concept to southern European countries or to countries with a more moderate climate. While keeping the same low-consumption objectives, the implemented technical solutions must certainly be modified with respect to the requirements of Table 6.22:

– The insulation thicknesses could certainly be reduced and less efficient windows in terms of U could be installed (from triple-glazing to double-glazing with low-emission layer).

– However, masks, sunscreens, inertia, and night ventilation will be more important to control the solar inputs.

– The indoor heat production by appliances and multimedia during the summer should also be controlled, by developing, for example, low-cost solar-cooling systems using the solar energy which is available in the summer.

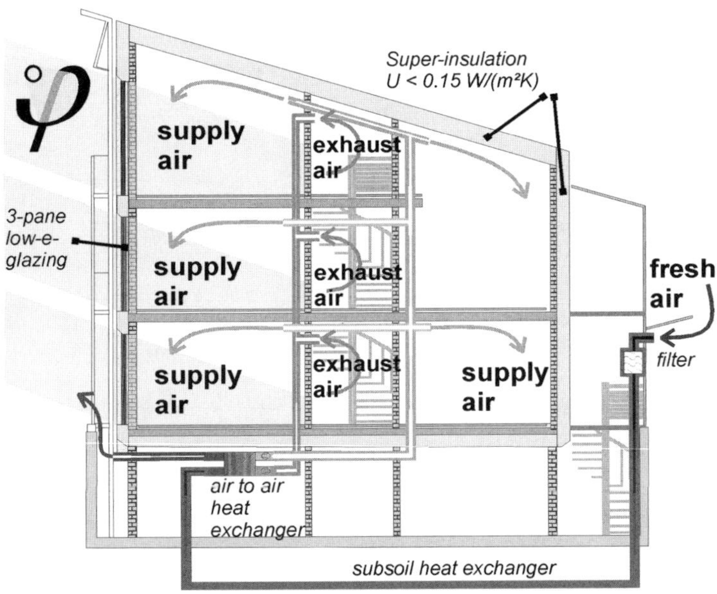

Figure 6.28. *Principle of a passive house (source: http://www.passivhaustagung.de)*

As far as the cost–benefit analysis of a passive house is concerned, it may be different from one project to another. The investigations carried out in Germany show, however, that if a global approach is applied at the design stage, the global cost (additional investment in construction and consumptions over 30 years) should not exceed, on average, the cost of a new house. This is mainly due to the elimination of the costs of a conventional heating system, unnecessary in a passive house, and to the corresponding reduction in energy bills over 30 years.

Moreover, the embedded energy, which is needed to build a passive house, should not exceed that of a traditional house, but could be higher than the exploitation energy during the house's life span (between 30 and 50 years).

The major requirements and recommendations necessary for achieving a passive house are summarized in Table 6.22.

Requirements	Value	Unit
Annual heating energy consumption (per net floor area)	15	$kWh/m^2.y$
Total primary energy consumption for all uses	42	$kWh/m^2.y$
Airtightness at 50 Pa over and under pressure [b]	120	$kWh/m^2.y$
Recommendations	$n_{50} < 0.6$	h^{-1}
Building opaque envelope		
Windows, including the frame	$U < 0.15$	$W/m^2.K$
Glass *g*-value	$U < 0.8$	$W/m^2.K$
Efficiency of ventilation heat recovery	$\psi < 0.01$	W/m.K
Annual heating energy consumption (per net floor area)	$g > 0.5$	%
Total primary energy consumption for all uses	$\eta > 75$	%
Notes:		
1. Net heated floor area is the sum of all the individual room and circulation floor areas, i.e. the carpet area if all rooms were carpeted (floor under interior walls is not counted).		
2. Total electric consumption, default value	End energy ≤ 18 Primary energy ≤ 55	$kWh/m^2.year$
3. Primary energy factors	Electricity : 3.0 (2.98) Natural gas: 1.07 Heating oil: 1.08	
4. Daily hot water consumption, default value	25 l/person at 60°C	

[b]: Hourly air changes per room volume

Table 6.22. *Requirements and recommendations for a passive house*

Figure 6.29 shows two individual passive houses: one in Germany, the other in Ireland. The Irish house is a 230 m^2 standard model marketed at the price of 1,130 €/m^2.

Figure 6.29. *Individual passive houses in Germany (top) and in Ireland (bottom) (source: www.scanhome.ie)*

In Figure 6.30 a group of passive houses, called "houses without heating" is shown. These houses are in Lindas, Sweden.

Figure 6.30. *Houses without heating – Lindas, Sweden*

6.6. The zero-energy houses: zero-energy house – zero-energy home (ZEH) – zero-energy buildings (ZEB)

A priori, a ZEH does not seek a reduction of it energy consumption, but simply an annual energy balance between energy consumption and on-site energy production [NRE 06, TOR 06]. Unlike the Minergie and Passivhaus approaches, which mainly consist of reducing thermal losses (HVAC, DHW) in order to limit, or even suppress, the additional thermal contributions, the zero-energy approach does not entirely meet this goal. The house, in order to find its equilibrium, uses on-site energy production, which is usually a photovoltaic system and which may consist of a complete roof, as proposed in Japan by many house builders. Wind-energy generation is also used if the site is suitable.

In practice, a zero-energy house is not completely autonomous all day long: when it generates a surplus of electricity (during daytime or in summer), this electricity is exported to the grid, whereas if the needs are not satisfied by the on-site production, it buys electricity from the grid. For this reason, up to now, there is no accurate description and clear criteria to define a zero-energy house. Several definitions are possible:

1) "Autonomous" house, which generates on-site the energy it needs by using different local energies: solar, biomass, wind, hydraulics, etc. Storage systems must be implemented in order to meet the demand all year round.

2) "Zero-energy" house, which reaches annual energy equilibrium in the generation of electricity. The surplus is sold, and when local generation is insufficient, electricity is bought from the grid. The energy balance calculation must be made in primary energy and must take into account the efficiencies of the equipment to transform the useful energy into final energy.

3) "Zero-balance" house, which looks for a financial equilibrium throughout the year, while generating photovoltaic electricity and taking advantage of the tariff differences in the cost of electricity between the peak periods and normal periods. Today, it is this kind of approach that seems to be the most frequently used, particularly in Japan where electricity is four times more expensive between 7 am and 11 pm than during the night time hours.

4) "Zero-emission" house, which presents a more global and environmental approach that takes into account greenhouse gas emissions.

As for definition 2, the zero-energy house, the calculation method to estimate the environmental impact of all available energy sources (oil, coal, gas, uranium, biomass, solar, etc.) including all the costs (operating, distribution, equipment manufacturing, waste recycling, etc.) still needs to be worked out.

However, in practice, the Building America program strongly advises that when building a zero-energy house, the demand for energy be reduced (to about 70%) and that the remaining energy that is required (30%) is produced on-site.

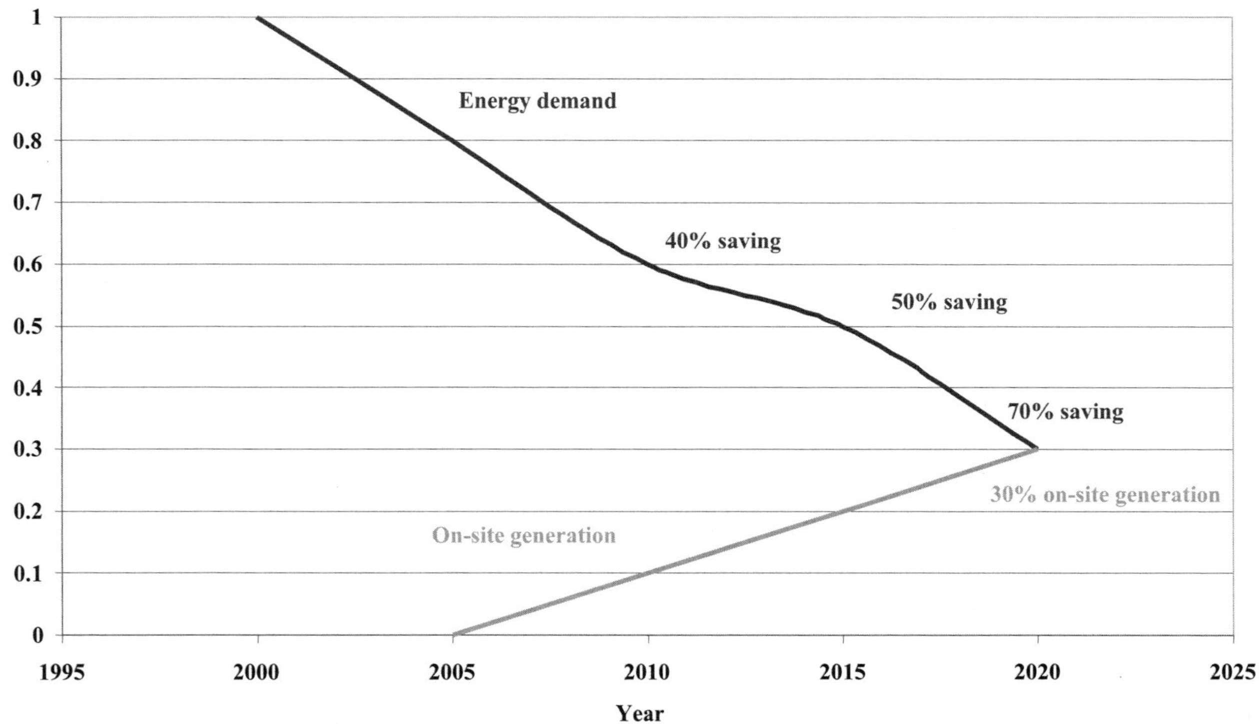

Figure 6.31. *The Building America program approach to zero-energy houses (source: www.eere.energy.gov/buildings/building_america)*

The success of a zero-energy house implies a set of requirements which start at the design stage, continue at the construction stage, and finish with the homeowners' behavior. Unlike the Minergie trademark and the PassivHaus label, these requirements are not quantified accurately, but it is recommended to follow the seven steps which reflect the main criteria or requirements of Minergie and PassivHaus houses.

1) Decrease HVAC needs by improving the insulation. To achieve this goal, the actions described in the previous sections are recommended, with some adaptations for the local climate: orientation of the house toward south, low surface walls to the west with screens and porches, reinforced insulation of the basement, walls and ceilings, low U-value for windows (double-Glazing + low emissions in all climates, but with a low SHGC g in warm climates (sunscreen included), airtightness of all leaks, cracks, and defaults on the envelope.

2) Use high-energy-efficient HVAC equipment: condensing or low-temperature boiler, heat pump, high-performance air-conditioning system, etc. It is recommended to buy the equipment with the best efficiency: price ratio, based on the local weather, in order to optimize the pipe network (short distance, insulation, etc.) and to ensure good seals. The ground or water/air heat pump must be taken into consideration when the place and the costs are suitable. Finally, when the climate is suitable, alternative cooling systems, such as natural ventilation, night cooling or evaporation systems, must be considered.

3) Implement a solar water-heating (for DHW), a preheating system, and an efficient additional system. The distribution networks must be efficient. Narrow-diameter pipes parallel to the hot water outlets are recommended along with reduced flow-rate systems. It is advised that solar systems certified by the Solar Rating and Certification Corporation (SRCC) be used, and that installation be by well-known and competent companies.

4) Install efficient lighting: compact fluorescent lamps and equipment with the energy star label are recommended.

5) Use efficient domestic appliances. For appliances (refrigerator, dishwasher, washing machine, etc.) consider products with the energy star label and compare their performances.

6) Install a well-designed photovoltaic system. In order to install this type of system, which is a fundamental component for zero-energy houses, the PVWatts program must be used for a quick estimation, a well-known company in the field of photovoltaics must be used, and a certified installer must do the installation. The typical power is around 2 kWc. For financing, tax credits, grants, and other incentives, measures must be taken into consideration.

7) Switch off computers and electric equipments when not in use.

In the United States, the Department of Energy (DOE) implemented the Zero Net Energy Buildings program in January 2000. This action plan aims to build 100,000 ZEH before 2020.

Since 2000, two significant programs have been launched in order to reach this goal. The first one concerns the establishment of six teams to design, build, and monitor individual homes or dwelling projects (Figure 6.28).

Figure 6.32. *Premier ZEHs Premier Gardens, Sacramento, California (source: www.smud.org/residential/saving/zeroenergyhomes.html)*

The second program is the creation of a competition, called Solar Decathlon, where universities design zero-energy houses. This initiative plays a very important role in that it makes architecture and engineering students sensitive to the problems of energy efficiency in buildings and to the needs to make buildings attractive to the young despite buildings not having a high-tech image.

In the early 1990s in Japan, the government implemented several programs to support the installation of photovoltaic systems, in particular in the residential areas. One program was the New Sunshine Project research program started in 1993. Another was the Residential PV System Dissemination Program and its previous Residential PV System Monitoring Program. According to a 2003 survey, in the framework of the International Energy Agency (AIE), about the photovoltaic systems, the subsidies per kilowatt were US$871 in 2003. During the same year, Japanese builders contributed more than 60%, or 400 MW, of the world production of photovoltaic modules, 55% of which was for the Japanese market [NOG 05]. The target was the residential area, which represents 85% of the market and the number of houses equipped of PV systems is increasing in Japan [NOG 05]. Between 1994 and 2003, this number of houses has increased in a spectacular manner, changing from 539 to 52,863 (Table 6.23).

Year	Number of houses with PV systems	Installed power MW
1994	539	1.9
1995	1,065	3.9
1996	1,986	7.5
1997	5,654	19.5
1998	6,352	24.1
1999	15,879	57.7
2000	20,877	74.4
2001	25,151	91.0
2002	38,262	141.4
2003	52,863	201.4
TOTAL	168,628	622.8

Table 6.23. *Residential PV systems in Japan between 1994 and 2003 (source: [NOG 05])*

In Japan, a zero-energy house is more like a zero-balance house; it is also called zero utility-cost housing. This concept leads to zero service costs by the use of a generation photovoltaic system. The house has reinforced insulation; it is completely electrified with a photovoltaic system of about 6 kW, and connected to the grid which supplies all the electrical equipment in the house, e.g. reversible heat pump, lighting, electrical appliances, multimedia, etc. The output of the electricity generated, when addition to the consumption, generally exceeds the cost of the network electricity used, this bringing the annual service costs to zero, or even providing a credit.

Photovoltaic roofs have been developed by the major Japanese manufacturers of prefabricated Japanese houses, such as Misawa Home, Sekisui House, and Daiwa House, together with the photovoltaic cells and systems manufacturers.

The one-floor "Parfait-Ex" house from Sekisui has a metallic structure and a photovoltaic system with a capacity of 2 to 5 kWc and which is integrated in the roof. As far as the "Eco Sunny House" model from Daiwa is concerned, it consists of a prefabricated house with 3 kWc integrated into the roof.

Figure 6.33. *Zero-energy house from Misawa Home – Hybrid Z*

The Hybrid Z house from Misawa Homes has a fully photovoltaic roof of 12 kWc, integrated on both sides of the roof (Figure 6.33).

6.7. The energy-positive house

An energy-positive house is an extension of the zero-energy house since it generates more energy than it consumes.

This type of house of can be described as a combination of the passive house concept, which minimizes the energy demand, and the Japanese solar roof in order to transform the building into a decentralized power plant.

The building production is used for its own needs and the surplus of energy can be sold to the grid. Thus, the grid becomes a major energy-manager based on the intermittent generation periods and the consumption peaks. The development of these houses will largely depend on decreasing photovoltaic costs, on the grid management (smart grid), and on the feed-in tariff policies.

As an example, the energy-positive house built at Thening in Austria is presented in Figure 6.34.

Figure 6.34. *Positive-energy house at Thening, Austria*

The main features of this house are listed in Table 6.24.

Heating needs	12.8 kWh/m^2 year
Controlled mechanical ventilation	yes
Thermal solar collectors in front	17.4 m^2
Photovoltaic on roof	10.35 kWp
Water tank	6.5 m^3

Table 6.24. *Positive-energy house, Thening, Austria (source: www.plusenergiehaus.at)*

6.8. Comparison of the three types of houses: Minergie, PassivHaus and ZEH

Many investigations have been carried out to compare various buildings or houses built in accordance with the requirements or recommendations from the three main low-energy house labels.

The first investigation [HAS 04] presents the results obtained for five apartment blocks in Switzerland (CH) and Germany (DE), which were built according to the passive-house criteria. All these apartments have in common a high-performance envelope (insulation and windows), but they differ as to the percentage of south-oriented windows and to the equipment choice for space heating and DHW (wood, HP, gas, grid, solar, district, etc.). This shows that the requirements for the envelope are fairly general (airtightness, insulation, inertia, solar and internal energy contributions, solar protection, ventilation, etc.), whereas the energy choice strongly depends on available local resources. The data from this investigation are collected in Tables 3.25 and 3.26.

Social Housing Dwellings	**Wechsel Stans CH**	**Balance Wallisellen CH**	**Rychenberg Str. Winterthur CH**	**Vauban Freiburg DE**	**Marsbachhöhe Kassel DE**	**Averages**	**Unit**
Estimated space heating	14.5	14.4	22.4	12.6	17.1	**16.2**	kWh/m2.a
Heated floor area	997.5	1015	781	781	1802		m^2
Cost	387	309	367				€/m3
Number of apartments	7	4/5	6	19	23		
Compactness S/V	0.57	0.44	0.46	0.42	0.49	**0.48**	m^2/m^3
U-roof	0.11	0.16	0.12	0.18	0.11	**0.14**	$W/m^2.K$
U-ground	0.14	0.16	0.20	0.25	0.11	**0.17**	$W/m^2.K$
U-walls	0.13	0.17	0.14	0.20	0.13	**0.15**	$W/m^2.K$
U-window - U_w	0.85	0.92	0.75	1.10	0.80	**0.88**	$W/m^2.K$
U-glazing - U_g	0.70	0.70	0.50	0.70	0.60	**0.60**	$W/m^2.K$
SHGC *g*-value glass	0.50	0.45	0.42	0.60	0.60	**0.51**	%
Ratio Window/south facade	40	46	48	50	21	**41**	%

Table 6.25. *Comparison of the building envelope characteristics (average by apartment)*

Condominium Apartments	**Wechsel Stans CH**	**Balance Wallisellen CH**	**Rychenberg Str. Winterthur CH**	**Vauban Freiburg DE**	**Marsbachhöhe Kassel DE**	**Averages**	**Unit**
Mechanical ventilation	Plate (2 x)	Plate	Cross-flow	Cross-flow	Cross-flow		
Efficiency η (%)	80	60	78	82	83	**76.6**	
Electrical power (summer/winter)	250/600	140/190	520/520	250/600	575/575	**347/497**	W
Earth heat exchanger							
Diameter	200	150	180				mm
Length/apartment	25	25	30				m
Solar system	Water	None	Air	Water	None		
Collector surfaces	42		60	42			m^2
Storage	4500		1750	3650	800		liter
Heat source	Wood pellets	Heat pump	Wood pellets	Gas heat & power	District heat		
Heating power/floor area	9.6	5.1 [a]	19.2	10.15	10.15		W/m^2

[a]: *Electrical input, actual heating capacity dependent on the momentary coefficient of performance (COP) of the heat pump*

Table 6.26. *Comparison of equipment characteristics of the buildings (average by apartment)*

Figure 6.35. *Wechsel Stans building in Switzerland*

Figure 6.36. *Marsbachhöhe building in Kassel (Germany)*

Figures 3.35 and 3.36 highlight the building compactness, the importance of the south-oriented windows, and the self-supporting metal balconies that limit the thermal bridges and the "fin effect".

Figure 6.37. *Detail of a wall with window (source: IDEA)*

Figure 6.37 shows a section of a wall of the Marsbachhöhe building in Kassel (Figure 6.36). We notice the difference between the opaque part thickness (48 cm, of which 30 cm is polystyrene insulator) and the window thickness with its triple-glazing and its reinforced insulation frame.

In the second investigation [CHA 05], several projects were compared to the framework of the IEA Task 28. We immediately notice the differences between the European approaches, Minergie or PassivHaus, which favor the reduction of energy needs, while the Japanese approach mainly seeks to balance consumption and production.

Indeed, the insulation levels (U-value of the walls and the windows) are much closer for the European houses whereas there is a factor of 3 to 4 between the European houses and both Japanese houses. In contrast, several European houses do not have a photovoltaic system and for those that have one the power is limited.

Finally, we notice that the Austrian energy-positive house (Figure 6.34 and Table 6.27) combines both approaches.

Project	City	Countries	Wall U W/m².K	Roof U W/m².K	Windows U W/m².K	PV	Thermal solar collectors m²	Thermal storage liter	Notes
20 row-houses	Göteborg Lindas	Sweden	0.10	0.08	0.80	0	5	500	No space-heating system
Demo houses	Hannover-Kronsberg	Germany	0.13	0.10		0	yes	yes	District heating co-generation
Minergie-P	Dintikon	Switzerland	0.11	0.11	0.74	49.5 m²	4.5	2,320	Pre-heating of the first-floor ventilation
Budstikka 18	Kongsberg	Norway	0.16	0.10	0.85				Electric heating floor, wood heating for the few very cold days
Eco Sunny House	KankyoKobo	Japan	0.45	0.19	2.33	3.0 kW	yes	yes	Rain and waste water recovery
Hybrid-Z	Kanagawa	Japan	0.38	0.48	2.55	11.3 kW			Heat pump air/air
Plus energy house	Thening	Austria	0.11	0.11	0.79	10.4 kW	17 (en facade)		Pre-heating and pre-cooling of the first-floor ventilation
Passive house	Gaspoltshofen	Austria	0.11	0.09	0.78	2.8 kW	22	2,000	Rain water tank

Table 6.27. *Description of eight demonstration houses*

6.9. Beyond the positive-energy building

The above sections show how to move toward energy-positive buildings. It is technically possible, but the cost of photovoltaic roofs, despite the feed-in tariff (55 c€/kWh in France), remains significant. However, when PV or urban windmills are integrated in buildings, they become power plants. So, the question is now: what to do with the surplus energy in order to fulfill the occupants' energy needs? The usual practice is to sell the electricity to the grid, but another use of this energy is possible: it may be used for electric vehicles (EV), in the garage, which could be transformed into a new equipment for buildings, such as a refrigerator, washing machine, etc.

In fact, EV can fulfill three functions: mobility, energy storage, and emergency batteries [WEI 05].

Indeed, after water and food, housing and transportation, which are also primary needs, consume more than 70% of the energy and contribute to more than 40% of greenhouse-effect gases. Therefore, it is interesting to think about a possible synergy between energy-positive buildings and electric cars. It is certainly a development possibility that has already inspired the Japanese car-makers.

Figure 6.38. *Toyota dream house*

During the 2005 Aichi exhibition, Toyota presented the "dream house" concept which has a connection with the Prius, a hybrid vehicle made by Toyota and which might be used as an energy source in case of an electricity failure [TOY 06]. With its tank full, the vehicle could supply the house with electricity for a day and a half. And, in return, the photovoltaic cells of the house could recharge the car batteries (Figures 6.38 and 6.39).

Figure 6.39. *Connection point: house–car*

At the 2005 Tokyo Automobile Show, Honda presented a co-generation system based on a fuel cell (home Energy Station) which generates, using natural gas, hydrogen for the car, electricity for the house, and, of course, recovers waste heat for space heating and DHW (Figure 6.40).

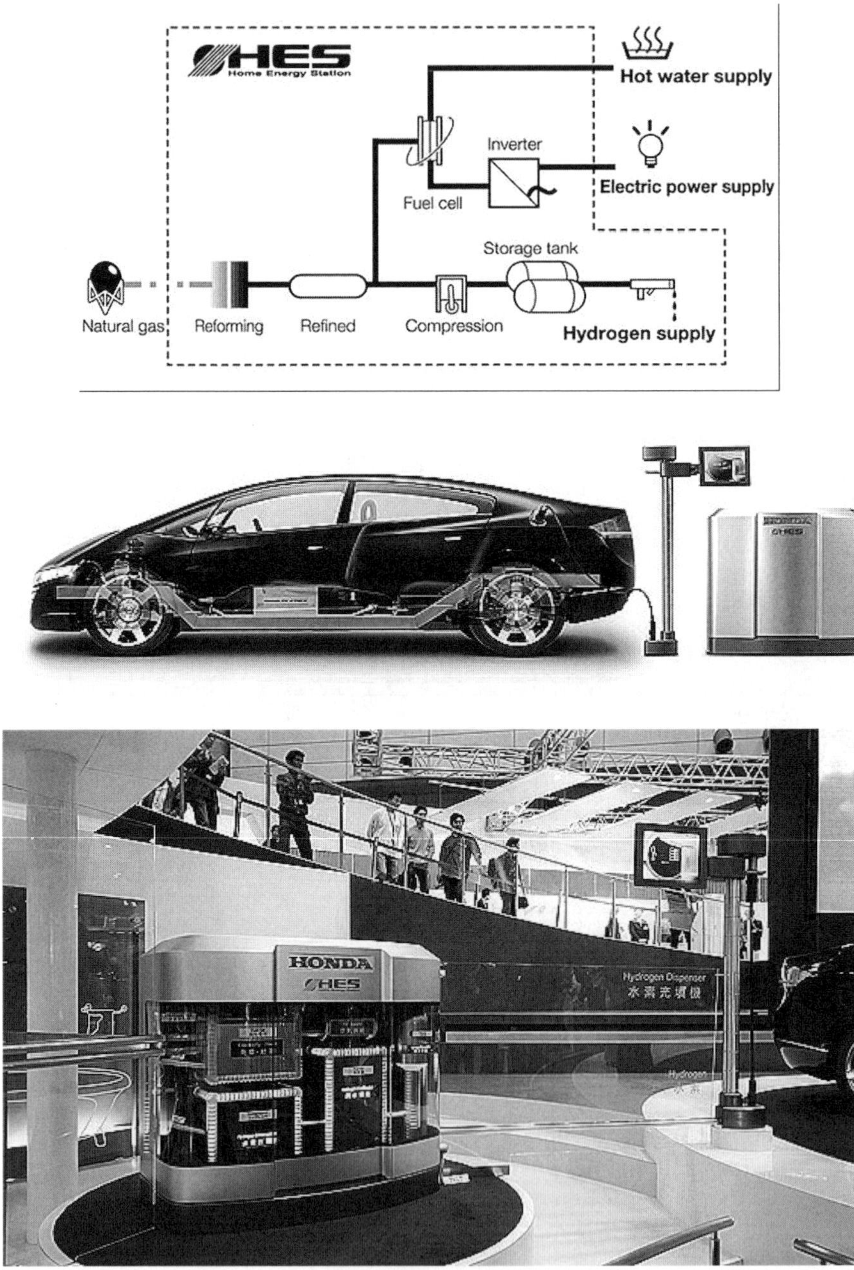

Figure 6.40. *Home energy system – Honda*

In the United States, there is also great interest in plug-in hybrid electric vehicles (PHEV) and the NREL (National Renewable Energy Laboratory) foresees that the connection of the PHEV to the ZEH could be one solution toward energy autonomy for housing and transportation, and toward what is called a renewable community [NRE 07].

Between individual dream and collective nightmare, the car and the single-family house are living spaces (90% of the time spent by people) overwhelmingly supported by consumers and considered to be without a future by the supporters of sustainable development. As a matter of fact, when separately considered, the car and the individual house are also viewed as predators (of space, of raw materials and of energy) and polluting agents (waste, CO_2, particles, etc.). However, when they are considered altogether, they can play a new role with a positive impact. Of course, this synergy will be all the more efficient if the houses are grouped, the displacements are limited, and that the user has respectively driving, and lodging licenses.

So tomorrow, the buildings, such as "habitat" and "power plants", coupled with electric vehicles as storage and emergency equipment, could reverse their negative image and contribute to curb climate change while meeting the needs of humankind for housing and transportation.

6.10. Bibliography

[ADE 04] ADEME, *Les chiffres clés du bâtiment - Données et Références*, ADEME, 2004.

[AEA 02] Austrian Energy Agency, "Buildings: a technology portrait – E.V.A.", *Arsenal Research*, Vienna, July 2002. Available at http://energytech.at.

[AHM 05] AHMAD, M., BONTEMPS, A., SALLÉE, H. and QUENARD, D., *Thermal testing and numerical simulation of a prototype cell using light wallboards, coupling vacuum isolation panels and phase change material*, Centre Scientifique et Technique du Bâtiment, CSTB, Grenoble Cedex, France, 2005.

[BUS 06] BUSCARLET, C. and CACCAVELLI, D., Les performances mesurées des chauffe-eau solaires en usage reel, Chaud-Froid-Plomberie-CFP, Catalogue des défauts des installations de chauffe-eau solaires individuels (CESI), Plan Soleil ADEME, number 695, 60–66, November 2006. Available at http://helios2006.cstb.fr/catalogue/index.htm.

[CAH 03] CAH, "Maîtriser les Consommations d'énergie et réduire les émissions de gaz à effet de serre dans l'habitat existant en France", *CAH Colloquium*, 2003. Available at www.cah.asso.fr.

[CAH 05] CAH, "Mieux répondre aux attentes des particuliers pour réduire la consommation d'énergie de leur logement", *CAH Colloquium*, 2005. Available at www.cah.asso.fr.

[CE 05] European Commission, *Comment consommer mieux avec moins – Livre vert sur l'efficacité énergétique*, Office des publications officielles des Communautés européennes, Luxembourg, 2005.

[CFP 02] "L'avenir du chauffage … sa disparition? L'exemple des maisons passives allemandes", *Chaud-Froid-Plomberie*, no. 651, November 2002.

[CHA 05] CHARRON, R., *Une revue de projets et mesures visant les maisons solaires à consommation nulle ou faible d'énergie*, Centre de la technologie de l'énergie de CANMET, Varennes, Canada, 2005.

[COU 06] COURGEY, S. and OLIVA, J.-P., *La conception bioclimatique des maisons confortables et économes*, Terre vivante, Mens, 2006.

[GUE 05] GUERRIAT, A., *La Maison Passive: Introduction pour les architectes et les futurs maîtres d'ouvrage*, Institut Supérieur d'Architecture de la Communauté Française, PHP, La Cambre, 2005. Available at www.maisonpassive.be.

[HAS 04] HASTINGS, R., "Breaking the 'heating barrier' – Learning from the first houses without conventional heating," *Energy and Buildings*, vol. 36, 373–380, 2004.

[HON 05] Home Hydrogen Refueling Technology – Advances with the Introduction of Honda's Experimental Home Energy Station, Honda, 2005.

[IEA 05] Vacuum Insulation Panels – Study on VIP-components and Panels for Service Life Prediction in Building Applications, Annex 39 HiPTI – High Performance Thermal Insulation of IEA/ECBCS-Implementing Agreement, Report on Subtask A, 2005, Vacuum Insulation in the Building Sector, Systems and Applications, Annex 39 HiPTI – High Performance Thermal Insulation of IEA/ECBCS-Implementing Agreement, Report on Subtask B, 2005.

[LEM 02] "Produire son énergie, c'est possible", *Le Moniteur Environnement*, special edition, April 2002.

[LEN 06] LENORMAND, P. and RIALHE, A., "Very Low Energy Houses", *AERE*, November 2006. Available at www.leonado-energy.org.

[MAU 05] MAUGARD, A., VISIER, J.-C. and QUENARD, D., "Le bâtiment à énergie positive", *Futuribles*, vol. 5, January 2005. Available at www.futuribles.com.

[MEA 05] Ecosystems and Human Well-Being – Millenium Ecosystem Assessment, 2005. Available at www.maweb.org.

[MIN 06] Le bilan énergétique de l'année 2005 en France, Bercy, April 20, 2006. Available at www.industrie.gouv.fr/infopres/presse/fichebilan.pdf.

[NIC 93] NICKLAS, M.H. and BAILEY, G.B., "Student Performance in Daylit Schools – Analysis of the performance of students in daylit schools", *Innovative Design*, North Carolina, 1993.

[NOG 05] NOGUCHI, M., *Habitations Solaires Photovoltaïques – Stratégie Japonaise de Commercialisation axée sur l'efficience*, Centre de la technologie de l'énergie de CANMET, Varennes, Canada, 2005.

[NRE 06] "The potential impact of zero-energy homes NREL", *NAHB*, February 2006. Available at www.toolbase.org.

[NRE 07] National Renewable Energy Laboratory, *Advanced Vehicles and Fuels Research Hybrid Electric and Fuel Cell Vehicles Plug-In Hybrid Electric Vehicles*, 2007. Available at www.nrel.gov.

[QUE 06] QUENARD, D., "Le bâtiment à énergie positive", *La Recherche*, supplément Ville et Mobilité Durable, vol. 398, June 2006.

[SAL 04] SALOMON, T. and AUBER, C., *Fraîcheur sans clim': Le guide des alternatives écologiques*, Terre vivante, Mens, 2004.

[SAL 99] SALOMON, T. and BEDEL, S., *La maison des "négawatts"*, Terre vivante, Mens, 1999.

[SID 00] SIDLER, O., Logements à Faibles Besoins en Energie, Guide de recommandations et d'aide à la conception, Région Rhône-Alpes ADEME, ODH 26 Conseil Général de Savoie, Cabinet Olivier Sidler, Enertech, 2000. Available at http://sidler.club.fr.

[SID 03] SIDLER, O., "L'Inertie thermique en climat méditerranéen" *Colloque Confort et consommations d'énergie*, ENERTECH, Montpellier, May 15, 2003.

[TOR 06] TORCELLINI P., PLESS S. and DERU M., "Zero Energy Buildings: A Critical Look at the Definition", *ACEEE Summer Study*, National Renewable Energy Laboratory, Pacific Grove, California, August 14–18, 2006.

[TOY 06] TOYOTA, A *New Future for People, Society and the Planet, Sustainability Report*, 2006.

[TRI 06] TRILLAT-BERDAL, V., ACHARD, G. and SOUYRI, B., "Les systèmes solaires combinés avec des pompes à chaleur géothermiques à échangeurs enterrés verticaux: étude de l'injection d'énergie solaire dans le sol", *IBPSA France 2006*, La Réunion, November 2–3, 2006.

[VIS 02] VISIER, J.-C. and ZIRNGIBL, J., "Vers l'autonomie énergétique des bâtiments", *CSTB Magazine*, vol 142, July–August 2002.

[WEI 05] VON WEIZSÄCKER, E.U., "Building Technology in the Vanguard of Eco-efficiency", *MP, The 2005 World Sustainable Building Conference*, Tokyo, September 27, 2005.

Chapter 7

Light Sources and Lighting: from Technology to Energy Savings

7.1. Lighting in the past and today

Man has always desired to carry on with his normal life after nightfall. Subject to the rhythms and hazards of natural light sources, he very quickly tried to replace the natural light with artificial sources.

For centuries, human society had to satisfy itself with the use of fire to generate light. Man thus, unconsciously, discovered one of the two processes of artificial light generation: incandescence. However, phenomena such as phosphorescence and fluorescence have shown that material was able to generate light without necessarily increasing its temperature. Man also desired to imitate other "natural" light sources such as fireflies, phosphorescent minerals or lightning. These sources generate light thanks to the luminescence[1] phenomena. This second process is much more efficient than incandescence, but was previously not exploitable because these sources were either hazardous or difficult to control. Furthermore, the energy efficiency was not up to date. However, during this period, man always had a dream of putting lightning into a bowl to give out light. It was only during the 19th century that this dream came true. The English chemist Sir Humphry Davy (1778–1829) discovered the electric discharge phenomena. In 1802, he tried to reproduce in his laboratory Franklin's experiment demonstrating the electric nature of lightning. He noticed the

---

Chapter written by Georges ZISSIS.

1. The German scientist Wiedemann suggested, in 1888, calling this cold light emission "luminescence" and to eventually clarifying its generation mode by means of a prefix (*bio*luminescence, *electro*luminescence, *thermo*luminescence, *chemi*luminescence, etc.).

appearance of an electric discharge emitting an intense light between two carbon electrodes connected to an electric battery. He did not use this new light source because it was too unstable and required a reliable energy source. Thanks to the development of chemical batteries between 1830 and 1840, experiments on the light generated by electric arcs were conducted, and in the period 1850–1870 many lamps appeared. However, chemical batteries were too expensive for a significant economic development. In the 1870s, less expensive and more practical batteries allowed the Russian Paul Jablochkoff (1847–1894), who was in Paris at that point, to patent the "electric candle" in 1876. He then put this on the market in France and the UK (Figure 7.1). It consisted of two vertical and parallel carbons, brought closer, and insulated except at their tips. The insulating layer thus burns at the same time as the carbon does, and a spark shines at the tips of the two pencils. This electric arc was used for the external lighting (urban) and the internal lighting of some factories. Furthermore, in order to be used in an easier way, a more compact form was given to these new artificial light sources known as "lamp".

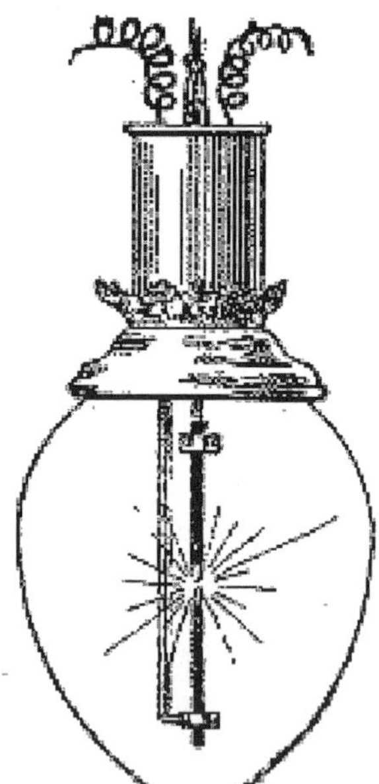

Figure 7.1. *The Jablochkoff "electric candle", an 1876 patent. This lamp lit up Paris, especially Opera Avenue, during the 1878 Exhibition Universelle*

At the same time, in the USA, the arc lamp was used for lighting shop windows and some small installations. The reliability and the efficiency of the electric arc were gradually improved, particularly due to the works of Rookes Evelyn Bell Crompton (1845–1940), by the use of high purity carbons and accurate regulators which reduced light fluctuations.

Finally, the introduction of enclosed space arcs in 1893 by the American engineer L.B. Marks represented a major development for discharge lamps. In a closed lamp, the arc was confined in glass bowls, which avoided excessive carbon consumption due to the air flow. The maintenance costs was then reduced, but the power required to the lamp operation was greater.

Arc lamps dominated the lighting market until the invention of the incandescence lamp by Thomas Edison in 1878[2]. Very quickly, the Edison Lamp replaced the arc lamp despite its more than mediocre efficiency, but once more, the energy efficiency was not the problem. Indeed, the first incandescent lamps generated a light quantity equivalent to 16 "candelas" whereas the arc lamps of the same period reached 2,000 to 4,000 candelas[3]. The success of the incandescent lamp is partly justified by three factors:

1) the incandescent lamp was well easier to mass produce;

2) its lifetime was much higher than that of the existing arc lamps;

3) it required very little maintenance (arc lamp electrodes had to be replaced very frequently).

However, the decisive factor for this invention coming to dominate the market was a forewarning of sustainable development: Edison very quickly noticed that to impose its invention, he had to "democratize" it. Thus, to reach this goal, electricity had to be brought to the individuals. Edison thus created the first electricity generation and distribution company. This is one of the first examples of a "systemic" vision of history. The Edison Company still exists to day; it is known under the name of the giant General Electric.

---

2. We recall that Edison "invented" and developed the incandescent lamp at the same time as his competitor Swan, but Edison knew how to market his invention, and above all he developed the electric distribution system. However, one of the true pioneers of the electric incandescent lamp is presumably Heinrich Gœbel who in 1854 first had the idea of making a carbon filament lamp, in order to illuminate his clock and watch shop in New York at night.

3. The ISO unit "candela" was afterwards defined as being the light intensity emitted by an "English standard candle". Let us recall that an "English standard candle" should be visible, in clear weather, at a distance of 27 km!

7.1.1. *Present-day lamp families*

Increasingly present in our activities, lamps and lighting have seen continuous development over the centuries. The use of electricity has been a revolution in this field. However, all the sources generating artificial light from electricity always use two basic properties, incandescence and luminescence, which had already been identified by prehistoric man.

Looking closer we find traditional source families (incandescent bulbs and discharge lamps), but also the electroluminescent diode families, which have entered the lighting world since over the last few years to a massive extent[4], along with some new products such as aggregate lamps and field emission lamps. Figure 7.2 presents, in a synthetic way, the different source families likely to be used for light generation or to provide sign post functions.

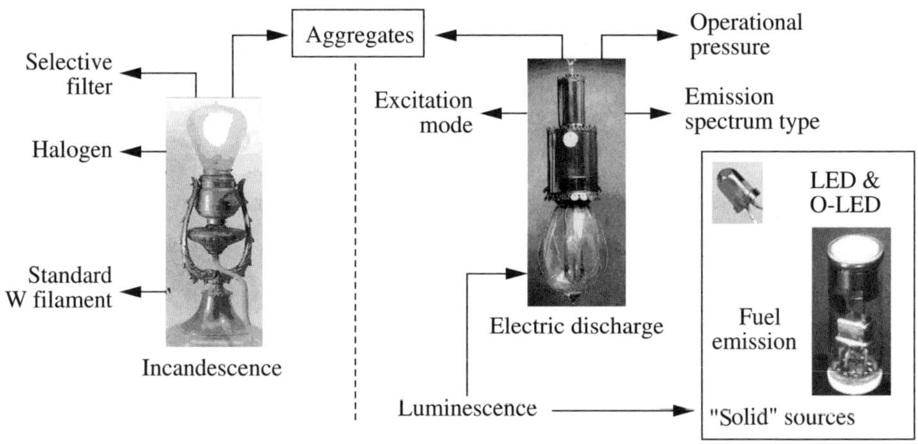

Figure 7.2. *The lamp families*

Today, the light sources represent a considerable market; the number of electric lamps operating on Earth is estimated at some 33 billion, whereas annual production exceeds 16 billion units. Table 7.1 outlines these data.

4. Today electroluminescent diodes are replacing the standard semiconductor diodes and their organic versions known as O-LEDs.

Lamp type	Annual requirements (billion units)	Lifetime (years)	Existing (billion units)
Incandescence	11.5	1	11.5
Halogen	0.8	2	1.6
Fluorescent	3.2	5	16.0
Compact Fluo	0.6	5	3.0
High pressure	0.2	3	0.6
Total	**16.3**		**32.7**

Table 7.1. *The lamp market in the world*

7.1.2. *The economic, environmental and energy impact of lamps*

From the energy point of view, lighting consumes more than 2,651 TWh electric energy per year. This represents approximately 19% of the total world electricity generation[5] [IAE 05]. It also represents an annual bill of more than 200 billion euros for the consumer [MIL 02]. The situation is different from one country to another. Thus, the energy consumption for the lighting of a western country varies between 7% and 15% of its energy generation (with a single case, the USA, where this is around 20%).

More precisely, France consumed 41 TWh for lighting in 1999. A little less than two thirds of this energy (60% of the total) is used by the tertiary area. Road and public lighting consumes 10% of the total, whereas 30% of the energy is absorbed by domestic lighting. However, it must be noted that this last area has seen its consumption increase threefold in the past two decades[6]. In the USA, the electric energy consumption for lighting in the year 2000 reached 659 TWh, which represents 19% of the electric energy generated in the country (in absolute value, this energy is equivalent to the combined electricity generation of France and Italy). Here again, tertiary lighting prevails in this consumption as shown by Figure 7.3. It must be made clear however that the major part of this energy is consumed by incandescent lamps: 46% [SCO 02]. Figure 7.3 compares the French and American situations.

5. This quantity of electricity represents the equivalent of 1-2% of the primary energy used on our planet during one year.
6. Domestic lighting energy consumption in France: 5 TWh in 1979, 14 TWh in 1999.

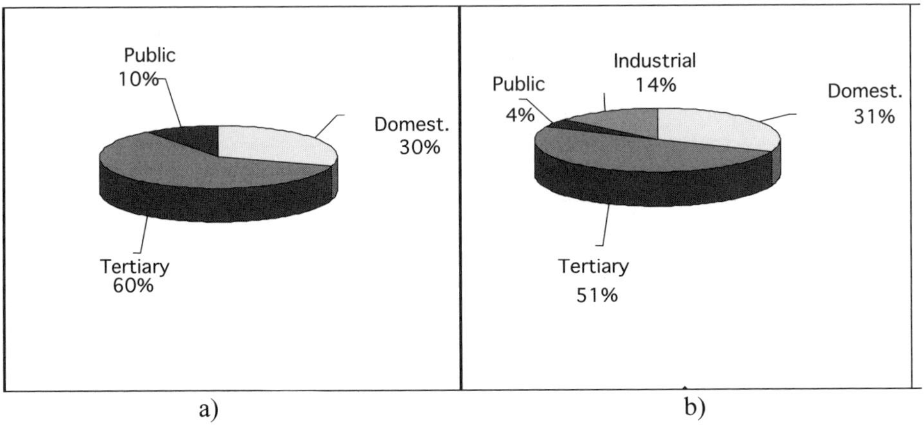

Figure 7.3. *Comparison of energy consumption dedicated to lighting in 1999 a) in France: 41 TWh and b) in the USA: 650 TWh*

As far as the developing countries are concerned, the situation is different. Lighting represents the major part of their electric consumption today: 30% for Tunisia, nearly 40% for Madagascar and up to 86% for Tanzania (champion of all categories in this field). Here domestic lighting prevails. This situation is easily understandable since light is a basic need for man, and, of course, as soon as electrification progresses, the population takes profit from it by first installing lamps. On the other hand, in these countries, the cost of lamps is a major factor in the choice of the type of light source. Thus, incandescent lamps dominate the market because "low consumption" lamps are often not available. Once again energy efficiency is neglected due to other considerations...

Pollutant	Fluorescent Lamp 36 W	Incandescent Lamp 60 W
Hg	0.64	4.45
Pb	19	136
Cu	26	185
No_x	21,700	152,000
SO_2	16,300	114,000

Table 7.2. *Emission of toxic substances due to the required energy generation following the production of 1 lumen during 1 hour, expressed in ng/(lm/h)*

The electric energy generation to satisfy the lighting needs of man has inevitably led to environmental pollution. It is thus estimated that each year, some 1,700 million metric tons of CO_2 accompany this energy generation, contributing to the greenhouse effect. At the same time, other toxic substances are also produced. Table 7.2 gives an estimation for two types of lamps generating the same light quantities (in lumens).

Furthermore, lamps contain rare and often toxic (Hg, Cd, rare earth, etc.) materials. They also contain radioactive materials (Th, ^{98}Kr), and their ferromagnetic ballasts contain lead. Thus, at the end of their lives the materials in the lamps may pour out into nature. For instance, the majority of lamps, except incandescent lamps, contain mercury. This is why, in the USA, to produce some 750 million fluorescent tubes each year, 2.5 tons of mercury are used. In other countries, such as France, about 80 tons of waste contaminated with mercury are collected each year. A new European regulation, effective since 1st January 1998, imposes that discharge lamps coming to the end of their life become "final" waste and that they are reprocessed before storage in special dumps (called "terminal" dumps). Storage in such a dump costs about 200 euros per ton of processed waste. Finally, regarding the quantity of toxic material in the lamps, there has been since 2003 a new European regulation, "RoH – *Removal of Hazardous wastes*", [EUP 03]. This regulation limits the amount of mercury in a discharge lamp.

The lamp industry is under development. The annual expansion of this area is around 0.9%. In absolute value, this rate seems low but it has remained stable for about the last 20 years, which makes the lamp industry one of the four industrial areas with the highest growth rate over the last two decades. This is why Japan, which represents today 20% of the world market, has experienced during the last five years a production growth of 3% to 8% per year and an annual sales increase of 9.7% (on average during the last ten years).

The total turnover of the light sources industry is about[7] 14 billion euros per year. However, the lamp industry and the lamp market have a true specificity: three major actors concentrate most of the activity: Philips Lighting, General Electric Lighting and Siemens-Osram. Some details are given below:

– Philips Lighting is the world number 1 in lighting. In particular, it holds half of the market share in Europe. The sales in 2005 rose up to 4.8 billion euros[8], an increase of 6.6% with respect to the preceding year. The company employs 45,650 people, with an increase of 1,500 people since 2004.

---

7. This turnover only concerns lamps and does not include either the electronic ballast area or the light fittings. The total turnover of the "lighting" area is probably 5 times higher.

8. Philips Lighting represented, at the first quarter of 2004, 15.7% of the Philips industry group turnover.

– The German company Osram, a subsidiary of the Siemens Group, is a company with 36,000 employees in 58 countries, with a turnover of 4.2 billion euros. With a profit of 10.8% of its turnover, in 2005 Osram reached the "highest level" in 5 years. Since its purchase of GTE-Sylvania (USA) in 1990, the USA, with 44% of the sales, has become the priority target for Osram, followed by Europe (37%) and Asia-Pacific (16%).

– General Electric Lighting is the American number 1 and represents 50% of the market share in the USA. Its turnover is about 3 billion euros. For the last 10 years GEL's preferred target has been Europe and its market. To reach this goal, GEL has made many acquisitions: the Hungarian Thungsram, the British Thorn EMI, the Italian Sivi, and the German Linder Licht. Investments over the last 6 years have exceeded 750 million dollars.

These three major companies represent by themselves a turnover of 12 billion dollars. This type of structure is called an "oligopoly". All the other companies, of which Sylvania Lighting International and Matshushita are the largest, represent a turnover of only 2 billion dollars (this represents for these companies a turnover 10 times lower than the "big three"). It remains that many national factories (which are small or middle size) conserve a market and know-how, either in the source area or in the equipment area. On the other hand, the "lighting" market (first installation and maintenance) still remains in the hands of "light designers" (engineers, architects, etc.) often linked to the principal parties such as the government or the communities.

7.2. Light sources and energy conversion

The previous sections largely illustrate the impact of lighting on the national and international energy level. On the other hand, we know today that the world population is growing at a steady rate and despite everything the quality of life of the population is progressing. Furthermore, as shown by Mills in his works [MIL 02], there exists a quasi-linear relationship between the internal gross product and the per capita lighting energy consumption. Moreover, electrification is progressing and new areas in the world are lit by means of electricity[9]. Taking into account these three statements, we know today for sure that the lighting needs of humans will continuously grow; a factor of about 2 is expected in the next two decades.

How can we satisfy this growth without the demand for energy increasing in the same proportion?

9. In 2004, more than 1.7 billion people on Earth did not have access to electric lighting (this was 2 billion in 1996, according to the World Bank).

A first answer to this question would be to put increased efficiency light sources onto the market. However, as we will explain later on, the answer is a lot more complex than this first approximation[10]. At this stage, however, we will voluntarily limit our discussion to the question of light source efficiency. For this discussion, it is important to separate the concept of "energy efficiency", η, expressed in %, from the concept of "light efficiency", η_{lum}, expressed in lumens per watt. Thus, for a light source absorbing an electric power P_{in} (expressed in watts), which emits a radiated power P_r (also expressed in watts), the energy efficiency is traditionally defined as the ratio of these two quantities:

$$\eta = \frac{P_r}{P_{in}} \times 100(\%) \qquad [7.1]$$

However, the energy emitted by the source presents a spectral distribution $P_r(\lambda)$, which characterizes, amongst other things, the emitter body and the emission[11] process. This distribution may extend to a very large wavelength range λ, but only the wavelengths of the "visible" electromagnetic spectrum (located between 360 nm and 780 nm) are perceptible by our visual system; the others do not contribute to the vision process. On the other hand, the sensitivity of our visual system depends on the wavelength and, as we will see later, it also depends on the light quantity. Figure 7.4 shows the standard response, $V(\lambda)$, of the human eye.

Under these conditions, instead of the total radiated power, light specialists use the notion of light flux, Φ, expressed in lumens (lm). This quantity better represents the "useful" light emitted by the source. The light flux is then defined as:

$$\Phi = k \int_{360}^{780} P_r(\lambda) V(\lambda) d\lambda \qquad [7.2]$$

in which κ is the normalizing factor of the curve $V(\lambda)$, defined by the International Lighting Commission (ILC) as being equal to 683 lm/W. Using the light flux defined above, the light efficiency of a light source is given by the ratio of the light flux (in lumens) over the absorbed power (in watts) and is expressed in lumens per watts (lm/W):

$$\eta_{lum} = \frac{\Phi}{P_{in}} \qquad [7.3]$$

10. Some studies, taking into account man's increased demand for lighting, show that by simply using the existing sources in a reasonable way, we could save between 10 and 15% of the consumed energy in the next decade; but others give opposite results…

11. Incandescence generates a continuous spectrum (close to a "black body" spectrum), whereas luminescence is characterized by a discontinuous spectrum (rays or spectral band).

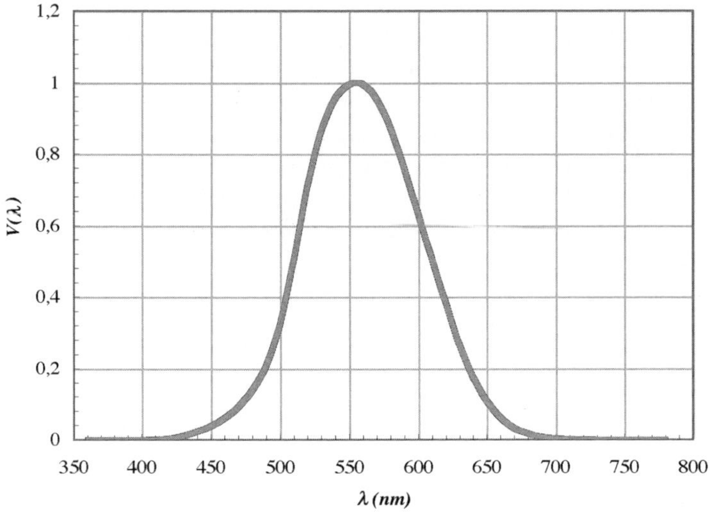

Figure 7.4. *Standard response of the human eye. This curve shows the "photopic" eye response. The standard coefficient is defined by the ILC and is equal to 683 lm/W*

7.2.1. *The thermodynamic limit of white light generation*

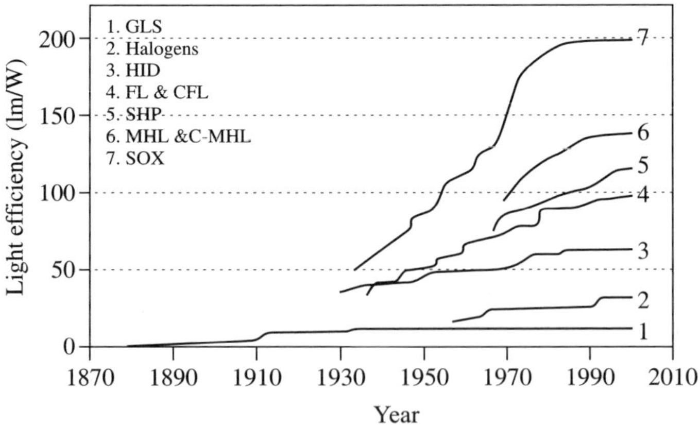

Figure 7.5. *Evolution of the efficiency of electric lamps used for general lighting. GLS: incandescent lamps, HID: mercury high pressure lamps, FL & CFL: compact and lineic fluorescent lamps, SHP: sodium high pressure lamps, MHL & C-MHL: metallic hydride high pressure lamps with a ceramic and silica envelope, SOX: sodium low pressure lamps (monochromatic yellow light)*

Currently, despite all the science and technological advances in the field of lamps, the maximum radiative efficiency of lamps for the production of white light has been stagnating for at least two decades, around 120 lm/W, as seen in Figure 7.5.

On the other hand, this graph clearly shows that the light efficiencies of "headlight" products[12] in the lamp industry reached their limit a long time ago. The question raised is the following: has the lamp industry reached a sort of "thermodynamic limit"?

To answer this question, let us try to determine what a reasonable higher limit of lamp efficiency generating white light of a "good quality" could be, with a color temperature, T_C, ranging between 3,000 K and 6,500 K and a good color rendering (color rendering index, CRI$\geq$80).

To obtain light perceived as "white", it is necessary to have, at least one emitter which allows a balanced excitation of the photo-receivers (more precisely, the three cone families each offering a maximum sensitivity centered around the red, green and blue colors) present in the fovea of the human eye[13]. Thus the desired "white" light can be obtained either by a radiation, the spectrum of which covers all the visible range of the spectrum (this is the case of the Sun or an incandescent body), or, thanks to the physiology of our visual system, by a combination of blue, green and red colors (this is the case of color TV) or else, by a combination of blue and yellow.

The Sun's radiation or incandescent radiation may be linked to "black body" radiation. They have similar properties, particularly regarding their dependency on the temperature of the emitting body. This is why the "black body" will be taken as the reference starting point. At temperature T_C, the radiation spectrum of the "black body" presents at wavelength λ, a spectrum brightness $L_{CN}(\lambda, T_C)$ given by the Planck relationship, which is the absolute higher limit for the radiation emitted around this wavelength. Integrating the Planck relationship, we can easily deduce the radiation fraction corresponding to the visible part of the visible spectrum. Thus, Figure 7.6 shows the "weight" of the visible spectrum in the total emission of the black body with respect to its surface temperature. We note that this fraction, in any

---

12. This graph does not include electroluminescent diodes (ELD). This omission is voluntary; electroluminescent diodes are not currently mature enough to be used in an intensive manner for general lighting; they are instead devoted to signaling and to decorative lighting. However, ELDs may not be ignored in the near future; a detailed discussion will be presented later in this chapter.

13. The fovea is the retina area located in vision direction; this area has a very limited surface area, but it presents the highest concentration of photo-receivers (cones essentially) and is reponsible for "accurate" and for "colored" vision.

case, cannot exceed half of the total radiated power. This is obtained for a surface temperature of about 5,000 K.

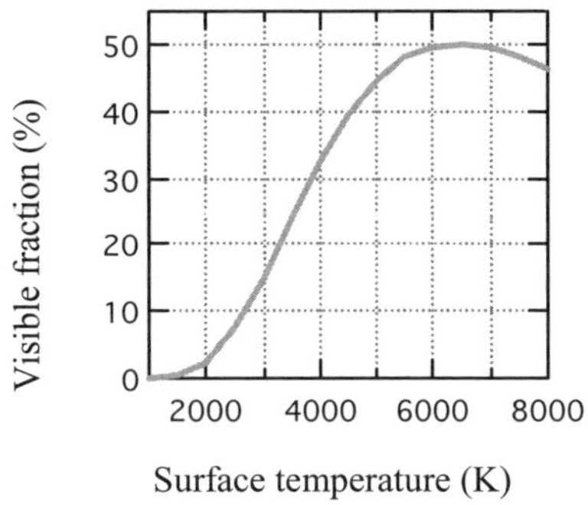

Figure 7.6. *The "weight" of the visible spectrum in the total emission of the black body with respect to its surface temperature*

However, the "black body" is only a theoretical concept. In fact, the brightness $L(T_C,\lambda)$ of a real body, when it can be characterized by a unique temperature, is linked to the black body emission by a quantity, the emissivity, which is always smaller than 1. The emissivity, $\varepsilon(T_C, \lambda)$, is specifically a function of the body characteristic temperature T_C and of the considered wavelength λ. We will then speak of "gray bodies", the brightness of which $L(T_C,\lambda)$ will be written:

$$L(T_c,\lambda) = \varepsilon(T_c,\lambda) L_{CN}(T_c,\lambda) \qquad [7.4]$$

Thus, for an incandescent lamp, the light quantity radiated in the narrow visible spectral band generally depends on the radiator temperature. The lamp filament is made of metal and is heated by the Joule effect when an electric current flows through it. The maximum temperature is then imposed by the fusion point of the metal used. Tungsten is the metal with the highest fusion temperature at 3,410°C, and this why it is used in lamp manufacturing. In practice, the mechanical and thermodynamic requirements limit the temperature to 2,700–3,000°C. At this temperature, Wien's law says that the major part of the radiation emitted by the filament is in the infrared portion of the spectrum (λ > 780 nm). Under these conditions, less than 25% of the total radiation is found in the visible range. This leads to a global energy efficiency of about 10% and a light efficiency of 14 lm/W. It seems difficult to make this any better with an incandescent body, except perhaps by replacing the metallic filament with a semiconductor (some attempts have been made, but the result is far from being exploitable). But, as already discussed, the emission of a continuous spectrum is not the only solution to obtain white light.

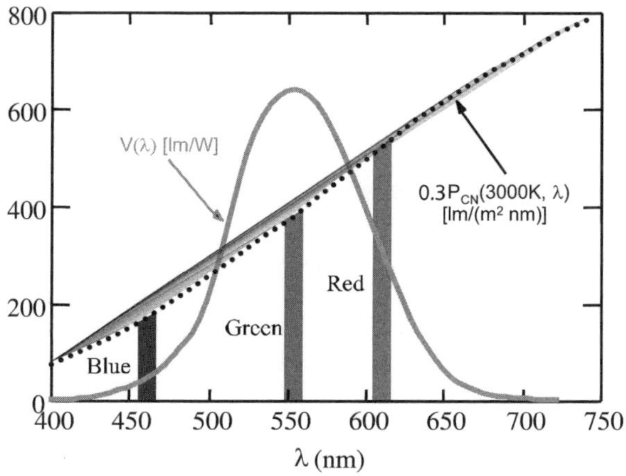

Figure 7.7. *White light can be obtained by combining three photons (blue, green and red). A lamp capable of emitting these three photons, such as a black body at 3000 K (black dashed line curve), would have a light efficiency of 300 lm/W!*

It seems that the best compromise between efficiency and quality would be the use of three emissions, blue, green and red, located in the respective sensitivity maximum of each three cone families which cover the fovea of our eye. Figure 7.7 shows the principle of a source which would radiate these three wavelengths only, such as a black body at 3,000 K[14].

The radiation efficiency can be computed using relations [4.2], [4.3] and Planck's law:

$$\eta_{ray}^{\max} = \frac{\kappa \sum_{i=1}^{3} P_{CN;i}\left(T_c = 3000K, \lambda_i\right) V\left(\lambda_i\right)}{P_{in}} \qquad \lambda_i = \begin{cases} 436 \text{ nm}\left(\text{blue}\right) \\ 546 \text{ nm}\left(\text{green}\right) \\ 700 \text{ nm}\left(\text{red}\right) \end{cases} \qquad [7.5]$$

$P_{CN;i}(T_C, \lambda_i)$ represents the power of each of the three photons according to Planck's law. A numerical application leads to 300 lm/W[15]. This value is three times higher than the best efficiency that can be achieved today. It thus seems realistic to declare that it would be possible, taking into account a global conversion efficiency

14. This lamp, despite its efficiency, would have a very bad color rendering because of its quite "poor" spectrum.
15. This value is the light efficiency of the electricity into white light conversion; it does not include the losses of the global system (mainly losses in the lamp supply circuit).

of the system, around 65%, to target light efficiencies of about 200 lm/W. The challenge is significant, and the associated energy savings (at constant light flux) are in proportion!

7.2.2. *Lamp technology and associated challenges*

How can we achieve the goal of the previous section? Today the answer is not known, but we are sure this is not the only answer. The efficiency and quality of each of the lamp families may be improved. Figure 7.5 shows hope for an evolution, each family having its own objectives and challenges. However a "revolution" in the lighting field is not impossible: electroluminescent (mineral and organic) diodes may be one "revolution", but discharge lamps have not had their last word... This section is dedicated to the description of these challenges.

7.2.2.1. *Incandescent lamps*

In today's incandescent lamps, the tungsten radiator (filament) is heated by the Joule effect. It is a very dense medium, the power density of which is high, and the (surface) emission presents a strong brightness. Its radiation is continuous and approximately follows (in the visible) the "gray" body law at the filament temperature. However, it is not possible to raise the temperature too much because the local filament evaporation and its recrystallization reduce the lamp lifetime very quickly by breaking the filament. This lifetime mainly depends, through these two phenomena, on the voltage, switching on and off, the vibrations, the temperature field around the lamp, etc. The other consequence of evaporation is the darkening of the bulb. This evaporation is controlled by a rare gas which reduces the speed of evaporation. However, the large size of the bulbs limits the rare gas pressure to a value close to atmospheric pressure. The tungsten-halogen cycle makes it possible, to eliminate the tungsten deposit on the glass surface in halogen lamps. The dimensions of the bulbs may then be reduced and the rare gas pressure may be considerably increased, which will limit the evaporation. This saving in the evaporation can be used directly, in terms of lifetime, or it may be transformed, at a constant lifetime, into an increase in temperature and thus in light efficiency. These trade-offs are made according to the use. For halogen lamps in general lighting, the color temperature has been raised to 3,000 K for a 2,000 h lifetime. Under these conditions, the light efficiency is limited to 20 to 25 lm/W, which is however higher than the 12 to 14 lm/W of the ordinary incandescent lamps. There have been attempts to change the emissivity of the radiator, but so far these have proved unsuccessful.

Attempts to replace the tungsten filament with semiconductor materials with a high fusion point are currently being studied, but the technology seems to be very complex. Light efficiency may also be increased by reflecting part of the infrared radiation generated by the lamp towards the filament. This can be done by laying (approx. 46) CVD thin layers onto the bulb surface. They behave like an interferential filter. The results are promising, but this method remains expensive at the present time.

However, even though this family seems to have exhausted its potential for evolution, taking into account the range of use of the lamps, their low cost and their excellent color rendering index (they are given by definition CRI $\approx$ 100), are still attractive. However, many countries have decided to ban them from the market over the next few years.

7.2.2.2. *Discharge lamps*

Although the electric discharge lamp, made for the first time in 1814 by Sir H. Davy and M. Faraday, was the first plasma application, it took more than half a century before a commercial achievement. Today, thanks to a better understanding of the physical processes that govern their operation, discharge lamps have considerably improved since their first appearance.

A discharge lamp is made of a sealed and transparent (or translucent) burner that contains the discharge gaseous medium, letting the radiation they want to use escape at the same time. This container, the shape and dimension of which varies greatly, contains a gas mixture and/or metal vapors, capable of radiating in the range of the desired wavelength. The excitation of the mixture is performed either by the flow of an electric current between two electrodes, or by injecting a high frequency electromagnetic radiation (radiofrequencies, microwaves, etc.) into the bulb. However, since the current-voltage characteristic of a discharge is negative or weakly positive (in the case of high pressure lamps), an impedance is necessary to stabilize the discharge current.

The active element pressure may lie between a few dozen and a few million Pascal. We tend talk about low pressure or high pressure (LP and HP respectively) lamps. This parameter will thus be used to discuss the different lamp families and the associated challenges.

7.2.2.2.1. *"High pressure" discharge lamps*

The "high pressure" lamps operate under pressures from 1 to several hundred bars (200 bars for the UHP™ lamp used in video projectors). It must be noted that most of the current HP lamps use mercury to control the electronic mobility in the plasma.

For these lamps, we generally consider that the thermodynamic equilibrium is approximately obtained locally. Under these conditions, for a given discharge, the medium is characterized well enough by its central temperature T_0 and its radial distribution. In the majority of the cases we are interested in, the radial distribution of the temperature follows a monotonous profile of the "parabolic" type:

$$T(r) = T_0 - (T_0 - T_W)\left(\frac{r}{R}\right)^{\beta} \quad [7.6]$$

in which T_w is the bulb temperature, R the lamp radius and β the "profile" parameter.

Under these conditions, the radiation brightness approaches the black body brightness at the temperature T_C in the spectral regions where the optical thickness is greater than a few units. We are now in situations which meet the above principles but with central temperatures between 4,000 K and 6,000 K and with a certain possibility of acting on the emissivity by selecting the radiating bodies (active elements: mercury, sodium, thallium, dysprosium, etc.). The light efficiencies obtained are currently between 50 lm/W and 110 lm/W. On the other hand, the medium, also less dense than a solid, still allows significant volume emissions (10^4 lm/cm^3).

The best known "high pressure" lamps are:

– high pressure mercury lamps ($T_C \approx$ 3,000–4,500 K, CRI $\approx$ 50, $\eta_{rad} \approx$ 50 lm/W), used for lighting large areas (warehouses) and for urban lighting;

– high pressure sodium lamps ($T_C \approx$ 2,100–2,300 K, CRI $\approx$ 15–60, $\eta_{rad} \approx$ 50–120 lm/W), devoted to road and urban lighting and to emphasize monuments;

– metallic halogen lamps with a polycrystalline aluminum oxide or silica envelope ($T_C \approx$ 3 000–6 000 K, CRI $\approx$ 60–95, $\eta_{rad} \approx$ 50–110 lm/W) for stadiums, esplanades and urban lighting.

It can be possible to change the spectrum through the supply mode, on the power balances. Thus, a commercial lamp using "high pressure" sodium, operating under this principle, has been manufactured. It can go from a color temperature of 2,700 K (yellow), corresponding to a good efficiency but to a poor color rendering, to a value of 3,100 K (white) with a better color rendering but with a lower efficiency.

As expected, because of the high temperatures reached in the high pressure lamps, their energy balance shows thermal losses that represent 25 to 50% of the electric power injected into the lamp. The reduction of the thermal losses is thus a major target in the research for an improved light efficiency. The thermal losses being proportional to the temperature gradient and to the thermal transfer coefficient, both directions must be investigated. However, it must be noted that the material transport phenomena often play an important role and that their control is necessary. The discovery of bulbs that are more thermally resistant could also reduce the temperature gradient while making the evaporation of some less volatile compounds easier. The ceramic metallic halogen lamps (C-MHL) are an example of such a principle (Figure 7.8).

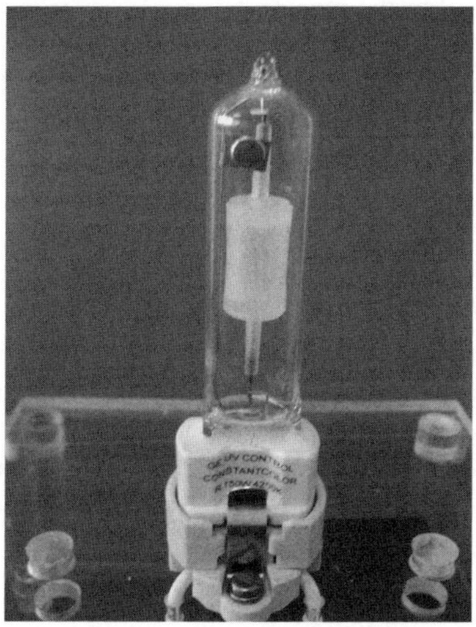

Figure 7.8. *Ceramic metallic iodide lamp (150 W) (source SIP-CPAT)*

These lamps generate a good quality white light (CRI in the order of 95) and they have a good efficiency (exceeding 100 lm/W). They have a long lifetime (more than 12,000 h) and an excellent chromatic stability. They are compact and are available in many nominal powers, from 20 W to 400 W. If the high power versions (from 150 watts) are invading the urban lighting market, the low power lamps will soon tackle the decorative and interior lighting market, replacing incandescent lamps.

Finally, in this family the UHP™ (Ultra High Precision) lamp must be mentioned. This lamp, which operates under 200 bars (mercury vapor) in a few cubic millimeters, is devoted to projection video systems. It really is a revolutionary lamp because the mercury plasma is so dense that it is close to the conditions of a liquid. On the other hand, during the operation, the electrodes are so hot that the tungsten fusion temperature is largely exceeded. Since the electrode is liquid during the operation of the lamp, there is thus a "liquid-plasma" interface. The physics which govern this particular discharge and its electrode are still not perfectly controlled, although the lamp, which is widely marketed, is operates correctly for more than 1,000 hours.

7.2.2.2.2. "Low pressure" discharge lamps

Fluorescent lamps are the best known lamps in this family. Fluorescent lamps[16] are technologically mature nowadays. Furthermore, it is the type of discharge lamp that is the most widespread in the world.

It is true that thanks to a better understanding of the physical processes which govern their operation and also because of the important improvement made in the field of materials, fluorescent lamps have been considerably enhanced since they first appeared. Moreover, the discovery of luminophores resistant to 185 nm ultraviolet bombing, led to a reduction in the diameter of fluorescent lamps from 38 to 24 mm and then to even further reduced diameters. These successive reductions of diameters have led, thanks to the change in the discharge characteristics, to more efficient lamps and particularly to more efficient lamp-luminary systems. On the other hand, taking into account the phenomena at the electrodes and the interactions between the plasma and its supply has allowed the choice of frequencies and wave shapes. These improvements have favored the generation of light and the modification of its properties. They also have offered to the user the possibility of better control of lighting installations.

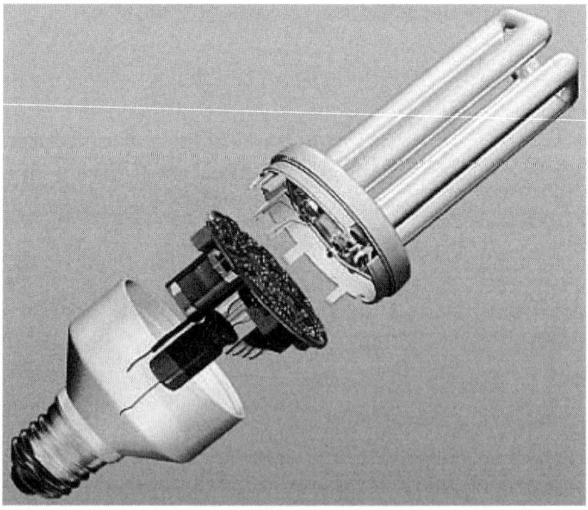

Figure 7.9. *Compact fluorescent lamp (CFL) with its integrated electronic ballast (with the permission of Osram GmbH)*

16. Fluorescent lamps are incorrectly called "neon" lights. They do not contain this gas.

Compact fluorescent lamps[17] (CFL) have appeared in parallel. The principle of operation is the same as the lineic fluorescent lamps, but the tube is "folded" here many times to guarantee the compactness of the lamp. On the other hand, to make their integration easy in the luminary, "integrated electronic ballast", which is light and compact, has been developed and included in the lamp base. This thus defines two types of compact fluorescent lamps: the lamps with a ballast integrated into the base, which are interchangeable with standard incandescent lamps (Figure 7.9), and lamps without the integrated ballast, which require specially-adapted luminaries (and which include this ballast necessary for the operation of the lamp).

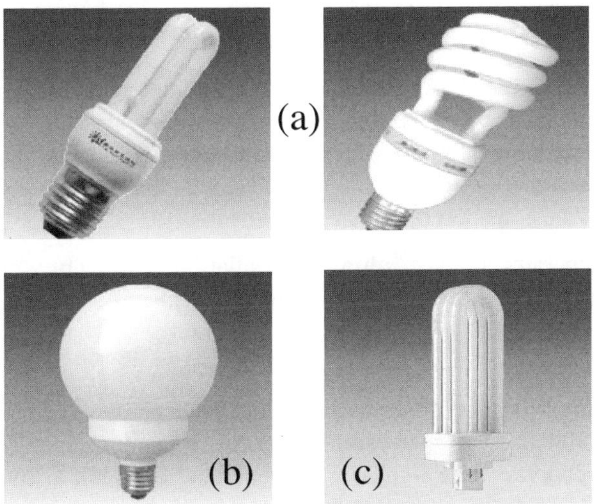

Figure 7.10. *Compact fluorescent lamps: (a) with ballast integrated into the base; (b) look-alike and (c) without integrated ballast*

CFLs today represent a market of about 8% in the residential area. The advantages are well known: they present a good efficiency (40 to 80 lumen/W), they are cheap when used under good conditions (use in rooms where lighting is durable), their lifetime is from 6,000 to 12,000 hours (estimation based on a single lighting per day). Let us point out that the significant efficiency range of the CFL (40 to 80 lumen/W) may be explained by many reasons: for the same type of lamp (shape and power) the efficiency strongly depends on the quality of the manufacturing line: the majority of the "made in China" lamps have a lower efficiency. In the same way, this also strongly depends on the quality of the electronic ballast integrated into the base of the lamp. On the other hand, the shape also has an impact of the efficiency, especially for the look-alike lamps (Figure 7.10) because a second spherical

17. The CFLs are commonly known under the name LCL (Low Consumption Lamp).

envelope must be added to achieve the shape of an incandescent lamp; this of course will also have an influence on the general efficiency of the lamp.

Despite all these improvements, there are still nowadays many scientific challenges in the field of fluorescent lamps:

– How can we avoid 50% of losses due to the luminophores? The answer probably lies in using luminophores able to generate two visible photons for each incident UV photon. This solution is possible but still expensive and less efficient. Another solution would consist of generating visible light directly, using volatile molecules or other gases. The research into new emissive bodies meets the second challenge described in the case of high pressure lamps.

– How can we avoid the use of mercury? Mercury, which is liquid at ambient temperatures and easily vaporizable, also has interesting atomic properties and seems to be unavoidable. However, mercury generates, especially at low pressures, a hard ultraviolet radiation and it is considered a highly toxic element. Many possibilities may be considered for its replacement:

- xenon generates a “hard” ultraviolet radiation at 147 nm which destroys the current luminophores and the generation efficiency is low (2.5 times lower than that for mercury). However, the xenon low pressure discharge generates light instantaneously;

- neon, one of the first gases used in the discharge field, generates red light at low pressures (100 Pa). By increasing the pressure to more than 104 Pa and with adapted luminophores, white light can be generated but with a low efficiency (20 lm/W);

- molecular emitters. Unfortunately, the presence of metallic electrodes strongly reduces the choice of usable gases. Some tests showed that hydroxyl (OH-) could play the role of the emitter gas but its use is very complex.

– How can we miniaturize fluorescent lamps? Although CFLs are more “compact” than linear fluorescent lamps, the efforts to miniaturize the source still continue.

– How can we control the color? Color control of the low pressure tubes is very important for applications such as signaling and display advertising. It is possible to control the discharge color with a convenient choice of the gaseous mixture (with or without mercury) and the associated luminophores. Another easier method consists of controlling the selective excitation of the emissive bodies by changing the characteristics of the discharge electric supply (wave shape, duty cycle, frequency). Thus, a Hg-Ne lamp may switch from a blue color (Hg) with sinusoidal excitation, to red (Ne) when the lamp is excited in pulse mode.

– How can we reduce the starting time? Let us note that this starting time idea is difficult; indeed, we are faced with important discrepancies according to the measurement mode. The available data show a delay of about 10 min whereas measurements performed by others give 3 min instead. It is not possible to conclude, but we must say here that this problem of starting time is important for the CFL user who, used to incandescent lamps, desires instantaneous lighting. The development of robust electronic ballast could partially solve this problem.

7.2.2.2.3. Discharge lamps without electrodes or molecular emitters

Almost all the discharge lamps used for lighting are equipped with electrodes to inject electric energy into the discharge and to guarantee a transition from the metal conduction to the gaseous conduction. Although the current electrode technology is better controlled, the presence of these metallic elements in the plasma seriously limits the lifetime of the lamp and also reduces the choice of the emitter gases. The elimination of the electrodes solves many problems and also simplifies the manufacture of the lamp. It is indeed possible to inject the electric power in the discharge by coupling, in an inductive or capacity way, the plasma with the supply circuit. The use of microwaves is also an alternative. Today, three different technologies exist: induction fluorescent lamps, dielectric barrier lamps using excimer radiation and microwave lamps.

The best known lamps without electrodes are "induction" lamps. The operation principle is similar to that of CFL. Today, there are products in this family on the market, for example, the QL™ (Philips), the Genura™ (GE) and the Icetron™ (Osram). The target applications are mainly devoted to ambient public lighting, since the price is still very high and despite the very long lifetime of these lamps, the depreciation is very long.

Dielectric barrier lamps use the emission of excimer molecules. These molecules, which do not exist in the normal conditions (the fundamental level is anti-bounding), may exist for very short times at a radiative excited state. Today, we can control the production of these unstable molecules very well in an electric discharge at atmospheric pressure far from local thermodynamic equilibrium conditions.

An important advantage of the excimer lamps is that any geometric shape of the tube is possible, due to the absence of electrodes. Furthermore, the lifetime, as in the case of induction lamps, is thus very long[18]. Moreover, a large choice of wavelengths is possible because there is no limitation in the choice of emissive body for the electrodes. For instance, we can mention a Xe-SF_6 lamp for the generation of

18. Lifetimes from 60,000 h to more than 100,000 h have been announced. As a comparison, an electrode lamp rarely exceeds 20,000 h and an incandescent lamp 2,000 h.

a quasi-monochromatic radiation of around 370 nm produced by Xe*F. We can also mention another lamp with Xe-Cl_2 which generates a 308 nm radiation because of the Xe*F. Regardless of these advantages, over the last decade this technology could only be used for industrial applications (photo polymerization, surface treatment, depollution, etc.).

However, attempts in using excimer lamps for general lighting and signaling have been made for many years. Osram has thus put onto the market a flat lamp called Planon™ (Figure 7.11). This lamp uses xenon as an active gas (up to a pressure of a few hundred mbars) to generate a radiation at 147 nm coming from the excimer molecule Xe_2^*. This radiation is then transformed into visible light by means of luminophores, as in an ordinary fluorescent lamp. The Planon™ lamp is a very interesting concept, but in the first version its efficiency was 27 lm/W. It was too weak a value to compete against a fluorescent lamp. This flat lamp may be used for general lighting and for signaling. Indeed, it is possible to combine many Planon™ lamps to have a large display panel visible from a long distance.

(a) (b) (c)

Figure 7.11. *a) The Planon™ lamp from Osram; b) its "planar" electrode structure and c) the "conical" micro discharges which generate the excimer radiation*

Another attempt at making an electrodeless lamp devoted to general lighting consists of using molecular emissions (spectral bands) and of course, it also consists of supplying energy to the discharge by means of microwave sources. This concept allows the generation of white visible light without a luminophore. The first attempt in this field was the sulfur lamp. Fusion (USA) developed this for public and monument lighting. Microwaves supply this lamp and its efficiency is about 100 lm/W. Figure 7.12 shows this lamp and its main characteristics.

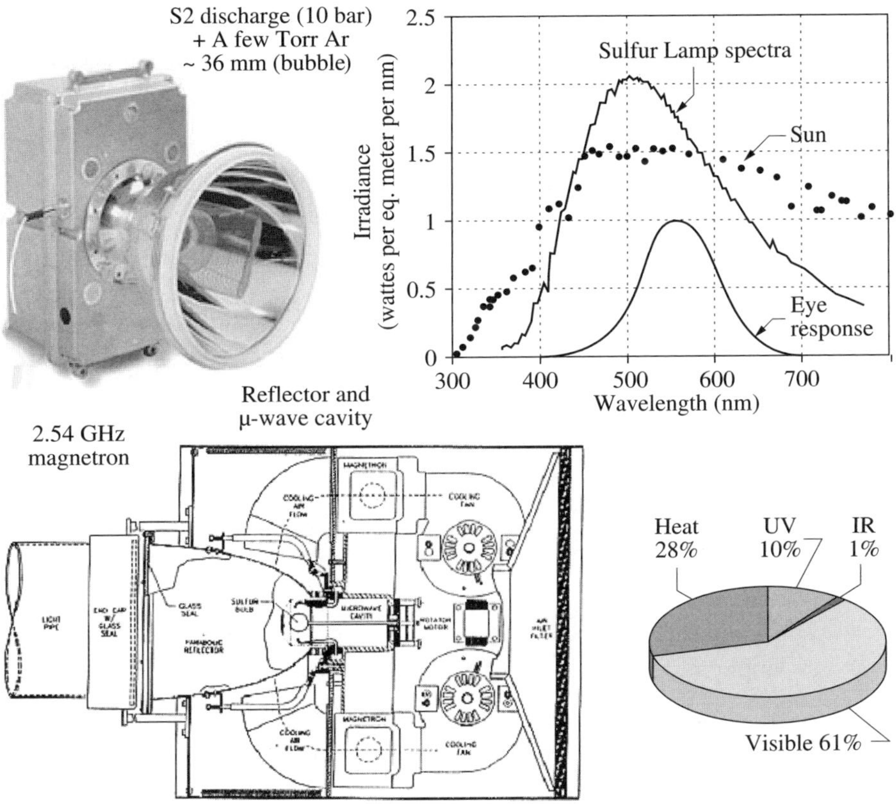

Figure 7.12. *The sulfur lamp, its spectrum and its energy balance*

Unfortunately, this lamp generates a light of relatively poor quality in terms of colors and its application is, at the moment, very limited. On the other hand, the system requires permanent cooling so the spherical bulb is rapidly rotating in the system. However, despite its poor commercial success due to these gaps, this lamp may be considered *a revolution in the field of light sources*. We are today convinced that the use of molecular emissions, combined with the electrodeless lamp technology, may result, within a few years, in the development of new products, that are more efficient and generate a light of better quality and in large quantity, with a long lifetime. Will this be the light source of the future?

7.2.2.3. *Electroluminescent diodes*

There is *also a revolution* in the field of light sources. In 1907, H. J. Round announced the first light emission of a semiconductor [ROU 07]. With no clear explanation, this discovery was rapidly forgotten. We had to wait until 1962 before Nick Holonyak and S.F. Bevacqua from General Electric sign the "official birth certificate" of the red electroluminescent diode [HO 01]. Since then, events have accelerated. In 1968, the first electroluminescent diode (or LED, for *Light-Emitting Diode*) on the market generated a red light of just 0.001 lm. Today, white LED lamps exist on the market and generate more than 1,000 lumens. This quick historical reminder shows that even if large difficulties remain and still make a rapid and large lighting market penetration impossible, it is realistic to think that LEDs have a future in this area.

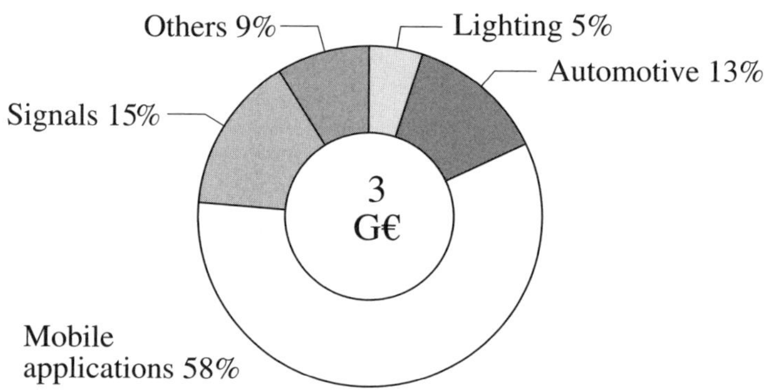

Figure 7.13. *LED market in 2004*

Today, according to Russ Arensman [ARE 05], the world LED market is valued at around 3 billion euros and its annual growth is 14%. If this rate is confirmed and maintained during the five coming years, this area would represent in 2009 about 6 billion euros and would exceed 10 billion in the following decade. However, today, the penetration of LEDs in the field of general lighting remains about 5% (Figure 7.13). Will this portion increase in the next few years? Despite the general optimism, supported by the spectacular scientific progress loudly announced by the manufacturers, the question today has no clear answer.

The simplified operating principle of an LED is as follows: the basic element of an electroluminescent diode is a semiconductor with two regions of different conductivity (p- and n-type) and a region of radiative recombination of the "n" (electron) and "p" (hole) carriers. Under the effect of a potential difference applied between the two layers, electrons penetrate the n-type layer. This is equivalent to an injection of holes in the p-type layer. At the interface of both zones, electrons and holes recombine and give a photon thereby generating light (Figure 7.14).

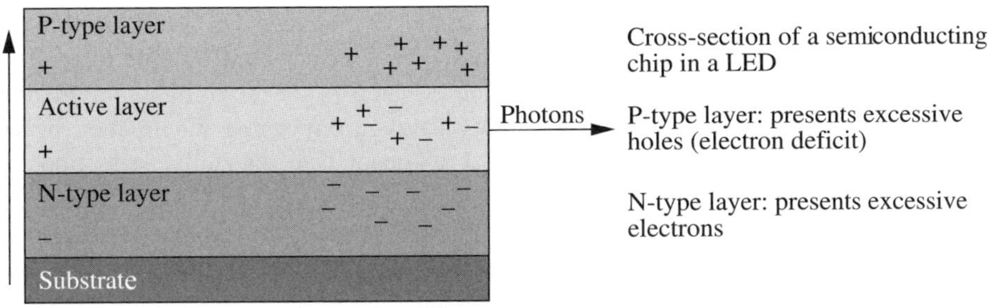

Figure 7.14. *Simplified operating principle of LEDs*

At the equilibrium state, the majority carriers of each region diffuse to the other zone: the electrons of the "n" region tend to diffuse to the "p" region, holes move in the opposite direction. These spontaneous motions disturb the local electric neutrality of the system and are at the origin of a "space charge" field, which, in turn, opposes these motions and the system balances itself. Consequently, the number of minority carriers in each region (electrons in the p region and holes in the n region) stays very weak and the radiative recombination probability is almost zero. Now, when applying a direct polarization voltage V to the junction, the potential barrier decreases by a value equal to eV. Consequently, the diffusion current of the majority carriers at each side of the barrier to the opposite region increase by a factor proportional to $exp\ (eV/k_BT)$, where T is the junction temperature and k_B is the Boltzmann constant. This diffusion current increase unbalances the system and leads to a population growth of the minority carriers in each region. If the direct polarization voltage is high enough, the radiative recombination probability is no longer negligible and the junction generates photons. The principle discussed above corresponds to a so-called "homo-junction" electroluminescent diode. The classical LEDs still use this structure today. However, the new high intensity diodes use many complex junctions, called "hetero-junctions". They operate on the space variation of the semiconductor composition. These structures are combinations of different chemical composition semiconductors also with different energy gaps. Today, we use simple (SH) or double (DH) hetero-structures.

The LEDs carry many hopes for lighting. In a white paper ordered by the US Senate, R. Haitz *et al.* have developed two scenarios [HAI 02]: the first, "evolutionary" scenario only foresees a white diode of 50 lm/W for 2010, whereas, the second, "revolutionary" scenario targets a 250 lm/W system from now to 2025. According to the first scenario, the 150 lm/W seems to be a realistic objective, which can be reached without considerable effort. In this context, LEDS may very well replace from now to 2025 20 to 30% of the incandescent and halogen lamps. This would result in an energy saving of about 400 TWh/year (corresponding to a generation capability reduction of 50 GW). In the second scenario, a 200 lm/W goal clearly requires a considerable effort. In that case, LEDs would be able to replace the fluorescent lamps and could represent more than 50% of the light source market share from now to 2025–2030. The energy savings envisaged would then be of about 1,000 TWh/year (corresponding to a generation capacity reduction of 120 GW).

Furthermore, LEDs offer major advantages:

– a strong development potential of their efficiency (in theory 100% of electricity into light transformation);

– a long lifetime: the foreseen objective, on which there is consensus, is to make a diode with a lifetime from 40,000 to 50,000 hours from now to 2025 (rather than the 100,000 hours announced a few years ago);

– high quality saturated colors, providing a good light perception by the user;

– an easy supply, a low energy demand;

– high brightness.

However, there are also well known drawbacks. The aspects that are not currently solved are as follows:

– the above scheme clearly shows that LEDs can only generate light "nibs": light is at the same time directional and monochromatic. Obviously, this is a huge drawback for general lighting (where a diffuse and non-blinding light is required). To solve this problem, heavy optics is necessary, which is expensive and damages the light efficiency. Because of these characteristics, the present applications are limited to spots and additional lighting;

– aging is erratic; the color "changes with time". The flux generated by white LEDs decreases quite linearly from the first hour of operation. Flux drop measurements have been made on a 1 W white diode. They show that the flux is reduced by half in 6,000 hours (this is data from 2000, we do much better today…). Advances in this field are thus also necessary for electroluminescent diodes to dominate the market one day;

– there are problems for mass production reproducibility for white LEDs only (non-identical color temperature);

– LEDs are limited to low powers: currently less than 5 W. It is mainly a thermal management problem of the junction since the heat generated by the diode operation changes its performances;

– the energy consumption is low. This contributes to give the diode an "environmental" image, but the production technique is, itself, less "green" (semiconductor industry). This point must be compared to the present-day lamp industry, which is very clean (few toxic wastes generated).

These problems are a snapshot, made at mid-2006; they might make us smile in the next few years, but today, nobody can say at what speed they will be solved.

One of the major problems with LEDs is because the light emitted is monochromatic. For this reason, one of the basic problems for the research consists of progressing towards a "white LED" with more than 100 lm/W of efficiency. Today, the commercial junctions only offer 25 lm/W of efficiency.

Three methods exist today to make an electroluminescent diode that generates white light:

– combining one diode emitting a short wavelength λ_1 with a luminophore (phosphorus) emitting a complementary wavelength λ_2 (long);

– using a diode emitting in the near ultraviolet range, coupled with one or more luminophores;

– using (at least) three diodes emitting visible wavelengths which combine to give a white light.

Each of these methods has advantages and drawbacks.

The first method is based on the fact that two photons of complementary wavelengths (λ_1 "short" and λ_2 "long") simultaneously reaching the human eye cause a white light feeling. The main idea is then to use a short wavelength diode covered with phosphorus, which absorbs some short photons, and transforms them into longer wavelengths. In principle, an infinity of combinations is possible. Practically, some manufacturers (Nishia, CREE, Lumileds, etc.) have a mass production of high intensity white diodes (up to 5 W today) using a blue diode (InGaN) combined with a yellow phosphor (YAG-Ce). In the short run, this technology will probably dominate the white diode market. However, it presents two drawbacks:

– The diode has a halo effect, which may be disturbing for some applications. This is due to the directional light generated by the white diode, whereas phosphorus has a Lambertian behavior (diffuse emission).

– The yellow phosphorus does not easily absorb the blue photons. This problem is raised because the present-day yellow phosphorus has a very low absorption in the blue because of the rare earths used as doping elements. The solution to the problem seems complex and requires a lot of research.

To reach an efficiency of 200 lm/W for the generation of white light in this technology associated with existing phosphorus would require a diode with a total conversion efficiency of 60%. Today, the best-known efficiency is 45% at a wavelength of 610 nm.

The second method consists of using a short wavelength diode (in the near ultraviolet or in the violet range) coupled with one (or many) phosphor(s) that transform the UV light into visible light. This technique is used today in fluorescent lamps. It presents the advantage of generating a good quality white light (good color rendering index). To use this technology to obtain 200 lm/W, as previously discussed, requires a UV diode with a total efficiency of 70%. Diodes generating 400 nm photons (violet) exist but the efficiency is almost 21%. To increase this efficiency up to 60–70% is a fantastic challenge but probably unrealistic in the present day state of art. However, we can note today a huge research activity in this field. The development of efficient UV diodes with a long lifetime and sufficient power represents an important challenge especially for lighting but, also, for other equipment using UV radiation in some applications (dermatology, surface treatment, decontamination, etc.). If this technology appears to be efficient, and if in the future, research on the luminophores brings onto the market good quality and reliable white phosphorus, this technique may then become the leading technology for white light generated by LEDs.

Finally, the third method consists of using three electroluminescent diodes, one for each main color (red, green and blue). The synthesis of three colors leads to the generation of white light, the color and temperature color of which depend on the proportion of each of the three components. Since the dimensions of a diode are small, we can think about the generation of "composite" white lights to which the addition of a diffuser would considerably improve the homogenity of the source. This method has many advantages: avoiding the use of luminophores, allowing a good control of the light quality (color rendering index, source color temperature). However, this technology has a major drawback: every single diode, out of the many dozen LEDs that make the composite source (even several hundred in some cases), must be controlled "individually". We also have to take into account the fact that the diodes of each elementary color require a different voltage supply. A direct consequence of these drawbacks is the price increase of the light source. Finally compared to a standard lamp of the same light flux, the "surface" of the composite source is considerable. To use this solution to reach the 200 lm/W target requires the use of electroluminescent diodes with 50% efficiency for each color…

Today, electroluminescent diodes, for the previously-mentioned reasons, have difficulties in proving their efficiency in the general lighting field. However, they have proven their efficiency in the field of signaling and they are penetrating the specialized lighting market. Figure 7.15 outlines the LED application fields according to color and flux [ZIS 04]. Signaling applications often require low light flux and a monochromatic light. For light, we need white light and important light flux.

On the other hand, if, in the 1970s, a single LED was used as an "indicator" (on/off, breakdown, level, etc.) in the control panel of electric appliances, more complex signaling systems (for instance traffic lights) use several dozen LEDs for a single function. The development of high intensity diodes will probably lead to systems dedicated to lighting that will operate with a limited number of diodes.

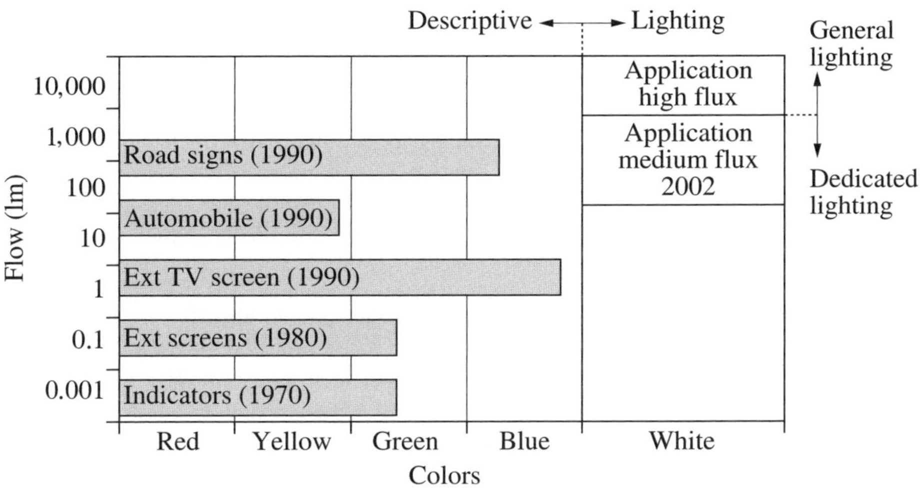

Figure 7.15. *Application field of electroluminescent diodes according to their radiated flux and their color*

The perspectives of research and development in the field of LEDs are important; the following list indicates some that are considered to be relevant [OID 02]:

– Fundamental research into materials and light generation: first of all a better understanding of the light generation mechanisms by semiconductors is necessary. This is why we must develop new diagnosis techniques adapted to LEDs and more sophisticated mathematical models than those available today. For the time being, the understanding of problems due to p-type doping faults and the addition of high indium and aluminum concentrations in the green diodes InGaN and AIGalnN, lead to the production of more efficient green diodes. As with the study of the carrier containment in phosphide-based LEDs, this seems to be the key for the achievement of more efficient red diodes.

– Substrate materials: today, for the development of GaN family diodes, which seem to be among the most promising, sapphire substrates, SiC or GaN must be used. Each of these three materials shows advantages and drawbacks. However, we are not able to produce large faultless surfaces with these materials. The solution to this problem must be considered as a priority in this field. On the other hand, it would be advisable to be able to reduce the substrate production cost to less than 70 €/cm^2 (for UV, green and blue diodes).

– Reactor technology for manufacturing diodes by epitaxy: the reactors used for mass production of the GaN family electroluminescent diodes are just machines developed for manufacturing GaAs diodes. If their reliability is good for the production of GaAs diodes, it is very bad for making modern GaN diodes. Thus, the diodes, mass-produced by these machines, quite systematically show efficiencies 1.5–2 times lower than those observed on the prototypes. On the other hand, in the same series produced by a machine, important characteristic dispersions may be noted. The ridiculous example is the manual selection, according to the "color", made at the end of the high intensity white diode production line… To go further in this field, we have to know the process chemistry and the nitride growth mechanism better in order to optimize the reactors.

– Luminophores: the problems of low blue light absorption by yellow phosphorus have already been mentioned. At first, luminophores able to absorb and efficiently convert photons with wavelengths between 370 and 470 nm must be developed, then it will be necessary to study other possible combinations of complementary colors and develop adequate phosphorus. On the other hand, as for fluorescent lamps, the development of "quantum-splitting phosphors"[19] will be a decisive step for LEDs.

– Light extraction and packaging: the improvements of the optical characteristics of transparent substrates and contacts, of the reflecting layers, etc., are important research directions for the near future. To improve the junction geometry in order to increase the light extraction efficiency, or even to look after complex structures such as photonic crystals are longer-term challenges. Finally, it must be recalled that the junction is only a small part of the diode and that the packaging[20] plays an essential role in its efficiency and reliability.

19. "Quantum-splitting phosphors" are a technology which makes it possible to obtain, for an initial UV photon with an energy E, two or three visible photons of energy E1, E2, E3 (E>E1+E2+E3). The ideal case, but one which has still not been reached to our knowledge, consists of using a blue, green and red photon to make white light. These phosphoresces are very expensive and need to be recycled at the end of the lamp lifetime. This technology will mainly be of benefit to fluorescent lamps with or without mercury and to LEDs radiating in the ultraviolet range.

20. "Packaging" is a term dedicated to the LED industry that designates the integration of a "semiconductor junction" in an LED (epoxy dome, phosphorus, electric connections, thermal radiators, etc.).

– Technological challenges must be envisaged in this field. For instance, for a high intensity diode, it would take a transparent material that is easy to machine with a refraction index greater than 1.6 and a transmittance of 80% from 440 to 650 nm able to stop completely short photons (λ<440 nm). This material must conserve its characteristics for at least 50,000 hours while being at a temperature of 150°C. Finally, the water concentration in this material must remain at about a few ppm and its mechanical characteristics must be compatible with the envisaged application.

– Luminaries and electronics: for the LEDs penetrating the lighting market, specific luminaries must be developed that take into account the strong light directivity, the small light source dimensions and especially its sensitivity to temperature, humidity and any other corroding factor encountered in its environment. Finally, because the generation of an enough light for general lighting requires a significant number of LEDs, a sophisticated electronic control is necessary. Let us note here that an LED color is very sensitive to both voltage and current supply. A small variation of one or the other of these quantities will provoke a significant color shift. Moreover, the behavior of every diode changes with time (flux is reduced, color shifts) more or les slowly, but because of lacks at the production level (see above), this variation is different from one diode to the next. They must be controlled individually.

These objectives, clearly announced by researchers for the 15 coming years, are ambitious stakes. Are they realistic? To answer this question by "yes" or "no" would be a speculation on the shape of a puzzle without the cornerstones!

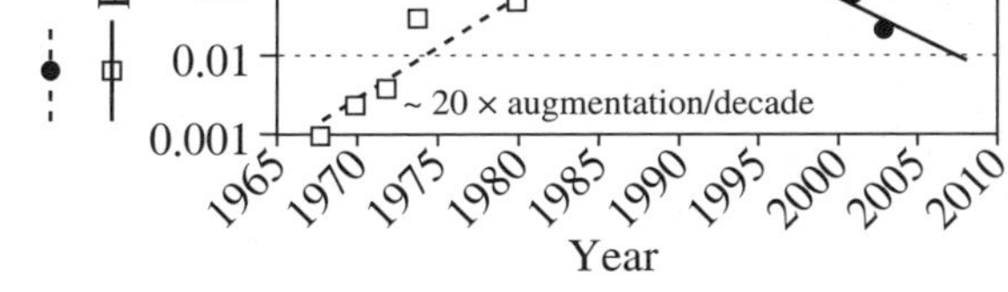

Figure 7.16. *Evolution of the flux light generated by LEDs and of their cost*

A key argument, often used by LED "supporters", is the rapid evolution of the light quantity (flux) generated by a diode. Indeed, as shown in Figure 7.16, the flux per diode is multiplied by 20 every 10 years. At the same time, the price decreases by a factor of 10 during the same period. Today, the researchers call these tendencies the "Craford laws".

These arguments are very encouraging. However, we must bear in mind that the quantities in Figure 7.16 concern not only white LEDs, but also the diodes of each color. On the other hand, nobody can guarantee the extrapolation of these curves until 2020–2030. Their evolution may remain steady, vanish or accelerate even more quickly.

7.2.2.3.1. Organic electroluminescent diodes (O-LED)

Organic electroluminescent diodes (O-LED) are the organic version of LEDs. This technology, patented in 1987 by the Kodak Company, was initially devoted to signaling applications and its first commercial application appeared around 1997. Since then, O-LEDs have evolved considerably, to obtain modules generating white light with an efficiency of 32 lm/W and Universal Display Corporation announced 102 lm/W in July 2008. In this context, and if the lifetime problems are solved, O-LEDs will be a serious competitor to standard LEDs, because contrary to LEDs, O-LEDs have an important and necessary "extension" to general light applications (Figure 7.17).

Figure 7.17. *Some O-LEDs prototype modules (top right the Novaled-Philips with a light efficiency of 32 lm/W; top left a prototype General Electric; at the bottom, a prototype luminary)*

Indeed, in an O-LED module, each diode, the thickness of which does not exceed one millimeter, contains three layers of an organic semiconductor (plastic film) surrounded by transparent electrodes (at least one of the two if we want light to "come out"). Every "light pixel" is made of three electroluminescent diodes placed side-by-side (one red, one green and one blue).

Here are some other advantages that O-LEDs seem to have in their favor:

– possible implementation on large surfaces (light signaling panel), which assumes a possibly very large lighting application and the associated scale savings;

– important potential efficiency: 102 lm/W (at the prototype stage 2008);

– simple supply (rectified grid, 10 V); possible light gradation, theoretically without problem;

– non-directional color or white light (large surface Lambertian emitters);

– light and flexible products (plastic).

The future will prove if this emerging technology will take market shares in the lighting field.

7.3. Energy savings in the lighting field: some typical case studies

Now, since we have seen the different technologies and the challenges of the light sources that dominate (or that will dominate) the general lighting field, we will discuss the possible energy savings in some key areas: residential and urban.

7.3.1. *Residential lighting*

The residential lighting area represents 30% of the total energy bill. In Europe for instance, if we could replace one 75 W incandescent lamp in our houses with one 15 W compact fluo lamp generating the same light, we could make huge savings. Each family would save an average of 150 kWh per year and since Europe represents about 150 million families, the yearly savings would represent more than 22 TWh or the equivalent of 7 million tons of oil (TEP). However, the behavior of the "residential" user is quite unpredictable and it is far more difficult to regulate through legislature. The solution would consist of using rather incentive means and especially in informing the population of some technology advantages, that are economical in energy. Thus, today, the penetration rate of energy-saving lamps remains very low in this area.

In the middle- or long-term, the arrival of lamps, which are marginalized at present, is possible. The price decrease of the metallic iodide lamps and the development of ballasts integrated into the lamps – making them interchangeable with the incandescent lamps – may let us imagine a breakthrough at the domestic level. For the same reason, regarding the electrodeless lamps, their cost reduction combined with a very long lifetime, make them competitive. On the other hand, it is not unthinkable that when LEDs come onto the residential lighting market, the latter will be always divided between incandescent lamps (GLS and halogens) and CFLs. The question of long-term distribution will then be raised between incandescent lamps, CFLs and LEDs. To try to come up with an answer, we imagined three scenarios; depending on the user behavior (we already showed that this point has a fundamental impact on the development of the different products):

– the "standard" scenario consists of extrapolating the future on the basis of what is known currently and without envisaging significant changes in behaviors;

– the "consumerist" scenario is the "pessimistic" option in which the recognition of the environmental aspects (and the energy savings) and of the advantages of "clean technologies" by the consumer, is the most limited;

– on the contrary, the "green" scenario represents an evolutionary model in which all these aspects (environmental impact, energy savings, "clean" energies, etc.) are massively developed.

This projection was made for the European Commission, in the framework of the EnERLIn[21] project. The analysis was based on the compilation of a significant number of disparate investigations on this subject. They are thus "averages" that are validated intuitively and by personal experience. The results of this projection, from now to the outlook in 2030, in the residential area, are described in Table 7.3.

Lamp types	Present distribution (percentage)	2030 projection "consumerist" scenario	2030 projection "standard" scenario	2030 projection "green" scenario
Incandescent	about 90%	75%	55%	15%
CFL	8%	25%	45%	75%
LED	0%	0%	0%	10%

Table 7.3. *Penetration rate of the different technologies in the residential area*

21. EnERLIn: Energy Efficient Residential Lighting Initiative project EIE-05–0176, http://www.enerlin.enea.it.

The different points to note are the following:

– In two scenarios out of the three, incandescent lamps remain the most popular in 2030 and only the "green" scenario allows a penetration of LEDs in the residential area. For the same reason, only this last scenario allows CFLs to be the most common.

– More globally, which also confirms this projection, the variation after 25 years, is great depending on the scenario that will be achieved (from 15% to 75% of CFL penetration). This will raise a comment: if the consumer perception of the usefulness of moving to energy-saving technologies is so important, it becomes very useful to communicate with the general public on this subject. Moreover, this phase may be followed by incentives, to help the advent of the "green" scenario in a concrete way.

– In the "standard" and the "green" cases, we assumed that the consumer considers the amount of light in a residence to be sufficient. In this context, the consumer does not try to increase beyond measure the number of lamps in his house. On the contrary, the "consumerist" scenario assumes that the user wants to increase the amount of light even if this is not necessary.

In this context, we can easily compute the energy impact of each scenario. Table 7.4 gives this calculation.

	Scenario	Standard	Consumerist	Green
GLS	%	55 %	70 %	15 %
	Number	15.4	26.6	4.2
	Energy (MWh)	0.5	0.8	0.1
	Flux (klm)	12.9	22.3	3.5
CFL	%	45 %	30 %	75 %
	Number	12.6	11.4	21
	Energy (MWh)	0.09	0.09	0.16
	Flux (klm)	11.9	10.8	19.8
LED	%	0 %	0 %	10 %
	Number	0	0	2.8
	Energy (MWh)	0	0	0.01
	Flux (klm)	0	0	1.4
Total	Energy (MWh)	0.6	0.9	0.3
	Variation %	21 %	-24 %	132 %
	Flux (klm)	24.84	33.12	24.77
	Variation %	13.9 %	35.4 %	13.6 %

Table 7.4. *2030 projection of the energy consumption of the residential area for lighting (European Union)*

Today, on average, a residence generally has 28 incandescent lamps (GLS) of about 60 W and few energy-saving lamps (CFL) of about 15 W. For an operation of 1,000 h per year, these sources consume 1.4 MWh/capita and generate 21.4 klm of light. In 2030, the efficiencies of the lamps of each type will be about 14 lm/W for the GLS, 63 lm/W for the CFLs and probably 100 lm/W for the LEDs. However, the penetration of each technology cannot be easily envisaged. Table 4.4 gives three possible scenarios: "standard", "consumerist" and "green".

We immediately note that, at the European Union level, according to the "standard scenario", we have an energy saving of 26% compared to the current situation. The "green" scenario leads to 141% of savings and at the opposite end, the "consumerist" scenario shows a consumption increase of 21%. From this simple computation, we first conclude that the advent of the "consumerist" scenario must be avoided. Then, we have to act in such a way that the number of economic lamps increases or, at least, stays the same. Unfortunately, the gamble is not paying off! To go further in that direction, it is necessary to continuously promote the technologies that save energy, to create a quality "label" for the best products, to understand the reaction and then to inform the consumer. Finally, it is also necessary to inform and to train the vendors of these products.

7.3.2. *Urban lighting*

Today, the urban lighting area is constantly expanding, and the corresponding energy consumption is growing rapidly. This area, today, makes up about 8–10% of the total energy bill[22]. It is estimated that an efficiency increase of urban lighting systems may allow, for Europe only, a reduction of 3 to 4 million tons of CO_2 waste in the atmosphere; which correspond to a 5% reduction as planned by the Kyoto agreement. On the other hand, cities are trying to implement lighting systems that guarantee a better quality of life and increased security for their citizens, without forgetting of course to highlight the city itself.

Research is very active in this field. We will present here the results obtained by a European project, which clearly shows the advances that may be achieved in terms of energy savings. Thus, the final objective of the European project NumeLiTe[23] was to conceive and to produce an innovative urban lighting system combining energy savings and quality of life, and to prove its potential in the city of Albi (Tarn, France). The project, particularly consisted of improving the light source efficiency (which is measured in lumens per consumed electric watt), and improving the color

---

22. In France, public lighting represents 46% of a village electricity bill.
23. NumeLiTe™: Numerical Light Technology (NNE-01–0282): project achieved in the "Energy program" framework of the European Union.

rendering, which means having a light which reproduces the colors of the surrounding world well.

The NumeLiTe™ system is based on arc lamps conceived in the framework of the project. It consists of metallic halogen lamps with a ceramic envelope (150 W) generating, with an excellent efficiency (98 lm/W), a white light of very good quality (IRC 97). Two versions exist of different color temperatures (3,000 K and 4,200 K). These lamps contain a mixture of different metals in their iodide form. In operation, the lamp generates a complex spectrum spread over the entire visible domain. The result is a white light of a very good quality. On the other hand, the ceramic envelope, combined with the overall design of the lamp (geometry, electrodes, etc.), guarantees excellent color stability and a long lifetime (estimated at more than 12,000 h). They work vertically as well as horizontally. The steady-state rising time is shorter than in a sodium lamp of equivalent power. The nominal voltage remains stable (+/-5%) during almost all their lifetime, whereas for most of the high-pressure sodium lamps this same voltage increases with the age of the lamp. These lamps are enclosed in luminaries, the reflectors of which are specially designed to better distribute the light on the road and to avoid the problems of light pollution. New electronic ballasts that allow the light gradation supply the lamps. Today, these ballasts allow the gradation in seven levels, but a continuous gradation will of course be possible in the future.

The lamps of the NumeLiTe project are specially conceived to support this gradation with no reduction in their performances or lifetime. Each light point can be controlled remotely (carrier current link or hertz connection in radiofrequencies) by a dedicated central station installed in the Albi city hall. This allows quality control of the light at all times. It also allows the scheduling of the system operation. Finally, the radiofrequencies used for programming the system offer many additional possibilities for the city (bus location, or public warnings, remote meter readings, etc.). We are thus moving towards an intelligent system for the user.

On the other hand, when the power is reduced, the spectrum radiated by the metallic iodide lamp is voluntarily shifted towards the smallest wavelengths (blue). This is done to take into account the peculiarities of the human eye, which is sensitive at night to wavelengths shifted to blue, in order to increase the energy gains without notably damaging comfort and uniformity. Figure 7.18 shows our visual system response in two extreme cases: the photopic vision and scotopic vision.

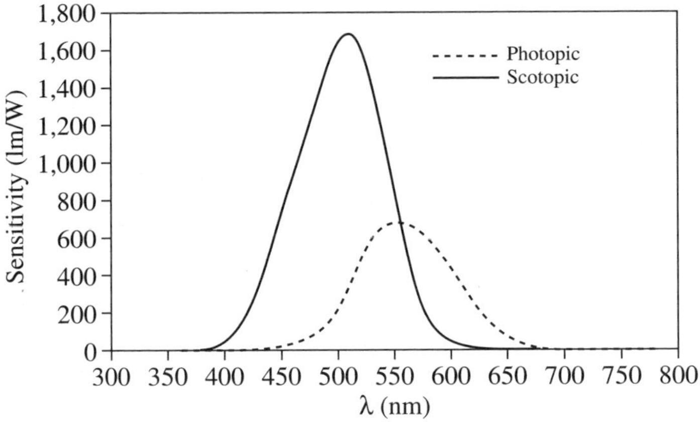

Figure 7.18. *Photopic and scotopic sensitivity of the human eye*

We note that when our visual system adapts itself to a dark environment, in modifying its sensitivity, our eye becomes more sensitive to the blue portion of the spectrum and its sensitivity increases. It happens that in cities, the available amount of light is often quite small and thus, our visual system works in an intermediate zone called the "mesopic" vision zone. Thus, an intelligent lighting system may benefit from this eye operation. Thus, by reducing the electric power delivered to the lamps during off hours, we save energy, but if the lamp spectrum shifts slightly to blue, our eye will continue to perceive a sufficient amount of light. Consequently, this system saves energy without damaging security conditions in the city. The photos in Figure 7.19 taken in the streets of Albi confirm this observation.

Figure 7.19. *Gambetta Avenue in Albi lit by the NumeLiTe system*

On the right of Figure 7.19, the system works at its nominal power, on the left, the electric power is reduced by 50%; the light quantity and quality are more than acceptable.

The first results obtained on the Albi site have demonstrated that at equal energy consumption, the new system generates twice as much light as the previous system. We can thus claim that this system will allow significant energy savings and will then contribute to durable development and land settlement. Our last estimates, based on the first results, show that the NumeLiTe concept may reach energy-saving achievements of about 35–40% with respect to a traditional system based on sodium lamps with ferromagnetic ballast. These energy savings would correspond to an annual reduction of 800 tons of CO_2 per 1,000 luminaries. In extrapolating these results to the European Union, we would save the equivalent of 1.3 million tons of oil in one year!

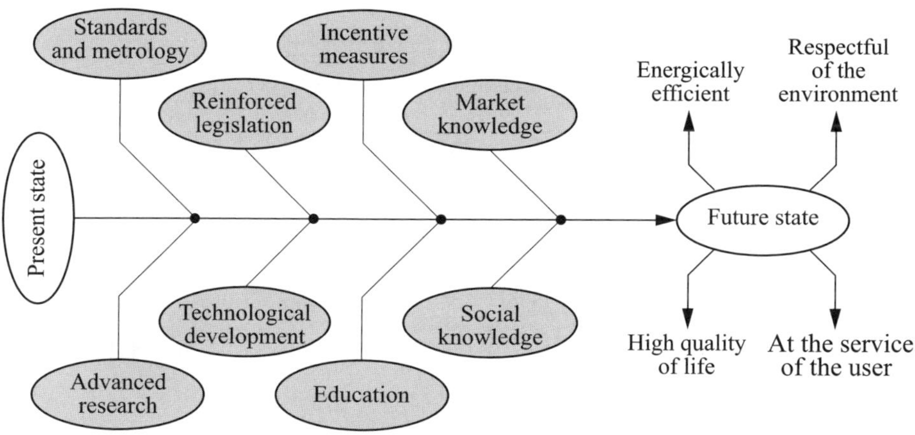

Figure 7.20. *How can we satisfy the lighting needs of the future without compromising energy and environmental resources?*

7.4. What is the future for light sources?

Today, we are certain that the lighting needs of humanity will keep growing; we can expect them to increase by a factor of 2 in the coming two decades. How can we satisfy this increase without the energy demand increasing in the same proportion? This question parties at the very core of sustainable development. Figure 7.20 illustrates in a synoptic way the means we have to find the answer to this question.

Any future lighting system resulting from this process must of course satisfy the light requirements, but it must also respect the energy and environmental resources.

It must contribute to the quality of life of man and finally it must be at the service of the user. The required qualities increasingly concern the sources' chromatic aspect, their lifetime and their stability (especially chromatic). They also concern their adapted operating conditions (instantaneous lighting and re-lighting, dimensions). The answers are probably, on one hand a better knowledge of the chemistry and the fluid motions in the source, on the other hand the control of the electronics (electrodes or coupling) and the steady-state regime. Finally, better modeling of the source-network interaction and of the ageing of lamps would probably reduce the management cost (in particular the maintenance) of urban, as well as building, lighting networks.

To increase the lifetime of the lamp, to improve its color rendering, to miniaturize the source and the system, to eliminate mercury and the other toxic elements without losing efficiency, etc., these are all interesting challenges for the future.

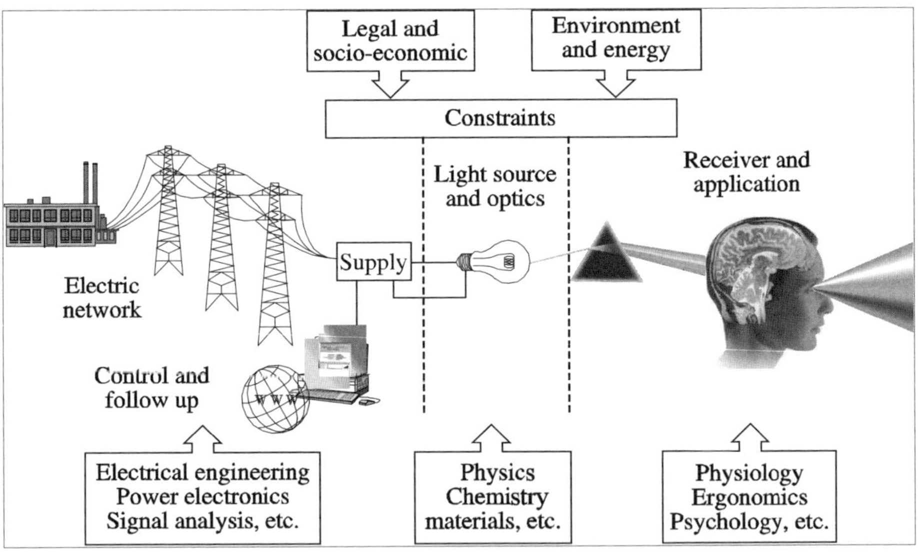

Figure 7.21. *The optimization of a light source cannot be realistic without taking into account the complete system from the power source to the photo receiver. It is a true multi-disciplinal investigation!*

However, as shown in Figure 7.21, if we want to win the gamble of a better and more economic light source, which respects the environment and which contributes to improve the quality and comfort of our life, we have to consider all the aspects of a complex system and this can only be done through a true a real multi-discipline study.

7.5. Bibliography

[ARE 05] ARENSMAN R., "LED market lights up", *Electronic Business*, Reed Business Information, a division of Reed Elsevier Inc, 4 January 2005.

[EUP 03] EUROPEAN PARLIAMANT AND COUNCIL, "On the restriction of the use of certain hazardous substances in electrical and electronic equipment", *Official Journal of the European Union*, Directive 2002/95/EC-27/01/2003, 37/19, 2003.

[HAI 02] HAITZ R., KISH F., TSAO J., NELSON J., "Another semiconductor revolution: this time it's lighting", *Comp. Semiconductor Magazine*, March 2002.

[IAE 05] Data of the International Association for Energy Efficient Lighting, http://www.iaeel.org/, 2005.

[MIL 02] MILLS P., "Why we're here: the $230-billion global lighting energy bill", *Proc. 5th Right Light Conference*, p. 369–385, Nice, May 2002.

[OID 02] OPTOELECTRONICS INDUSTRY DEVELOPMENT ASSOCIATION, "Light emitting diodes (LEDs) for general illumination", *OIDA Road Map Update*, 2002.

[ROU 07] ROUND H.J., "A note on carborundum", *Electrical World*, vol. 49, p. 309, 1907.

[SCO 02] SCHOLAND M., BRODRICK J., PETROW E., "Lighting energy consumption trends and conservation opportunities in US buildings", *Presentation to the 1st COST-529 meeting*, Toulouse, June 2002.

[ZIS 04] ZISSIS G., "Diodes Electroluminescentes pour l'éclairage", *Techniques de l'Ingénieur*, Cahiers Innovation IN-18, p. 1–13, May 2004.

Chapter 8

Distributed Generation: Impact and Solutions

8.1. Introduction: a threat or an opportunity?

For more than 20 years, the electricity sector of many industrialized countries has been subjected to many changes. Energy market deregulations, followed by many privatizations and vertical disintegration have brought a complete restructuring of the electricity sector leading to new forms of organization. The traditional schemes, based on monopoly, are threatened.

The opening of the energy markets and the technological development of the low- and medium-power generation devices, clearly emphasize this evolution. The central architecture of the current energy systems results from a historical evolution which reached its height during the last third of the 20th century. During the first half of the 20th century, industrialized countries organized their electricity sector according to a monopolistic model, which included generation, transport and distribution. In fact, for decades, the technical-economic characteristics of the electricity sector have been in favor of the creation of powerful vertically integrated monopolies. The main reason comes from the elevated costs of construction and maintenance of generation, transport and distribution devices. These costs indirectly implied the idea of network monopoly. Another reason comes from the "legal" specificities of the public service electricity supplier such as:

– supply commitment;

Chapter written by Raphaël CAIRE and Bertrand RAISON.

– network evolution according to consumer needs;

– consumer treatment equality;

– service continuity.

Network evolution was based on technical and economic principles, system security, and quality of the supplied energy [BOL 00]. This centralized structure enabled continuous monitoring and control of the network. The generation, in this case, was instantly and optimally adjusted to the consumption and losses. Reallocations took into account the injection of foreign energy across borders. This model remained dominant until the end of the 1980s. In Europe, before deregulation, the electricity system was well integrated on technical and functional levels [GLO 03]. This integration enabled useful energy exchanges, and was on a basis that was more collaborative than competitive. During the years 1980–1990, some countries questioned the traditional organizational form of the electricity sector. Many factors led to this questioning, including:

– the slowdown in growth of the demand for electric energy after the industry and services maturation in industrialized countries;

– the increasing importance given to environment protection;

– the expectancies based on the decentralized generation and the use of renewable energy sources and co-generations plants;

– the desire of consumers to choose their suppliers.

The evolution of the electricity market, leading to deregulation, led to different organizational forms in different countries. Some forms stand apart from the classic "monopolistic industrial model" of generation, transport, and distribution separation; these are the models that introduced competition between suppliers, and created pools and spot markets. The following sections of this chapter come from the thoughts of a research group from industry, Electricité de France (EDF) and Schneider Electric, and one from universities, namely the Grenoble Institute of Technology.

8.2. Deregulation

In a deregulation context, described at the European level by the guideline 96/92/CEE, a heavy influx of distributed suppliers (or distributed energy generation (DEG)) is expected at the medium-voltage level (MV, 20 kV mainly in France) and at the low-voltage level (LV 400/230 V, mainly in France).

The advantages of distributed generation (DG) concern both the electricity suppliers and the consumers. For the former:

– A generation plant may be installed near the consumer, hence leading to a cost reduction in transport and distribution, and sometimes to less loss of electricity.

– Sites for small generators may be more easily found.

– Distributed generation devices have smaller installation delays than centralized plants.

– The new technologies are often cleaner and quieter.

– Co-generation, one of the most widespread forms of DG, improves the energy efficiency of the installation.

But this massive arrival will also greatly modify the energy flows. In fact, a power flow inversion may appear on the distribution network, which could lead to questions about the architecture of the network in relation to a radial and descending distribution of energy from transmission to distribution grid.

8.3. New generation equipment

The next section presents a few technologies for the distributed generation of electric energy, which are currently used or are still at a research stage.

8.3.1. *Introduction to new generation equipment*

From an efficiency energy point of view, the different regulations promote operation in a co-generation mode. In fact, the idea is to recover all the wasted energies and make them worth something. In Figure 8.1, the energy needed to generate a given amount of thermal and electrical power is compared between a co-generation unit and two dedicated and separated units (thermal and electrical); see Chapter 3.

Thus, to generate the same level of power, both electrical and thermal, the gas consumption is 1.6 times higher on this specific example. If the economic payoff of tri-generation is examined, one fact is accepted: the viability of this technique. This technique operates in two distinct modes: one of them uses the wasted heat, and the other uses the electricity directly generated by the turbine or the co-generation engine. Tri-generation, although reliable and currently used, remains marginal in many countries. From the two refrigerant systems available today, the use of absorption groups only generate cold water from the extra heat of a co-generation.

The capital cost of an absorption group I, however, higher than simple tri-generation using a compression engine.

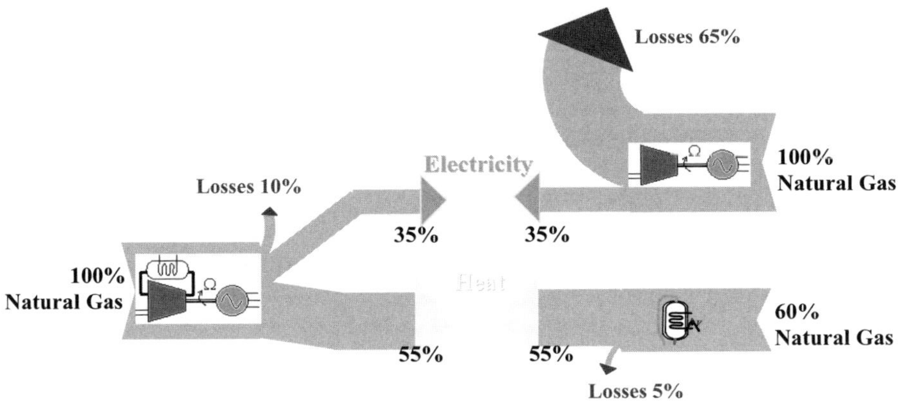

Figure 8.1. *Principle of a co-generation plant*

The use of compression refrigerant groups, which in fact are simple thermodynamic machines, is well known. This technique enables the cooling of a cold source. It is based on the compression of a refrigerant fluid, the choice of which is dependant upon the desired cold temperature level. The compressor used is supplied, in the case of tri-generation, either by the electricity directly generated by the co-generation or by the available mechanical energy. Such a system has a double advantage: an energy efficiency of about 87% and a reduction of CO_2 and SO_2 by 50% and 100% respectively in a standard power plant.

The tri-generation absorption machine does not use the electricity directly generated by the turbine or the co-generation engine. It is a real synergy and not, as in the case of compression engines, of a refrigerant generation using the electricity produced by co-generation. The absorption group tri-generation uses the waste heat from the turbine or from the engine of a co-generation. No electric supply system is required to do that. The absorption, which may be direct (using hot smokes coming from the generation units) or indirect (using hot water or steam), uses the affinity principle of steam with an aqueous solution of lithium bromide or ammoniac. The solution of lithium bromide-water is used for air-conditioning applications. It cannot generate negative cold, whereas the solution of ammoniac-water especially adapted to industrial cold applications can. In the case of a lithium bromide-water solution, the co-generation heat warms up the generator where there is a concentrated solution. Because of the heat effect, steam is pushed into a cooling tower where condensation and heat extraction is produced. Water then reacts with the aqueous solution of concentrated lithium bromide (62%). The endothermic reaction thus cools – at the level of an exchanger – the water to be used for air-conditioning.

In this chapter, we focus on the following distributed electricity generation technologies:

- small hydraulics;
- aerogenerator or wind generator;
- photovoltaic cells;
- gas or combustion turbines;
- combustion engines;
- Stirling engines;
- fuel cells.

These different technologies use many energy sources, such as wind, sun, water, standard fuels, biomass or waste.

8.3.2. *Small hydraulics*

The small electro-hydraulic plants, consisting of a generator and a turbine group, transform the hydraulic energy into electric energy (by using the water-driving force) [LHO 04]. Water, from a well or from a stream, is captured through a basic hydrant and is then oriented to a penstock and/or a channel toward a turbine located below. The water flow will activate the turbine which drives an electric generator. Finally, water goes back onto its course.

The turbine (low/medium/high waterfall) is coupled to a (synchronous) generator which generates alternative current. This current may then be injected in to the distribution network. Most often, the small plants are classified into three categories depending on their power:

- micro plants, < 100 kW;
- mini plants, from 100 to 1,000 kW;
- small plants, from 1 to 50 MW.

Figure 8.2 presents a synoptic of what may be found in each small hydroelectric plant. Many elements may be added, such as an upstream sand-removal system, a load chamber (upstream reservoir), an overflow release for the load chamber, flywheels, hydropneumatic dampers for water hammers, automatic discharge valves, etc.

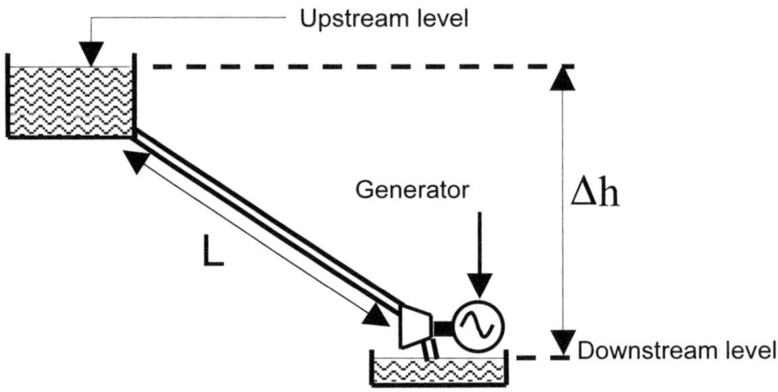

Figure 8.2. *Principle of a water duct power generation station*

In Figure 8.2, the principle of a small hydroelectric plant is shown. There are many types of civil engineering works: river power station without a reservoir (where the power is variable according to the available river flow) or with reservoir (which guarantees a larger capacity all year round).

Many types of turbine may be used, such as reaction turbines (Francis turbines) which are constant-pitch propellers, or turbo propellers (Kaplan turbines), for the low- or medium-height waterfalls. These immersed turbines work with the help of water pressure and kinetic energy. Likewise, pulse turbines, or Pelton turbines, are used for higher waterfalls. They use the kinetic energy of a high-speed water jet [SAB 09].

The key parameters for the design of water duct plants are the following:

– altitude difference (Δh in m);

– duct length (L in m);

– water volume used annually (V in m^3/year) gives a water quantity per year which is transformed into a water flow Q in (m^3/s).

In order to size a small hydraulic installation, the use of a spillway (using a triangular spillway, and for small flows the Thompson method is effective) may be necessary to measure the flow many times during the year. This enables the estimation of the mean flow in m^3/s. According to the river flows, which are classified by the local authorities, thus defining the legal "residual flow" all throughout the year, it is then possible to give the flow which may be used during the year. The maximum available electric power, in watts, is:

$$P_{elec} = \eta \cdot \rho \cdot Q \cdot g \cdot \Delta h \qquad [8.1]$$

with:

– ρ, the water mass density (ρ # 1 000 kg/m^3);

– Q, the water flow in m^3/s, calculated from the yearly usable water (taking into account the annual correlations or variations from year to year);

– g, the gravity acceleration, 9,81 m/s^2;

– Δh, the height difference of the installation in m;

– η is a global efficiency term of the hydroelectric installation. This efficiency takes into account (multiplication of the different terms):

– the efficiency of the duct (generally equal to 0.95) which gives the net waterfall height at full load;

– the efficiency of the turbine (typically: 0.9);

– the possible multiplier efficiency, if the speed must be adapted;

– the efficiency of the electric generator (typically: 0.93);

– the efficiency of the different electric elements before the delivery nodes (transformers and lines, etc.).

More exact calculations may include the real energy losses in the duct.

8.3.3. *Wind generators*

Blades rotate under the strength of the wind and enable the generation of electric energy when there is enough wind potential (speed and wind regularity all year round). A mechanical transmission transfers the power from the rotor and blades to the generator [LHO 04]. The electric generator then transforms the mechanical energy into electrical energy. A nacelle contains all the elements. It is located on the top of a mast and, by virtue of an orientation system, it orientates itself to face the wind. Some wind generators may also control the blade pitch, which enables us to obtain an electric power slightly decoupled from the wind speed which is often variable [SAB 09].

A wind-power plant may use a synchronous or an induction generator, each of which is connected to the grid in a different way through power-electronics-based devices. Up to 10 years ago, the generators used in a wind plant were of the asynchronous type as their cost was lower than the other types of generator. Today, there is a tendency toward plants with doubly-fed induction generators (DFIG) along with synchronous machines which cost at least 50% more than induction machines because they have additional power electronics. This extra cost is justified by the larger control latitude (wind speed versatility among other things). It is not possible to estimate the real lifetime of a wind generator because practical knowledge is not enough.

The output power increases with the swept area. The peak power of the wind generator, its type (constant or variable speed [LAV 05]), and the quantity of energy the wind plant is able to collect in one year is determined by the surface of the disk swept by the rotor and the wind speed.

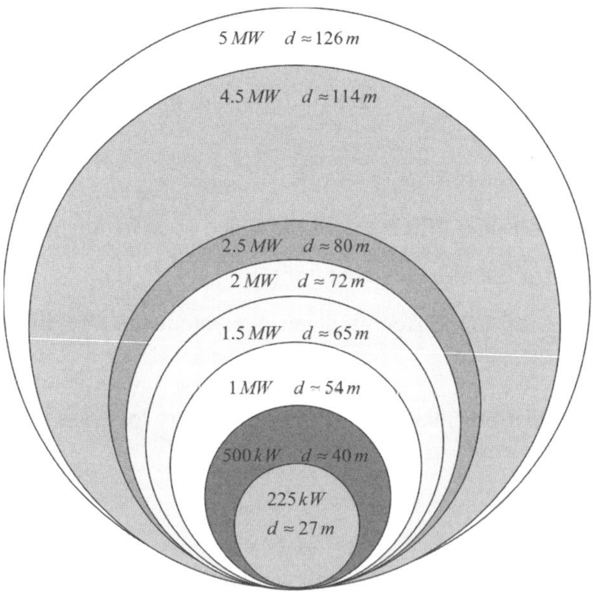

Figure 8.3. *Mean blade diameter vs. electric power*

During the preliminary design of a wind-plant construction project, a mast is often installed for a year or more to estimate the wind potential of the site. A wind plant starts to generate electric energy with a typical wind speed of 3 to 5 m/s; this is usual until a speed of 15 m/s is reached, which is its nominal power. The *onshore* (installed on the land) plants do not work, for safety reasons, with wind speeds of over 25 m/s. Some *offshore* (or at open sea) plants can operate with winds up to 34 m/s [SAB 09].

8.3.4. *Photovoltaic modules*

Sunlight may be directly transformed [TIW 03] into electricity without noise and pollution by photovoltaic panels. The photovoltaic module installed in the form of solar panels converts light into electricity. The energy thus obtained can be stored in batteries in order to be permanently available. A regulator protects the battery against overcharges and discharges.

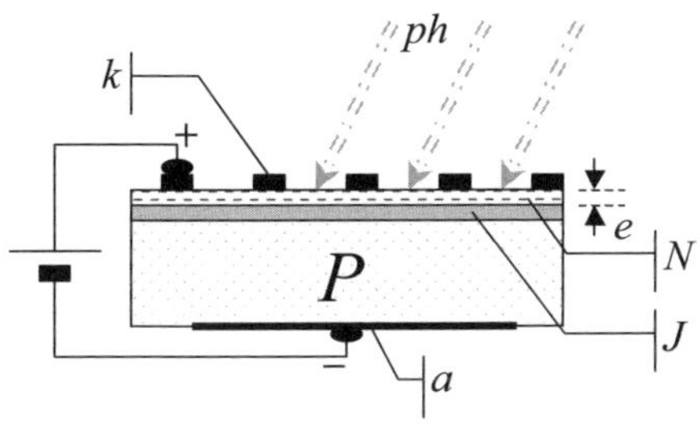

Figure 8.4. *Principle of a photovoltaic cell*

The photovoltaic cell is the basic element of solar panels which generate electricity. It is made of a semiconductor material which absorbs light energy and directly transforms it into electricity. The operation principle of this cell calls for radiation and semiconductor properties.

A P-doped semiconductor is covered with a very thin N-doped layer which thickness e is equal to a few thousandths of mm. Between both layers there is a J junction. The N layer is covered with a metallic grid which serves as a cathode “k” whereas a metallic plate “a” covers the other face of the semiconductor and plays the role of the anode. The total thickness of the crystal is in the order of the mm. A light ray (“ph”) which hits the device can penetrate the crystal through the grid and provoke the appearance of a differential voltage between cathode and anode. Generally, the basic semiconductor is mono-crystalline as is the one used in transistor manufacturing, but now we find more frequently polycrystalline silicon which is less expensive to manufacture.

The individual cell – the basic module of a photovoltaic system – generates a very small electric power, typically from 1 to 3 W with a voltage of less than 1 V. To generate more power, the cells are assembled to make a panel (or a module). The series connection of many cells increases the voltage for the same current, whereas the parallel connection increases the current for the same voltage. Most of the commercialized modules contain a series of 36 crystalline silicon cells connected for 12 volts applications. The output current, and thus the power, will be proportional to the module (or to the parallel connected modules) area. The interconnection of modules, in series or in parallel, in order to obtain an even greater power, defines the notion of the photovoltaic field. Large installations require about 40,000 m^2 per MW_{peak}. The panels must be oriented to face the sun with an optimal angle which depends on the installation latitude.

A photovoltaic generator, the useful area of which is 1 m^2, normally oriented to the sun rays and receiving a light power of about 1,000 W, will deliver only 60 to 200 "electric" W depending upon the technology used for its manufacturing process. The efficiency of a cell made from a mono-crystal is higher than 15%. If it is made with polycrystalline silicone, its efficiency will be in the order of 10% to 15%. With amorphous silicone, its efficiency will lie between 5% and 10%. The manufacturing costs are also regressive. A photovoltaic unit must operate for three (polycrystalline) to four years (amorphous) before starting to deliver an energy quantity higher than the one used in its manufacture. The energy used for the manufacturing process (the purification and the crystallization of silicone are the most energy-consuming parts of the process, after which it must be sliced and then assembled into modules) and the installation of photovoltaic panels using mono- or polycrystalline cells connected to the grid is about 600 kWh/m^2, and in the case of amorphous photovoltaic units, it is about 420 kWh/m^2.

The dissemination of photovoltaic inverters on the distribution network may however raise connection problems. In fact, the massive insertion of inverters, and in the case where the networks could become exporters, or at least at zero generation, the protection strategy may be invalidated. This could endanger the security of goods and people in the case where the generator is disconnected from the grid incorrectly.

8.3.5. *Gas turbines*

Gas turbines are also known as combustion turbines. They directly use the heat from the gas combustion, rather than steam, to drive a generator, thereby generating electricity. Regarding the turbines, the power ranges from 100 kW up to a few hundred MW; it is a mature technology.

They are used in centralized generation (about 250 MW) and in distributed generation (500 kW to 25 MW).

The mini and micro gas turbines are very small combustion turbines. They come from aeronautic applications (also called aero-derivative applications) such as helicopter turbine. Their power ranges from 30 kW to a few hundred kW. Their electric efficiency ranges from 20% to 30% according to their load.

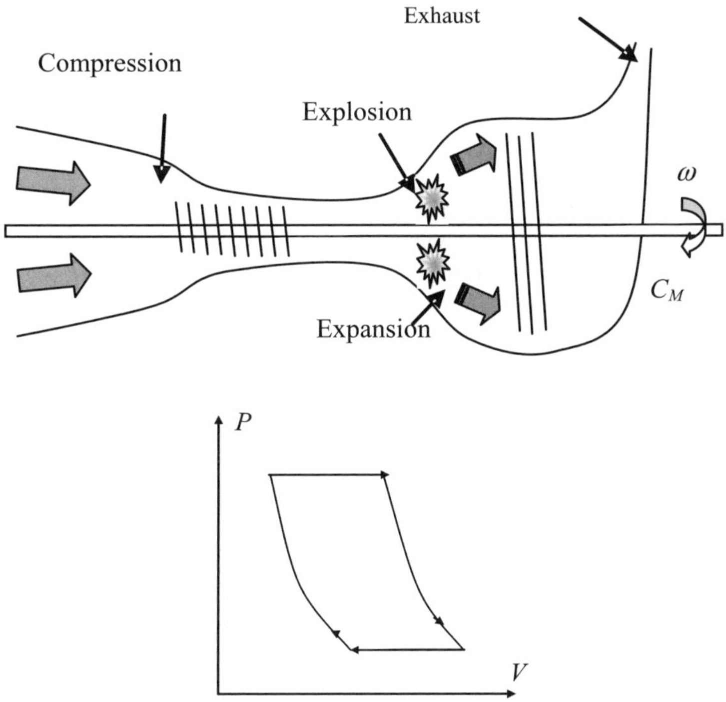

Figure 8.5. *Principle and thermodynamic cycle of a gas turbine (two adiabatic phases and two isobar phases)*

Although combustion turbines use sophisticated technologies, their operation principle is quite simple: inlet air is compressed in a compressor, mixed with fuel (introduced in gaseous or liquid form), and then sent, under pressure, to a combustion chamber where the combustion takes place. The hot gases from the combustion then expand in a turbine. The hot gas expansion generates a mechanical power on the turbine shaft, which drives the useful load (the alternator if electricity must be generated) and the compressor (which needs a good half of the total power to compress the input air). These power stations may operate with or without heat regeneration (co-generation). The applications without regeneration "heat" the ambient air in the same was as a helicopter turbine does.

Most of the micro turbines – at least those designed for high efficiency generators – take the air at the output of a scavenger (a system which uses the exhaust gases of the turbine to heat the input air), thereby increasing the internal temperature of the turbine. The scavenger is a type of radiator or heat exchanger which transfers the exhaust heat into the input air. In some micro turbines, the scavenger is a rotating porcelain disc with holes which allow the air to flow through. In other units, a heat exchanger is used. In heat exchangers, the exhaust gases flow into long metallic tubes, around which the input air circulates.

8.3.6. *Combustion engines*

A combustion engine turns the energy of a conventional fuel (diesel, natural gas) into mechanical energy. This mechanical energy is then used to make a shaft in the engine rotate. An electric generator is coupled with the engine to transform the rotating motion into electrical power. The cycle which characterizes the internal combustion engines, including the diesel engine, is the Beau de Rochas cycle which is characterized by four steps: admission, compression, explosion and detent (during these steps, fuel is injected in the combustion chamber where it lights by itself, that is, without a spark), and exhaust. The abscissa axis of Figure 8.6 represents the volume V of the combustion chamber; the ordinate axis represents the pressure P inside this chamber.

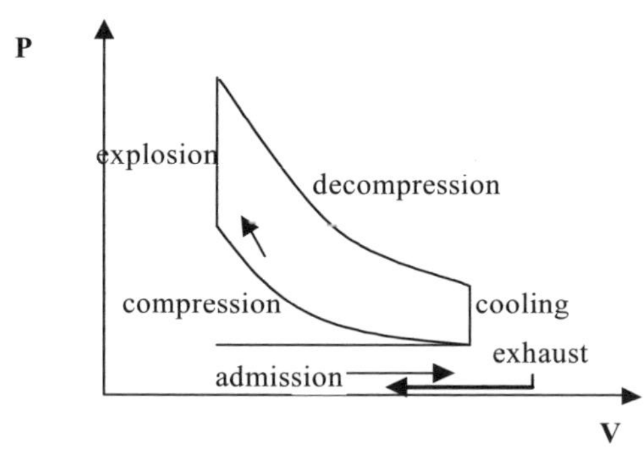

Figure 8.6. *Beau de Rochas cycle for a diesel engine*

The availability of combustion engines reaches 90%. Their lifespan is around 20 to 30 years. Their electric efficiency is about 35% and their thermal efficiency 42%.

Compared to combustion turbines, gas engines present greater operating flexibility: their efficiency is a little bit reduced on a large load range, typically between 40% and 100% of the load, and they are not sensitive to weather variations (temperature, pressure).

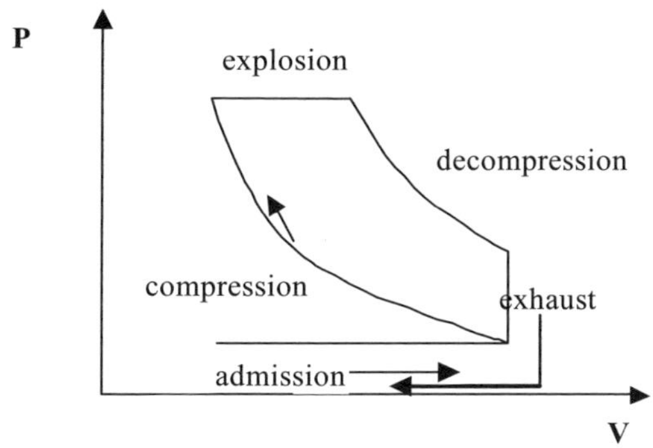

Figure 8.7. *Beau de Rochas cycle of a gasoline engine*

8.3.7. *Stirling engines*

A Stirling engine is be a small sized (from 1 to 20 kWelec) external combustion engine. It is manufactured in small quantities and for specific applications. The typical electric efficiencies for these small units range from 1% to 30% (Pelec less than 20 kWelec). The advantage of an external combustion is the easy regeneration of thermal power and the thermodynamic liquid flow in a closed circuit [SAB 09].

The thermodynamic cycle of a Stirling engine is quite simple; it has four phases. During each of these phases the used gas is subjected to the following transformations in a closed volume. First, constant volume heating (the heat source delivers energy). The pressure and the temperature of the gas increase during this phase. Then the gas is subjected to an isothermal detent (at constant temperature), and the volume increases whereas the pressure decreases. The driving power is generated during this phase. Finally, it experiences an isothermal compression; the gas pressure gradually increases as its volume decreases. A small amount of mechanical energy is delivered to the gas during this period.

The achievement of an engine such as the one described above raises difficulties: heating; cooling the gas – this is done by two inserted mechanical devices, a regenerator (or metallic matrix type thermal buffer) and a mover. This last device modifies neither the volume nor the pressure of the gas, but it pushes the gas toward either the heat source or the cold source. The regenerator is supposed to pick up calories from the gas in transit between the heat zone and the cold zone and inversely:

– during the constant volume heating phase: the volume remains constant, and the mover lets the gas flow from the cold source to the hot source, and it recovers calories at the regenerator level;

– during the isothermal detent: the mover follows the engine piston during the detent for the gas to stay in contact with the heat source;

– during the constant volume cooling: the volume remains constant, and the mover lets the gas flow from the hot source to the cold source; there is an exchange of calories between the gas and the regenerator;

– during the isothermal compression: the mover keeps its position for the gas to remain in contact with the cold source.

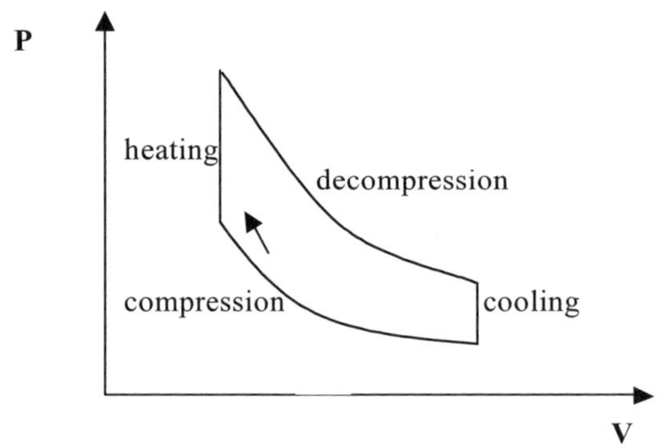

Figure 8.8. *Thermodynamic cycle of the Stirling engine*

The Stirling engine may also be used in co-generation for powers ranging from a few hundred Watts to 100 kW and investigations are being carried out to increase this power range between 1 and 1.5 MW. In co-generation, its global efficiency lies between 60% and 85%, 40% of which is electric efficiency. The advantage of external combustion engines is the possibility of using a standard external burner. The major drawback is the reliability of a waterproof system (where pressure is often very high) with mechanical parts moving inside.

8.3.8. *Fuel cells*

A fuel cell enables the direct conversion of chemical energy into electric energy [LAR 00].

Figure 8.9. *Heart of a fuel cell*

This technique is based on the inverse principle of electrolysis [JOH 93]. A fuel cell is made of an elementary cell which contains two electrodes separated by an electrolyte (see Chapter 5). Before obtaining reactions in domestic fuel cells (supplied with common gas or oil products), chemical reactions are necessary to "transform" natural gas or oil byproducts into hydrogen:

$$CH_4 + H_2O \rightarrow 3H_2 + CO \qquad [8.2]$$

Then the *shift* reaction, which must be thermally controlled, enables the elimination of the carbon monoxide:

$$CO + H_2O \rightarrow H_2 + CO_2 \qquad [8.3]$$

The reactions at the heart of the cell are shown in the following.

At the anode:

$$H_2 \rightarrow 2H^+ + 2e^- \qquad [8.4]$$

At the cathode, the hydroxides (issued from the water oxidation by the oxygen) react with the hydrogen:

$$\frac{1}{2}O_2 + 2H^+ + 2e^- \rightarrow H_2O \qquad [8.5]$$

There are many kinds of fuel cells which differ according to the electrolyte used and which are more or less interesting for co-generation because of their operating temperature:

– AFC (alkaline fuel cell). The operating temperature of an AFC is close to 90°C under a pressure of 4 bars. This kind of cell is present in mass applications in the stationary field (domestic and industrial applications, tertiary sector, assistance or security) and the automotive sector.

– PEMFC (polymer exchange membrane fuel cell). Its operating temperature lies between 60 and 90°C. For a temperature higher than 90°C the membranes do not retain water and cannot guarantee a convenient proton migration. Investigations are being carried out to increase these operating temperatures to around 160°C to 180°C. It may be used in co-generation.

– DMFC (direct methanol fuel cell). its operating temperature is around 60°C. Its applications are limited in the stationary field since it is in direct competition with the PEMFC.

– PAFC (phosphoric acid fuel cell). Its advantage is in the use of generated heat. It has an operating temperature around 85°C.

– MCFC (molten carbonate fuel cell). Its operating temperature is between 600°C and 660°C. It can be used in medium- and high-power co-generation.

– SOFC (solid oxide fuel cell). Its operating temperature is high, from 800°C to 1,000°C.

The fuel cells may be used for distributed generation, but this technology still requires much investigations; some cells are at the demonstration level, for example the MCFC and the SOFC, and they seem very promising for co-generation units. The small co-generation units (1 to 5 kVA) using fuel cells have typical electric efficiencies in the order of 25% to 30%. The most significant units have efficiencies ranging from 27% to 40%, depending on the load.

8.4. Impact of distributed generation on electric networks

In order to better understand the context of this study, along with its necessity and its difficulty, the following section presents the major characteristics of distribution networks. Some concepts linked to the operation of distribution networks will also be defined in this section. The changes that have been noticed in recent years, at the distribution network level, will also be briefly presented in the second part of this chapter, section 1.5.

8.4.1. *Distribution networks*

8.4.1.1. *General structure*

The whole infrastructure which enables the flow of the electric energy from the generation sources to the final consumers (industrial or domestic) is an electric power network (or grid). The general function of this network is to transport the energy from the producer site and to distribute the electricity locally to the consumer. The real time balance between generation and consumption (plus the losses), is ensured by the use of many automatisms, a fine observation of some electrical data and an excellent forecast of the load, D. Between the generation centers and the final consumers, there are many voltage levels. Figure 8.10 gives the names of the different voltages in the French electric networks.

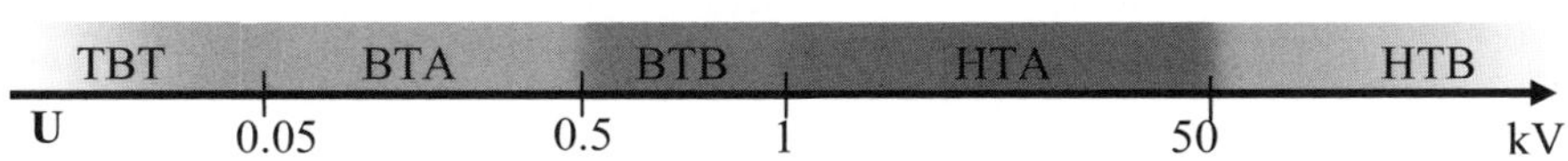

Figure 8.10. *Voltage levels in the French electric networks*

The distribution networks (the most commonly used voltages of which, in France, are 20 kV and 400 V phase-to-phase) are the supply networks for the small, residential consumers, along with the small industrial areas, the services, and agriculture [DOU 00]. They are the property of the local communities which concede the exploitation to a distributor. In France, these infrastructures are mainly exploited by EDF (Electricité de France) and Gaz de France. EDF Gaz de France Distribution (EGD) guarantees the maintenance. The control of the electricity network (all the energy-flow management decisions) is made by ERDF (EDF Distribution Network).

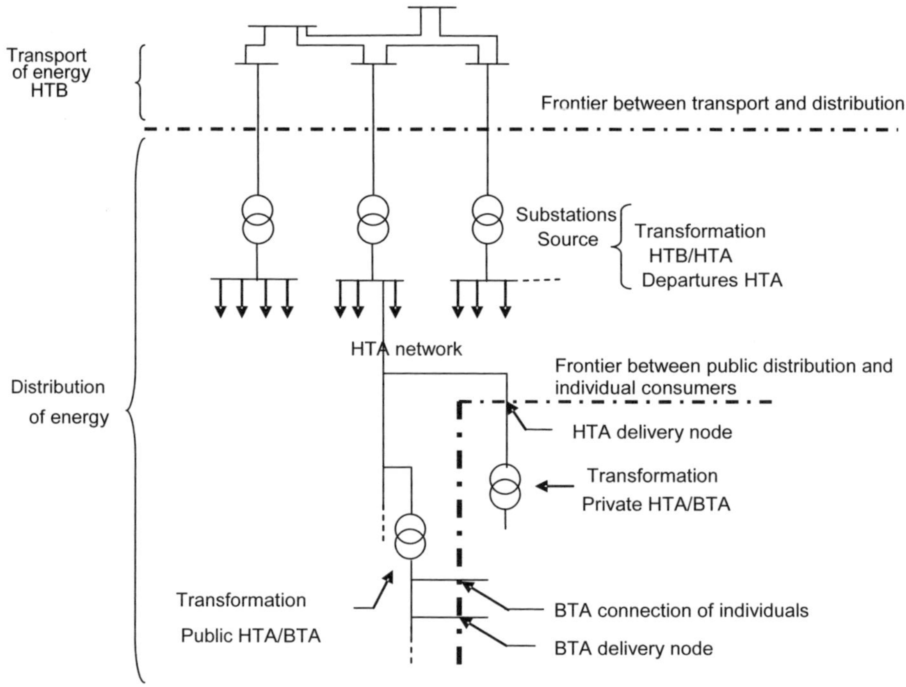

Figure 8.11. *General structure of the electricity network from the high voltage level*

As shown in Figure 8.11, the substations, high voltage B level/high voltage A level, also called "source substations" (in France often 220, 90 or 63 kV toward 20 kV) supply these distribution networks. A transformation substation is a key element of the electricity network. From a source substation, the energy is distributed through one or more cables or overhead lines, supplying a given part of the distribution network. These HTA conductors are connected to the bus bars of the source substation.

Most domestic customers are connected to the low voltage lines supplied through HTA/BT transformers. The other customers (industry and the self-employed, among others) that have significant consumption, are directly connected to HTB or HTA networks.

8.4.1.2. *Loads and energy generation at the distribution level*

Electric energy consumption is the fact of all economic life: industry, services, and families. It is present under different aspects: synchronous and induction motors, heating apparatus, etc. Unlike the generation groups, one cannot identify the consumption without an individual power meter. All the consumption behind a network node or a load node represent the "load" characterizing this node. On medium-voltage distribution networks there are two kinds of loads: the load/customers which have their connection at the 20 kV level, and the HTA/BT public distribution substations. If the easiest static model is taken into account, a load may be considered as consuming constant active and reactive power (PL, QL). This definition is historical and is linked to the metering process of the electric network manager by the customer or by the final consumer. Actually, each load could be more finely modeled (for example: a hot water tank could be modeled by a pure resistance).

The worrying ecological situation of our planet, the expected exhaustion of fossil energy sources, the many environment constraints, and the lure of subsides have, in recent years, encouraged research into alternative means of electricity generation. This is why distributed generators (DG) are being developed. These sources are often inserted in the distribution network because they have a lower power with respect to that of a standard centralized power generation station. It is in fact more and more difficult to undertake large-scale projects of centralized electric power generation units. The DG often give value to local and/or renewable energy sources, such as biomass, solar, or wind energy. They are considered as active and reactive (PQ) power injections. The presence of these energy sources has an impact on the operation and the exploitation of the network, which was originally designed for unidirectional energy fluxes (from the source station to the consumers). Much works has been done in order to study these impacts.

8.4.2. *Impact of distributed generation on the electric variables*

During the past decade, investigations into the impact of previously presented distribution networks have been carried out. Because of these investigations, [CAN 00] has been able to identify the impact of distributed generation connected at the HTA level on the HTA network itself. Electricité de France R&D [COR 99, FRA 01, LEM 99] have undertaken studies internally, and in other research groups [CAI 04] in England ([JEN 00, MAS 02] among others) and elsewhere in the world [MIA 01, SMI 01] have also done studies. The results of these studies has led to a modification of the distribution network operation. The following sections list some of the elements of the electric variables that need to be watched carefully. The network operator has put at the disposal of the suppliers all its connection rules and the key points which must be checked at every connection request [EDF].

8.4.2.1. *Load flow*

Every power transit on a conductor will develop, because of its impedance, a voltage drop which can be approximated by the following equation:

$$\frac{\Delta U}{U} = \frac{R.P + X.Q}{U^2} \quad [8.6]$$

The voltage drop ΔU is thus linked to the P and Q powers which flow through line impedance R, X. This equation is valid in the consumer convention. In this equation, P is the algebraic difference of the active powers, consumed and generated, and Q is the algebraic difference of the reactive powers, consumed and generated. Every customer will have the tendency to reduce the voltage, whereas a supplier will tend to increase it. Every overvoltage will generate an accelerated deterioration of the equipment. Overvoltages will, in any case, occur when coupling (dynamic). An under-voltage may cause equipment to malfunction.

The French standard (NF-EN 50–160) set out the model of the voltage shape of the distribution networks. There are all the voltage quality problems along with the associated levels (rms, tolerance, and the different harmonic levels). Therefore, the maximum values for the voltage delivered to the customers is defined. In France, these limits are 230 V – 10% to + 6% on the low voltage side and ± 5% at the HTA. The insertion study of a supplier (connecting node and injected power) must be done to verify that the voltage margins in the entire network are respected.

8.4.2.2. *Voltage quality*

In order to be accepted as a distributed electricity supplier, candidates must fulfill requirements on, for example, harmonics, rapid voltage variations or flicker, unbalances generated by single-phase connections, voltage sags, and overvoltages. The regulation is clear on quality. In order to check the contractual commitments of the delivery devices for quality measurement (counting at least long and short power cuts may be installed) on the request of the network or the supplier manager, introducing an independent supplier on a distribution network can modify the voltage quality. For instance, if the supplier has a power electronic interface, it can add perturbation harmonics on the network. Another problem is if the supplier dynamic is such that when it is connected to the network, flicker phenomena appear, affecting the quality of the energy delivered to the consumer. Likewise, unbalances and voltage sags and overvoltages may appear.

8.4.2.3. *Dynamic and transient stability*

In order to guarantee stability, the coupling must be made at the synchronism. Moreover, other recommendations suggest that, during the coupling and in order to aid stability while accepting discrepancies with respect to the synchronism, transient limits are defined for the voltages (±10%), the frequency (±0.1 Hz), and for the phase (±10 degrees) difference. It is also specified that the voltage variations during induction machine connection must remain below 6%. Finally, the load take-on and take-off speeds must not exceed several MW/minute.

Generally speaking, the stability is characterized by the power variations crossing the network and is measured by the time variations of the associated voltages and frequencies. Two types of stabilities may be considered:

– transient "large signals" (or how the generators react after rapid frequency variations due to a fault or a DG connection);

– dynamic "small signals" (load connection, among other things).

8.4.2.4. *Short circuit power and currents*

Regulations require that the installation connection does not lead the short circuit current to exceed the assigned limits for HTA or HTB equipment. The chosen verification method is the one recommended by IEC 60–909, with the short-circuit-stage greater or equal to 250 ms. If we consider a network without DEG, with a given short-circuit power, this is equivalent to dimensioning parameters of the network equipment. The network is characterized by its sensitivity to any kind of perturbations. Indeed, with an infinite short-circuit power, the network impedance is zero, and the network is therefore considered as an ideal voltage source. This short-circuit-current knowledge enables the calculation of the electrodynamic forces on the conductors. A distributed supplier will then contribute to the fault, which can result in a significant increase of the current in the conductors. Norm IEC 909, applicable to all networks, meshed and radial, up to 230 kV, gives a short-circuit calculation method. It uses Thevenin's theorem to have the equivalent circuit of the sources at the short circuit point, then it uses the superposition theorem to add the contributions of the suppliers.

This modification of the short-circuit currents may cause problems for the protection plan. Indeed, in order to prevent inopportune tripping (circuit breaker opening without a real fault) or blinding (the protection does not see the fault), it must be verified that the distributed suppliers do not modify the short-circuit currents to any great degree. A significant modification of the short-circuit powers may endanger some network elements. Indeed, the distribution network elements are often designed to accept a short-circuit current greater than 8 kA over one second.

Setting of the protection may change. If not, some undesired or even blinded tripping may occur.

8.4.2.5. *Tariff signals*

The distributed suppliers may perturb the economic signals, of the central remote control at musical frequency. This remote control emits low rate carrier current signals at a given frequency (in France often 175 Hz). As the network impedance is being modified because of a synchronous machine connection, this signal may sometimes be amplified again. This amplification may be charged to the supplier.

8.4.2.6. *Over currents due to the connection of coupling transformers*

Investigations on inrush transformer currents are also conducted during the supplier connection demands. Care must be taken to ensure that the transients thus generated do not trigger the network protection or cause transient voltage sags all along the distribution network.

8.5. Solution elements

Many different solutions exist to control some of the impact of the distributed generation on the distribution networks:

– coordination of the distributed suppliers (system service distributor type);

– active control of the distribution networks.

8.5.1. *Distributed suppliers coordination: the virtual plant*

In order to limit the impact of the distributed generation on distribution networks, coordination between suppliers needs to be considered. This coordination would maximize the distributed energy generation rate in the distribution networks, while respecting the legal constraints.

Many investigations have proved that the acceptance capability of the generation follows the evolution capability of the load: the network may accept non-negligible generation quantities. However, this comment needs explanation. Indeed, it must be taken into account:

– "weak" networks, or networks with a low short-circuit power;

– possible change in the design requirements of the distribution network. Indeed, it can be imagined that, following the market opening and the unbundling of the large monopolistic companies, a modification in these requirements in order to obtain cost reductions in the infrastructure and then weaker networks will be required.

An optimal coordination to minimize some of the impact of the generation on the network and thus to indirectly increase the connectable quantity is then conceivable. This optimization would allow the character, both discrete and continuous, of the network control variables to be taken into account. Multi-objective functions with strong technical and economical connotations (reduction of the adjustment costs by reducing the number of the operating adjustment means) must be implemented. Taking into account real-time simulation methods or *in situ* tests [FEN], coordination strategies, media and communication protocols must be validated.

New distribution concepts must be addressed, especially the assembling of diverse generation means which may be associated inside virtual plants to guarantee, for instance, a given energy supply to customers or energy stipends. This coordinated management would enable the maximum rate of connected distributed generation to be increased, while respecting the technical constraints and the security of goods and people.

When proposing a new dispatching and coordination strategy for distributed generations, the virtual plant would offer new perspectives for the evaluation of the generation means. The major principle of this "plant" would be to aggregate the generations, to control them, and even to guarantee an active power supply while offering services to the distribution network. These services must keep the network parameters (voltage, among other things) within acceptable limits. The two major functions of this aggregation would thus be linked to the evaluation of the generation on the energy market and to the services delivered to the network. The *virtual power plant* could then be defined as an energy cooperative. It would be then a set of distributed generation means, but also loads coordinated by one or several supervisory control centers, using a telecommunication and information system. These distributed resources could be connected to one or many nodes of the distribution and/or transport electric network.

The concept of assembling many actors is equivalent to simultaneously considering several investigation areas: economy, multi-infrastructure modeling, computer science, and telecommunication, all of which open new fields of innovation. The supervisory control center must collect information regarding the operating costs of the many electric power plants, the energy needs (electrical or thermal) of the customers or of the energy suppliers, the needs in terms of services to the distribution network and the prices of the different electric markets. This "virtual plant", which is based on a common telecommunication network, becomes, in essence, the favored key contact between these local resources and the other actors of the distribution network [SUR 06].

8.5.1.1. *Reminder of the possible functionalities*

To guarantee the power generation, control the active or reactive power reserves. The development of the virtual power plant could be done by an offer–demand process. A strategy to upgrade uncertain energies, such as renewable energy, consists of aggregating them with more secure means (co-generation with associated or emergency storage). The possibility to have many generation means, the possibly of different technologies (heterogenous plant); enables, through an optimized management, a better guarantee of the whole operation. In order to better manage this association, the operation of the virtual plant operation under electric, and possibly thermal, power guarantee constraints must be considered.

To encourage and to develop renewable generation means and distributed generation. A non-negligible part of the distributed generation is often said to be "fateful" or to be of the intermittent/renewable type. It is thus difficult to predict it *a priori*, despite recent advances in terms of weather forecasting, among other things. Renewable generation means may then be put together with fossil-type generation means. Thus real fuel savings may be made.

To sustain the grid (voltage, overload management of the distribution grid component). The introduction of distributed generation in distribution networks leads to weaknesses which are usually treated by network reinforcement. At the same time, the load increase requires investment: the network has to be designed to face extreme situations. With a control of the "virtual plant" type, it may be possible to avoid network constraints. The generation thus delivers a service (generation guarantee) which enables the distribution network manager to make it work without overloads while limiting network investments.

The coordination concept of control means with the associated telecommunication network must be characterized in a technical and economic way: the demonstration of their technical feasibility must validate two major points:

– the control command aspects of the distributed generators in order to determine the architecture and the coordination methods that are the best adapted to the technical and economic environment and to the technical requirements;

– the communication infrastructure which can be based on medias (cellular phones, telephone lines, local networks, high frequencies, among other things) and existing (IP, GPRS, LAN among others) or yet-to-be-developed protocols.

8.5.2. *Active control of the distribution networks*

8.5.2.1. *What is expected for future networks?*

What follows are some synthesized thoughts on the possible evolution of distribution networks. Four elements characterize what is expected:

– accessibility: the networks must host all the suppliers who wish to be connected;

– economics: the focus will be on an exploitation very close to the margins;

– flexibility: the redundancy of the circuits will be increased with respect to the construction of new components in order to optimize the energy efficiency of the existing circuits;

– reliability: in order to ensure and to increase security and supply quality.

Many areas of investigation are therefore open following the expected modifications of network operations:

– more frequent grid re-configuration in order to be very close to the supplier exploitation margins;

– adaptation and evolution of monitoring tools to this new context (protection, detection, and localization of the faults);

– development of communication channels for information exchange and the intelligence "distribution" in the feeders in order to improve the adaptation ability to context changes.

Three evolutionary steps may be considered:

– Over five years: improvement of the exploitation performance through more systematic information reporting on the network state and the use of remote breaker apparatus in order to perform more frequent reconfigurations. This is consistent with load control practice with a commercial evaluation and a network evaluation.

– Over 10 years: it is expected that the exploitation and control teams will become closer. It is also expected that information from intelligent customer meters will also be taken into account.

– Over 15 years and longer: it is assumed that transmission and distribution operators will become closer, following the new exploitation configuration (meshed, interconnected, neutral regime change), and a greater number of distributed suppliers.

8.5.2.2. *Toward a meshed and flexible distribution network*

Let us ask a serious question of the exploitation, of its control and even of the distribution network structure. Indeed, the distribution network structure may change considerably, first into a partly meshed network, and then into an interconnected one. A middle step between both exploitation modes will consist of introducing greater flexibility in the distribution networks in order to optimize the power flows, for example, and to control the network taking into account the constraints of its exploitation (voltage plane, admissible power flows in the conductors, losses to be minimized, influence of the energy spot market).

At the same time, the structure modification will also significantly change the network defense plan and will require a complete modification of the exploitation practices as far as the network structure choice and the use of new protection functions are concerned. These tools must be able to adapt themselves to the flow inversion of the active and reactive power specific to the exploitation configuration changes.

8.6. Conclusion: a challenge and a development opportunity for the electricity sector

Until now, in France the electric energy transmission and distribution tasks were considered as a "public service" to transport the energies generated in large power plants through a transport network, near the consumption centers and, finally, to distribute these energies by means of local distribution networks. The organization and the hierarchy of the distribution were clear and the energy flows were foreseeable. With a strong distributed energy injection, generated by many distributed generation means, the characteristics and the properties of the distribution networks may change in a significant manner.

The result may be a bidirectional energy flow. A huge injection of electric energy on the distribution network will also, in the long run, reduce the number of large, national power generation plants. These plants currently guarantee the efficient operation of the electricity system because of the existing energy reserves (under their inertial or other forms). The loss of these plants, and consequently of the associated reserves, may render the system more fragile. It is obvious that small suppliers will have to participate in the network adjustments (voltage and frequency) and thus in the power flows. This is called the "ancillary services".

New distribution concepts must now be addressed, especially:

– The aggregation of diverse generation which may be associated with virtual plants to guarantee a given energy supply to customers or network services. This coordinated management would, in the long run, enable the increase of the maximum rate of the implemented distributed generation while respecting the technical constraints and the security of people and goods.

– The active management of the distribution network devices, which would allow a smart management of the energy flows due to the local apparatus such as the breakers and other possible protections, distributed in the network.

Many research directions are needed on the future networks in order to overcome the scientific obstacles to a massive injection of distributed energy generation. Therefore, a safe management must be created for the future networks. This safety implies watching for the network stability and then initiating the conditions of a strong development of the distributed generation from renewable energies allowing a citizen behavior for sustainable development.

The long-term research must be in a multidisciplinary context (economy, sociology, industrial strategies) which addresses the transformation and the storage of energy (chemistry, physics, mechanics, thermodynamics, electrotechnology), the management, and the safe control of the energy flows on the networks (electrotechnology, new communication and information technologies, electronics, automatic control), and the development of these technologies and their integration into the electric system.

8.7. Bibliography

[BOL 00] MATH, H.J., *Understanding Power Quality Problems*, IEEE Press, 2000.

[CAI 04] CAIRE, R., Impact de production décentralisée sur les réseaux de distribution, PhD Thesis, Institut National Polytechnique de Grenoble, 2 April 2004.

[CAN 00] CANARD, J.-F., Impact de la génération d'énergie dispersée dans les réseaux de distribution, PhD Thesis, Institut National Polytechnique de Grenoble, 2000.

[CAR 03] CARRIVE, P., "Réseau de distribution structure et planification", *Techniques de l'Ingénieur*, traité Génie Electrique, D4210, 2003.

[COR 99] CORTINAS, P.J., "Assessing the impact of dispersed generation on medium voltage networks: analysis methods", *PowerTech Budapest 99, International Conference on Electric Power Engineering*, 29 August–2 September 1999.

[DOU 00] DOULET, A., "Réseaux de distribution-exploitation", *Techniques de l'Ingénieur*, traité Génie Electrique, D4230, 2000.

[EDF 07] Le référentiel technique de la distribution, Electricité de France, Réseau de Distribution, 2007. Available at: http://www.edfdistribution.fr/130350i/le-distributeur-EDF/publications/referentiel-technique.html.

[FEN 07] FENIX, Projet Européen Flexible Electricity Networks to Integrate the Expected "Energy Evolution", 2007. Available at: http://www.fenix-project.org/.

[FRA 01] FRAISSE, J.L., BOULANGER, F., JUSTON, P., LEMERLE, P. and JEANNIN, O., "Praogen, a tool for studying connection of generating plant onto the medium-voltage network", *16th International Conference and Exhibition on Electricity Distribution*, Part 1: Contributions, CIRED, vol. 4, 18–21 June 2001.

[GAI 03] GAIN, E., "Réseaux de distribution – Conception et dimensionnement", *Techniques de l'Ingénieur*, traité Génie Electrique, D4220, 2003.

[GLO 03] Global Chance "Petit Mémento Energétique", Les Cahiers de Global Chance. Available at: http://ww.agora21.org/global-chance/memento_nrj2003.pdf/. January 2003.

[HAM 03] HAMILTON, S.L., *Microturbine Generator Handbook*, PennWell,, 2003.

[JEN 00] JENKINS, N., KIRSCHEN, D., STRBAC, G., ALLAN R. and CROSSLEY P., *Embedded Generation*, The Institution of Electrical Engineers, 2000.

[JOH 93] JOHANSSON, T.B., KELLY, H., REDDY, A.K.N. and WILLIAMS, R.H., *Renewable Energy – Sources for Fuels and Electricity*, Island Press, 1993.

[LAR 00] LARMINIE, J. AND DICKS, A., *Fuel Cell Systems Explained*, John Wiley & Sons, 2000.

[LAV 05] LAVERDURE, N., Sur l'intégration des générateurs éoliens dans les réseaux faibles ou insulaires, PhD Thesis, Institut National Polytechnique de Grenoble, 2005.

[LEM 99] LEMERLE, P., CORTINAS, D., VITET, S., MEYER J.-L. and FRAISSE J.-L., "Insertion de la production décentralisée en BT", *15th International Conference and Exhibition on Electricity Distribution*, Part 1: Contributions, CIRED, vol. 4, Nice, 2–4 June 1999.

[LHO 04] LHOMME, J.-C., *Les énergies renouvelables*, 2nd ed., Delachaux et Niestlé, 2004.

[MAS 02] MASTERS, C.L., "Voltage rise: the big issue when connecting embedded generation to long 11 kV overhead lines", *Power Engineering Journal*, vol. 16, issue 1, February 2002.

[MIA 01] MIAO, Z., KLEIN, R.L. and CHOUDHRY M.A., "Case Study of Distributed Generation on the Power Distribution System", *Proc. of NIAPS*, Texas A&M University, 15–16 October 2001.

[SAB 09] SABONNADIÈRE, J.-C., *Renewable Energy Technologies*, ISTE, 2009.

[SMI 01] SMITH, J.W. and BROOKS, D.L., "Voltage Impacts of Distributed Wind Generation on Rural Distribution Feeders", *Transmission and Distribution Conference and Exposition*, vol. 1, IEEE/PES, 28 October–2 November 2001.

[SUR 06] SURDU, C. *et al.*, "On the interest of the virtual power plant concept in the distribution networks", *Conférence Internationale des Grands Réseaux Electriques, CIGRE 2006*, papier C6–301, 2006.

[TIW 03] TIWARI, G.N., *Solar Energy – Fundamentals, Design, Modelling and Applications*, Narosa Publishing House, 2002.

Chapter 9

Control of the Energy Demand: Network Load Shedding

9.1. Nomenclature

We would like to start this chapter by defining some abbreviations and acronyms; see Table 9.1.

CMC	Communication, measurement, control
DSB	Demand Side Bidding
DSM	Demand-side management
EMS	Energy Management System
DEG	Distributed Energy Generator
GPRS	General Packet Radio Service
GSM	Global System for Mobile communication
HVAC	Heating, Ventilating and Air Conditioning
NCIT	New communication and information technologies
DER	Distributed Energy Resources
RTE	Transmission System Operator (in France)
UCTE	Union for the Coordination of Transmission of Electricity

Table 9.1. *List of abbreviations and acronyms used in this chapter*

Chapter written by Guillaume VERNEAU.

9.2. Introduction

The continuous increase in consumption and the potential integration of distributed production will lead electric networks to support more and more constraints in the coming years. At the present time, the reliability and security of the networks are guaranteed by the ancillary services, supplied by the centralized production groups, in the form of active or reactive powers, with specific dynamics in normal operations or during unexpected events.

With the advent of new technologies, especially those related to information and communication (NCIT), it can be considered that the distributed energy resources, including distributed energy generators (DEG) and loads, will no longer be "passive" but will have an important role in the good operation of the network. The "controllable" loads will be thus be able to deliver, directly or indirectly, services to electric systems, and will thus be able to contribute to the ancillary services. The advantage of using loads is justified, all the more as these loads are naturally connected in the current distribution networks, contrary to the DEG, for which implementation is still limited.

Behind the terminology of energy control, there are many actions. The first concerns the user choices in terms of equipment and behavior: indeed, a customer may choose "low consumption" equipment, such as eponym lamps or class A electronic domestic operators, for economic reasons or simply because of self conviction. Moreover, this customer may also choose to have, in his way of using energy resources, an eco-consumer behavior, for example in using preferably, on his own initiative, the major electronic appliances during off-peak hours. Doing so, the main winner of this behavior is the final user who will see his electricity bill decreasing.

A second action concerning the energy control is to accept that his habits or consumption are guided according to financial requirements coming from one or many actors of the electric system, such as the distributor or the supplier. In this case, the volume and the proliferation of customers allow large-scale actions such as peak consumption reduction. By delaying consumption actions during the less critical periods for the networks or when the electricity is less expensive on the market, the security and reliability of the systems are increased. These two approaches to energy control can be summarized by the following expressions: the first aims at "lower consumption", the second at "better consumption".

This chapter is devoted to energy control, in terms of "better consumption", while examining the general problem of load control through a series of questions, considering examples of already existing load management programs. First of all, questions on the reasons, the needs and the requirements of load control in the electric system, are detailed. Then, questions on the kind of loads to control, as well as the way to control them from a practical and material point of view, will be handled. Finally, the question of load control from its concept aspects will be tackled.

9.3. Stakes of the load control

As mentioned in the introduction to this chapter, at the present time, evolution perspectives linked to deregulation, to the possible implementation of the DEG, and to the increase of the loads are such that the network will support increasingly severe constraints. The increase in these constraints may endanger the security, reliability and good operation of the network. This last point is guaranteed by the ancillary services, intrinsically linked to the energy market: the consumption peaks lead to a high energy price. A better load management would reduce these peaks, and the prices at the same time, and consequently the constraints on the network.

Load control has many advantages. First of all, for the system, an intelligent load control could facilitate the management of the strong constraint periods on the network. Moreover, in the economic field, aggregations of consumers could become actors in the energy market by increasing the demand elasticity, while being controlled (reduction and/or delay of the demand). These consumers would become a new source of reliability for the network. Finally, contrary to the DEG, also able to guarantee the same services in the future, the loads are already connected to the network.

9.3.1. *System aspect*

To guarantee its reliability and security, the electric network requires ancillary services. Generally speaking, the ancillary services may participate in frequency control, voltage control and load power cuts during the most critical periods, and to restore the network following a black out [BEL 05]. These supplier services are usually located at the transport network level, because their size makes it possible for them to efficiently offer these services; the network operators pay them for these services. These contributions mainly consist of a production (or consumption) of active and/or reactive power during specific periods, with some time margins. This power, kept in reserve is currently generated by the generators, but can also be delivered by the load.

Reserve Categories		Period	Volume
Primary		30 s	750 MW
Secondary		800 s (or 133 s in exceptional situations)	500 to 1,000 MW
Tertiary	*1*	15 min	> 1,000 MW
	2	30 min	> 500 MW
	3	> 30 min	undefined

Table 9.2. *Active power reserve categories in France*

In France nowadays, the disconnection of loads is a last resort [RTE 04], brought about in case of voltage and frequency collapse, where entire line is disconnected. A more precise disconnection, punctual and located, would be more suitable for an active contribution of the loads to the ancillary services. But the development of new communication and information technologies should allow the implementation of these intelligent load control strategies, increasing the reliability and stability of the network, during critical periods of consumption peaks. Their implementation requires certain things:

– upgrading standard disconnection plans for finer plans, where loads are considered individually and temporarily;

– implementing load supervision strategies which may be controlled, including, as far as their responses are concerned, probabilistic and economic criteria;

– being able to implement algorithms able to adapt themselves to the risks;

– being able to detect the disturbances which necessitate the disconnection.

The aim of the disconnection is the good operation of the network: to maintain the production/consumption balance, to maintain the voltage plan, etc. It then contributes by definition to the ancillary services. Assuming that all the control, measurement and communication systems are implemented, it would be easier in the case of production/consumption imbalance to reduce the demand, rather than increase the generation. Load reduction may be performed almost instantly, whereas the increase of a generation is not always possible, or requires at least a continuous increase (constraint due to the group starting); moreover, the resulting overload could lead to new disturbances. Finally, the possible local action and the significant load distribution on the network are undeniable advantages for their control.

In the USA, especially in California, network weakness explains why load control has already been in existence for more than about ten years. Two program categories exist: those related to network reliability, and those in response to the energy prices. The first aims at keeping the network safe, the second has a financial goal [CER 03, CER 04]. Table 2.3 shows some management demand programs that have been used in the USA for a few years.

Category			Examples
Special price options	Real time pricing (RTP)		Real Time Pricing (Georgia Power)
	Time of use (TOU)		CPP (SDG&E)
Programs offering compensation for loads reduction	Request decided according to the needs of the program manager	Voluntary	Emergency Demand Response (NYISO)
		Mandatory	Real Time Demand Response (01–03, ISO-NE)
		Developed programs	Early advantage (ComEd)
	Request criteria linked to market prices		Real Time Price Response (ISO-NE)
	Without notification		Scheduled Load Reduction (SDG&E)
Programs offering compensation for the distributed generations			Rolling Blackout Reduction (SDG&E)
Programs offering services for services participation			Blackout Protection (PacifiCorp)
Programs promoting direct participation to the markets			Day Ahead Demand Response (ISO-NE)
Other programs			Energy Smart Thermostat Program (SCE)
			Real Time Profiled Response Program (ISO-NE)

Table 9.3. *Examples of US Demand-Side Management programs [EUD 04]*

9.3.2. *Energy market aspects, Demand-Side Bidding and Demand-Side Management programs*

The loads control programs described above enable the management of emergency situations due to the load reactivity, and are in keeping with a contribution to the ancillary services. These services enter into a tariff mechanism. The customers may thus become new actors on the demand side and participate in the energy market (D-1 or real time), while offering to control the loads themselves or to let an operator control them.

Customers, or groups of customers, may subscribe to market-oriented programs, in which they answer to tariff calls, while reducing their consumption. These are Demand-Side Bidding (DSB) programs. These programs offer customers the possibility of actively participating in the commercial electricity exchanges, by offering to undertake modifications in their consumption habits, and receiving financial compensation. These financial compensations may be in the form of a reduction in the electricity price, through a direct payment for unused electricity, or through a payment according to an availability guarantee to make a consumption modification at a specific time depending upon an agreement [EUD 04]. These programs are quite close to Demand-Side Management (DSM) programs, but they have several differences. The latter essentially concern customers over the long term, whereas DSB programs involve them in more punctual changes, as described in Figure 9.1.

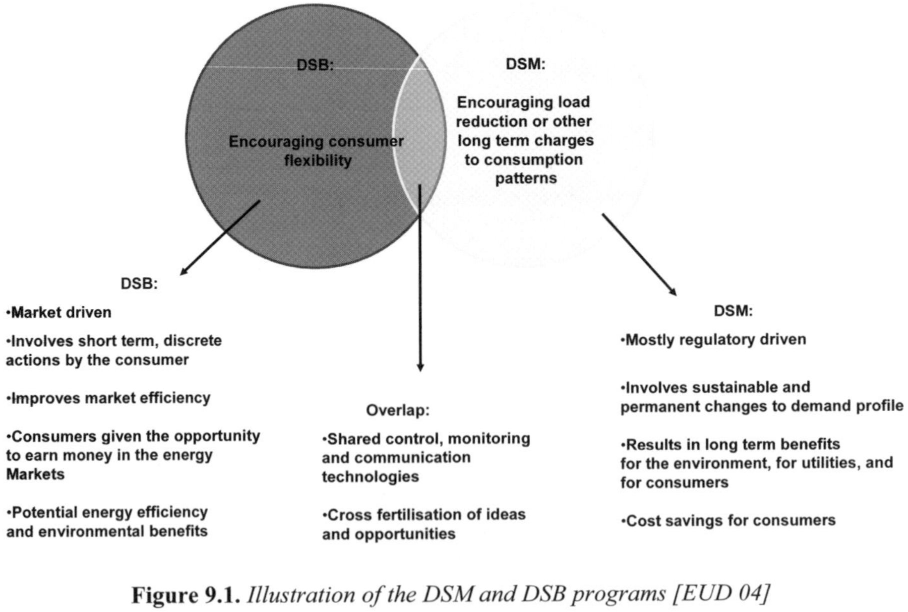

Figure 9.1. *Illustration of the DSM and DSB programs [EUD 04]*

In these DSB programs, the offer essentially aims at a reduction of the load. The offer may concern either an individual consumer (important industrial consumer), or an aggregation of small consumers grouped together in order to represent a non-negligible load at the network level (residential customer group). Consequently, there are five categories of DSB programs [IEA 01]:

– programs which aim at maintaining the supply quality;

– programs concerning the network constraints;

– programs for the electric balance;

– programs to reach market prices;

– programs for price management.

DSM programs aim at reducing and/or temporarily postponing the electricity consumption, without necessarily reducing the total consumption. These consumption modifications are established in response to the price and/or consumption peaks. DSM programs have many advantages: reduction of consumption peaks and network constraints, delay on the investments to be made in the network, better control of the generation losses (especially with intermittent generation), reduction in greenhouse gas emissions, etc. Moreover, as previously discussed, these programs make it possible to open up the energy markets and the ancillary services to the loads, with a participation which may prove efficient and useful, especially as far as the frequency control and reserve participation are concerned.

There are three types of DSM programs:

– those aiming at postponing consumptions in order to balance costs and/or energy;

– those aiming at obtaining economical benefits;

– those aiming at selling services on the energy market.

From the supplier point of view, these DSM programs may be an interesting alternative to network reinforcement. In reducing the consumption peaks, the network is subject to fewer constraints, and the generators have a better efficiency. The costs for the supplier and the consumers are thus reduced [KRE 05].

9.3.3. *Implication of the customer as a market and system actor*

Load control will be of interest both at the system service level as well as the electricity market level. Whatever its aim, it needs customers to have adequate communication, measurement and control (CMC) equipment at their disposal. These pieces of equipment will allow the customers to receive security or economic calls and to answer them by reducing, postponing or canceling their consumption.

At the same time, this equipment will enable the aggregation of small customers to have a non-negligible weight on the network. This load accumulation could be controlled by an aggregator. The new control and communication technologies would allow this aggregator to know his load state and to transmit the general state of his stock to the supplier, in order to gain better management of the energy. All this equipment will make the decision making of the consumer easier with respect to external calls. However, the comfort aspect of the customer should be taken into account: loads should not be disconnected for too long and disconnection proposals should not be too restrictive.

Finally, the DSB and DSM programs enable their participants to actively contribute to the energy markets and to the electricity commercial exchanges. One of the consequences for the customer of this active participation in the electricity commercial exchanges is that he can reduce his electricity bill by controlling his loads. Indeed, the customer receives financial compensations for erasing or postponing his consumption. However, it is obvious that for small consumers this contribution will require a load aggregation in order to obtain a non-negligible power at a network scale.

9.4. Choice of loads to control

There are many kinds of loads on the network. They differ in terms of power, rate of use or use criticality. A multi-criteria classification of these loads will thus be necessary for the implementation of load control programs. This classification must take into account:

– the ability of the loads to be controlled;

– the load power;

– the type of load (motor, convector, etc.);

– the way of controlling loads (all or nothing, or in a continuous manner);

– the control dynamics;

– the time margin allowed for the control;

– the area concerned (residential, commercial, agriculture), and the investigation factor (building, residential area, industrial site, town, area, etc.).

This classification will make it possible to define useful modeling requirements according to the loads and the targeted control applications (system or market).

9.4.1. *Load classification according to their controllability*

The controllability of a load is defined as its ability to not be supplied or under-supplied during a certain time, in response to an external call, which is linked to an economic or security factor. However, the load control must not affect the customer or the linked application critically. Each load has different characteristics at its level of operation. Moreover, some loads are subject to external and temporal factors (weather, season). All this leads to different controllability levels. The major load classes, according to their controllability [CRI 04], are as follows:

– *Control of the thermal comfort of a building*: the thermal comfort-type loads (convectors, air conditioning, ventilators, etc.) represent between 30 and 50% of the energy consumption of services and residential buildings and the electrical energy part represents a non-negligible portion [OH 06]. One idea for the control of these loads is to use the thermal inertia of the buildings to disconnect the loads during a certain time without altering customer comfort.

We can also pre-heat (or pre-cool) the building during the night when the price of energy is low [CRI 04]. The control of this type of load in the context of a deregulated market and DEG insertion is an important challenge. This type of load is one of the most flexible and is a major candidate for intelligent load control.

– *Rotating equipment*: these are ventilators and electro-hydraulic pumps. In California, there is a load control program concerning this type of equipment in the industrial area; a network of pumping stations generates a reserve by opening and closing gates in order to store energy [KIR 03].

– *Heating and cooling industrial equipment:* as for the thermal comfort loads, we can profit from the thermal inertia of these operators. Currently, this equipment works with an on/off system with immediate power generation when temperature lies outside a certain range. The new equipment should have a continuously controlled mechanism in which tariff signals can control the amount of power to be delivered during a given period. Combined with storage equipment (of heat or cold), they could reduce the system load [CRI 04].

– *Combined loads*: this kind of load could be the association of an electric heat pump, an air conditioning operator and a storage element. The operating periods may be cut or temporarily delayed, without altering the associated thermal comfort. The heat pump could also be used for cooling [CRI 04].

– *Light applications:* these loads are generally not interruptible because of their operation and because they strongly influence visual comfort. An investigation conducted in the tertiary sector has shown that a light reduction of about 30% in offices did not change the working conditions of the employees [LIG 03]. Other equipment is under investigation (lighting in public areas, supermarkets, etc.).

– *Hot water generation and heat storage*: electric heat water tanks are easily controllable loads. They have been under investigation in France since the 1980s [ANG 04, CAS 95, ORP 99], but also for a few years in the American DSM programs [CRI 04, ENE 04]. The integration of new communication, measurement and controlled technologies should allow the adjustment of the settings according to tariffs or security signals.

– *Energy storage*: the implementation of energy storage means will be a way to act directly on the electricity market signals. Energy storage will prove to be interesting in cases of over or under production or breakdown. Research efforts into energy storage, in different forms from milliseconds (super condensers) to several hours (hydraulic pumps), have been carried out [CRI 04, ISE 02].

– *Domestic equipment:* washing machines and dryers represent an opportunity for load control as operation of this equipment may be delayed in time. DSM programs use these loads to erase consumption peaks. Moreover, manufacturers of domestic equipment are improving products by implementing intelligent systems (temperature control, washing cycles, etc.) which could be used for load control [CAL 03].

– *Computer, audio and video equipment*: generally these items of equipment are not interruptible, except in the case of computer applications equipped with emergency invertors, where a disconnection is possible.

– *Non-controllable loads*: very low power applications (wireless telephone bases) present little interest to be controlled; critical applications (security equipment, hospitals) must not be controlled.

9.4.2. *Area classification*

Four important areas of activity must be considered: industrial, residential, agricultural and commercial. The electric usages vary according to these different areas [EUD 04]:

– *residential area*: five types of housing have been considered according to the main technology used for heating, knowing that a second technology is possible. For the sub-areas, we consider different housing classes, for which the energy usages are then listed. The electric equipment whose aim is not linked to thermal aspects are listed in this table as a separate section below, independent of the housing type (see Table 9.3);

– *industrial area:* nine industrial areas have been considered, each having its own sub-areas; for each of them, the applications using the energy are illustrated (note: the energies used are not necessarily of electrical origin for all the listed applications) (see Table 9.4);

– *agricultural area*: for the agricultural sub-areas, the different energy usages groups are defined here and are then subdivided according to the technologies (see Table 9.5);

– *commercial and tertiary area*: here different energy usages are defined, and are then subdivided according to the technologies (see Table 9.6).

Areas	Housing Types	Energy usages
Housing	Housing with central heating which may also use gas connected to the central heating boiler (apartments) Housing with electric heating (no partial heating) Housing with direct gas heating (apartments or private houses) Housing connected to urban heating network Partially heated housing and agricultural buildings	Heating Cooking Hot Water Air conditioning
Electrical equipment		Washing machine Tumble dryer Dishwasher Lighting Refrigerating equipment Video, audio and computer equipment

Table 9.3. *Residential area [EUD 04]*

Areas	Sub areas	Energy usages	
Iron and Steel	Electric arc Iron and Steel intcgrated	Air compressors Blast furnace Electric arc Electric process Foundries Lighting	Low enthalpy heat Motor drives Process furnace Rolled steel Sinter making Steam and high enthalpy heat
Non-ferrous metal production	Primary aluminum production Secondary aluminum production Copper production Zinc production Lead production	Air compressors Lighting Motor drives Electric furnace	Electric furnace Low enthalpy heat Steam and high enthalpy heat
Chemicals production	Fertilizers Petrochemical Inorganic chemicals Low enthalpy chemicals	Air compressors Low enthalpy heat Lighting Motor drives Electric processes	Steam and high enthalpy heat Thermal processes Energy use as raw material
Building materials production	Cement dry Ceramics and bricks Glass basic production Glass recycled production Other building materials production	Electric kilns Cement kilns Air compressors Lighting Motor drives Glass annealing electric	Low enthalpy heat Glass annealing thermal Material kilns Drying and separation Tunnel kilns
Paper and pulp production	Chemical paper Mechanical pulp and paper	Lighting Motor drives Pulping electric Refining electric	Steam and high enthalpy heat Low enthalpy heat Pulping steam Refining steam Drying and separation

Food, drink and tobacco production	Food, drinks and tobacco foods	Air compressors Cooling and refrigeration Lighting Motor drives Drying and separation electric	Steam and high enthalpy heat Low enthalpy heat Space heating Drying and separation thermal Specific heat Direct heat
Engineering	Engineering goods	Air compressors Lighting Motor drives Drying and separation electric Machinery Coating electric Foundries electric	Steam and high enthalpy heat Low enthalpy heat Space heating Drying and separation thermal Thermal foundries Direct heat
Textiles production	Textiles goods	Air compressors Cooling and refrigeration Lighting Motor drives Drying and separation electric	Steam and high enthalpy heat Low enthalpy heat Space heating Drying and separation thermal Direct heat
Other industrial areas	Other industrial areas goods	Air compressors Lighting Motor drives Drying and separation electric	Steam and high enthalpy heat Low enthalpy heat Space heating Drying and separation thermal Direct heat

Table 9.4. *Industrial area [EUD 04]*

Areas	**Energy usages**	**Technologies**
Agricultural Area	Lighting Heating Different electric needs Pumping Energy for motors	Lighting Heating/cooling Greenhouses Pumping Motor generation and operation

Table 9.5. *Agricultural area [EUD 04]*

Areas	Energy Usages	Technologies
Offices	Lighting Heating Air conditioning Different electric needs Hot water	Lighting Electric heating/cooling Gas heating/cooling Boiling electric/cooling Urban heating Different electric equipment
Commerce	Lighting Heating Air conditioning Steam requirements Different electric needs Hot water	Lighting Electric heating/cooling Gas heating/cooling Boiling electric/cooling Urban heating Different electric equipment
Public services	Lighting Heating Air conditioning Steam requirements Different electric needs Hot water	Lighting Electric heating/cooling Gas heating/cooling Boiling electric/cooling Urban heating Different electric equipment

Table 9.6. *Commercial and tertiary [EUD 04]*

9.4.3. *Most promising loads*

The most promising applications dealing with load control are obviously those linked to thermal control (HVAC). The use of thermal inertia of this equipment (building, hot water, cold chamber, etc.) should temporarily allow the interruption or reduction of the load supply. These loads have already been used in the DSM programs, and taking into account the part represented by this equipment in the tertiary and residential area, they have a strong possibility of control, thus contributing to the ancillary services. However, some parameters should be taken into account: thermal inertia of the buildings, consumption curves of the customers, current regulations (for refrigeration rooms). Some applications may operate under a reduced regime without modifying the good operation of other applications (fans, pumping systems, etc.).

The light control of the tertiary and commercial areas may also be interesting, in so far as the light intensity level seems to be too intense and so a reduction in the level will not alter customer comfort.

As far as electronic domestic equipment is concerned (washing machines, dryers, etc.), they would require more energy control than load control. The consumers should be encouraged to delay their usages over time in order to smooth out their consumption peaks.

In addition to HVAC-type equipment, a certain potential, which is specific to the industrial area, seems to exist (slow down or delay of the industrial processes). However, the possible applications will be taken case by case, according to the processes and operating equipment.

Beyond these classifications, an inquiry among the consumers will be necessary in order to know the exact possibilities of each load. This consumer inquiry will determine their acceptability factor, preference, requirements and the price they are ready to pay in order to delay or cancel their consumption. This inquiry will also make them sensitive to the energy consumption.

9.5. Needs in communications, measurements and monitoring to control the loads

9.5.1. *Introduction*

The information on energy flows and the system state are fundamental to the management, exploitation and distribution of the energy resources. The major information sources are the measurements. But the controllable loads are not the only ones to have measurement needs: the DEG and production in general as well as the storage also have many communication, measurement and control (CMC) needs. These needs include offers, exchanges, equipment protection, authority regulations, monitoring and response verification, etc.

In the USA, where the demand management measurement operators are developed enough, four large manufacturers dominate the market: ABB, General Electric, Schlumberger and Siemens. The equipment offered by these companies includes data collection and a communication interface [CER 03]. This equipment attempts to gather many requirements into a single instrument. This possibility will improve the data availability, the fault tolerance and will reduce the measurement cost. It will also make the system integration easier. The regulations concerning data format play a key role in their accessibility, as well as in the investment permanence made on the measurement equipment.

9.5.2. *Various measurement requirements*

These requirements are the offers, the load modeling, the response verification, the final usage control, the resource management (loads and DEG), the energy exchanges, the distribution control, the power quality and the protection. This requires different measurement conditions, as listed below.

Invoicing and payment: the measurements depend on the applied pricing, as well as the corresponding time resolution (quarter of an hour, half an hour, hour). The measurement system must simultaneously measure the consumption and the active (or reactive) power generation but they must be separated since the rates are generally different. Data must be accurate and reliable. We have enough time to transfer them (days, weeks). Detection of measurement breakdown or incorrect measurement is useful, as well as the detection of service interruption or wrong power delivery measurement. Compensation for these delivery quality failures may be taken into consideration.

Load modeling: it must allow the understanding of the load behavior and a precise time resolution (a few minutes) is necessary. At the same time, the response verification requires the same time accuracy in the time resolution. On the contrary, the responses cannot be different from the other load variations.

Load monitoring: non-invasive monitoring means that the measurement of a load set must make it possible to take into account the individual loads without modifying their correct operation. The required time resolution depends on the loads to be monitored. The data transfer conditions mainly depend on site measurement.

Control of final electricity usage: electric management equipment controls and monitors the loads. A time resolution of one minute is required and long delays are not acceptable, especially in the case of online monitoring [EUD 04].

Loads and DEG: the controls, management, and monitoring of the resource state are typical requirements. Information on breakdown and stops due to maintenance are useful for commercial energy exchanges. Distance access measurements are necessary for cost reduction. The control of these distributed resources also requires measurement and load forecast and power flows. This requires measuring active and reactive powers, currents and voltages in order to have a more accurate estimation of the system. Other non-electrical measurements are necessary, such as the temperature, weather forecast, voltage in the distributed group batteries, etc. Indeed, the control at the distribution network level does not generally have enough measurements at the customer connection nodes because the cost is too high. In this type of use, long delays and a lack of date are not penalizing. Another important date for the good behavior of the network is related to power quality. IEEE 1547–2003 standard says that each DEG of at least 250 kVA power should be able to deliver at its connection node information on its active and reactive power as well as its voltage.

Power quality: measurements of the power quality are required mainly at the generation level. These measurements are more important for the customer (consumer or supplier) than for the network itself. Measurements related to the voltage level, the voltage interruption and the voltage sags must preferably be permanent and the DEG protection systems must adapt their response. On the island-type networks, the frequency control is subject to fine monitoring. There are two types of control: continuous and temporary monitoring. Invoicing systems include a permanent control of the power quality; the data are stored and analyzed separately (rates and power quality) [EUD 04, ROM 02]. The large scale monitoring will be based on data storage; these data will be analyzed only if a warning signal indicates bad quality. The measurement intervals and the interpolations of the power quality are mentioned in the IEC 61000–4-30–2003 standard. The intervals are of 200 ms, 3 s, 10 min, 2 hours and 1 week. However, in order to be compatible with the EN 50160–1999 standard, many recording intervals are used simultaneously, depending on the type of measurement.

Protections: many measurements related to the power quality are used to trigger protections (frequency, voltage). The responses must be fast and a high reliability of the systems is recommended. To avoid information redundancies, it is common that the protection equipment delivers its result to the other systems.

Measurement equipment with self-diagnosis: modern measurement equipment often has means to detect the problems on transmission lines and to establish their diagnosis. This information aims at helping the operator to take adequate decisions quickly in order to limit the zone affected by the incident [CRI 04].

Regulation authorities and research centers: in the future, these organizations will need other specific measurements that are difficult to estimate now.

NOTE: Other measurements, not dependent on electrical quantities, but linked to environmental parameters are also necessary. These measurements may be the temperature, pressure, humidity, luminosity, etc.

9.5.3. *Required conditions for data measurement and communication*

Generic multifunctional equipment, implemented in an electric pricing system, will have to [EUD 04]:

– register invoicing measurements every hour;

– measure the voltage characteristics;

– deliver data following a specific request;

– send warnings which can identify their sources;

– record simultaneously data with different time intervals;

– read data sent by other equipment;

– have inputs to read other measurements;

– and have a continuous real-time measurement output.

Four other options could be included in an additional unit implemented only in case of necessity in order to:

– have outputs for control signals (of the tariff type or weather signals);

– have the possibility of compressing and encrypting data;

– control the measurement equipment access (by the network or by the data port);

– and to filter the warnings in order to avoid the pollution of the communication systems for some significant warnings.

The redundancy of the measurement equipment communicating each other allows the detection of the measurement fault and the limitation of the communication breakdowns.

The measurement reading system must also store invoice data separated from the other data, deliver measurements to other users, control the data access and efficiently filter the warnings in order to cancel the redundancies. The data transfer could be guaranteed by the standard mobile phones technologies (GSM, GPRS), or via the Internet (e-mail, website, etc.). The warning transfer should be fast, which will be made easier by means allowing a continuous communication. Finally, it is still not possible to say whether all the functions could be implemented in a single piece of equipment. However, the implementation of separate units could raise maintenance and space problems and could be expensive. The equipment should then be reduced.

9.5.4. *Choice between common or separated measurement system*

	Separate measurement systems	Single measurement system
Component properties	Cheap and simple	Complex
Required conditions for the system	Simple	Complex
Data management	Relatively simple	Complex
Data availability	A few	Good
Number of system interfaces	A few	High
Cables and communication	Many in parallel	A single system
Implementation	Different installations in parallel	A single system
Access management, coding	Parallel	Divided but more complex
Organization, maintenance	Parallel	Divided
Possibilities to improve and to develop the system	Limited	Fewer constraints
Application development	Need to learn different simple systems	All data in a single system
Regulation control	Low	Important

Table 9.7. *Comparison between separated and unique measurement systems for applications requiring measured data [EUD 04]*

Although the measure is a generic action, the traditional systems are generally dedicated to a single objective. Indeed, the required measurement conditions are different depending on the goals. Moreover, the information management is then easier and less critical. On the other hand, in the absence of regulations, the equipment manufacturers use their own standards. Finally, separate systems are preferred because they are less expensive and because their objectives are different. Table 9.7 outlines the arguments for and against separate equipment versus single systems.

Possibilities exist to improve the current measurement systems. Usually, a single instrument is dedicated to one measurement and many pieces of equipment are thus used in parallel. Consequently, a huge number of cables and installations are implemented, which is expensive and less efficient in the long run. The electric consumption measurement instruments measure voltage, current etc. and indicate the breakdowns. The results are sent only by a pulse connection. Other information (power quality, statistical data) would be necessary. This is why this transfer may require a data bus.

Deregulation is still in progress and the investments related to electric system developments are at risk. An intelligent regulation should impose measurement standards, standard interface, and competition in the measurement area.

Finally, the electric market rules are different depending on the country. But in order to have the best technological solutions and their large-scale diffusion, stable and precise regulations for large areas should be established (UCTE, etc.). This will lead to production volumes large enough to reduce equipment price.

Other problems exist. The needs for a common measurement system are complex and sometimes not defined enough. The standardization for the reading, storage and data transfer is not ready at present. The short time business interests often limit the system development, the inertia of which has longer time constraints.

For the DEG, it is not possible to have separate units for invoicing, continuous power quality monitoring and control. However, separation is possible for the protections and relative measurements.

Standard interfaces and common data structures are necessary to allow information exchange with other systems (customer load energy management, distribution control, delivery quality monitoring, commercial exchanges, etc.). These standard interfaces are also useful because of the lifetimes of the different components, which are not the same.

Field	Norms
Reading of invoicing measurements	ANSI.C12.19:1997/IEEE Std 1377:1977, Standard for Utility Industry End Device Data Tables, 201 p. (IEC 61334–4-41:1996, Distribution automation using distribution line carrier systems – Part 4: Data communication protocols – Section 41: Application protocol – Distribution line message specification. IEC 61334–4-42:1996, Distribution automation using distribution line carrier systems – Part 4: Data communication protocols – Section 42: Application protocols – Application layer, 89 p. IEC 62056–51:1998, Electricity metering – Data exchange for meter reading, tariff and load control – Part 51: Application layer protocols, 69 p. IEC 62056–52:1998, Electricity metering – Data exchange for meter reading, tariff and load control – Part 52: Communication protocols management distribution line message specification (DLMS) server, 25 p.) IEC 62056–53:2002, Electricity metering – Data exchange for meter reading, tariff and load control – Part 53: COSEM application layer, 123 p. IEC 62056–61:2002, Electricity metering – Data exchange for meter reading, tariff and load control – Part 61: Object identification system (OBIS), 25 p. IEC 62056–62:2002, Electricity metering – Data exchange for meter reading, tariff and load control – Part 62: Interface classes, 81 p.
Common data structures for the control of the distribution	IEC 61850–7-3:2003, Communication networks and systems in substations – Part 7–3: Basic communication structure for substation and feeder equipment – Common data classes, 64 p. IEC 61850–7-4:2003, Communication networks and systems in substations – Part 7–4: Basic communication structure for substation and feeder equipment – Compatible logical node classes and data classes, 104 p.
Norms proposals for common data structures for DEG	IEC CD 61400–25 Ed.1.0: 2003, Wind turbines – Part 25: Communication standard for remote control and monitoring of wind power plants, 199 p. Utility Communications Architecture (UCA) Object models for Distributed Energy Resources (UCADER), Version for IEC TC57. October 2003.
Generic presentation of the measured data	ISO/IEC DIS 19501–1: Information Technology -Unified Modeling Language (UML) – Part 1: Specification.
Norms related to the measurement of power quality	IEEE P1159.3/D9 Draft, Recommended practice for the transfer of power quality data.(PQDIF) 2002 IEEE, US, 129 p. IEC 60255–24 Electrical relays – Part 24: Common format for transient data exchange (COMTRADE) for power systems, 105 p.
Norms for the presentation of measurement data for electricity exchanges	Ediel UN/EDIFACT standard (Nordic countries) www.ediel.org ODEL – Object oriented Data model for ELectricity supply, Version 2.0, English description. SINTEF Energy Research TR A5207, Trondheim, October 2000, 50 p. ETSO Modelling Methodology for the Automation of Data Interchange of Business Processes (EMM), Version 1, Release 3, ETSO (European transmission system operators), 12 February 2003, 25 p.

Table 9.8. *Communication standards for the measured data [EUD 04]*

Standardization is also required between the measurement equipment and the data-reading equipment, between the data-reading equipment and the application using these data. The measured data should be easily available, the communication means should be flexible and the measurement data should be standardized.

9.5.5. *Control needs*

The distributed resources (loads or DEG) interact with the electric network. The latter may use these resources in view of good operation. The network can thus benefit from these resources, which must not affect its power quality or reliability. Good control of the loads should allow this goal.

Two types of applications exist for load control: standard control equipment and the EMS. A computer may be used as a control center.

9.5.5.1. *Standard control equipment*

For the first type of application, four important control equipment technologies exist: load control breakers, controllable thermostats, lighting controls and variable speed controls:

– *load control breakers*: this electronic control equipment includes a relay and a communication module. The communication module receives information from an aggregator or operator and the relay disconnects the load. The time during which the load is disconnected depends on a pre-programmed timer, included in the product. Manufacturers are: ABB, Grinteck, Cannon Technologies, etc.;

– *controllable thermostats*: these combine a communication module and a standard thermostat. The energy supplier, thanks to the communication system, can raise or reduce the threshold depending upon the network constraints. Manufacturers are Domosys, Delta Control Inc, etc.;

– *lighting control*: this mainly depends on a more efficient management of the lighting energy. The major manufacturers and distributors in Europe and in the USA are Barrington Systems, Watt Stopper, Lutron, etc.;

– *variable speed controls*: these electronic systems allow the induction motor to operate over a continuous speed range. As the load of the motorized applications varies with time, this electronic equipment allows the motors to meet the required conditions of operation while saving energy when the application is not at its maximum. The principle of this equipment is to transform the alternative current into direct current, then to generate an alternative current again but at a variable frequency. The speed of the motors is thus controlled. The loads using these systems are pumps, compressors and ventilators. The manufacturers are Emerson Controls Techniques, GE Industrial System, etc. [EUD 04].

9.5.5.2. *Energy Management Systems (EMS)*

EMS are digital management systems which control and program the energy demand of a building or of a larger industrial estate. If the system is correctly installed and used, the control is more efficient than if an individual performed it. The complexity of the system varies from the control of a single piece of equipment in a building up to the distributed control of an entire building complex; in every case, a central computer supervises the control. The system must allow a better distribution of the energy consumption (to program, to delay turn-on and turn-off, to limit the demand during certain periods, etc.). Most of the functions must not affect customer comfort as far as possible. Many companies offer EMS systems: ABB, Schneider Electric, Echelon, etc.

It is difficult to estimate the general cost of the implementation, use and maintenance of EMS. The energy saved must not be compensated for by logistic costs. Cheap, simple and well-used systems are often more efficient than more complex systems which may not be well installed and thus require additional maintenance costs. However, well installed systems and systems used correctly by qualified personnel may reduce the maintenance costs of the buildings in a significant way (about 15% according an American study) [EUD 04]. Finally, generally speaking, the maintenance of the EMS is low but the actuators and sensors still require periodical control.

9.5.6. *Conclusion*

Taking into account what has been discussed previously, it will be important in the future to reduce costs and to increase the quality and availability of the measured data. A realizable transition must be found to move from the current systems towards multiple goal measurement systems. These systems will not only take measurements but also communicate with an external operator and will also control the loads. In order to ensure an efficient interface, standards and measurement conditions must be clearly defined and implemented. The measurement, communication and control systems must be mature enough not to disturb the development of more adapted technologies.

The development of load and DEG control will be enforced by the abandonment of the interfaces and non-standardized communications between systems and by the abandonment of equipment in parallel without communication. Standardizing this equipment will be advantageous, mostly for the maintenance.

Future requirements will require common conditions for the measurement, communication, control, storage and exploitation of the measured data. They will also require different needs in terms of common standards satisfying measurement, with object-oriented protocols and a transition towards common communication platforms for the different measurements.

9.6. Model and algorithm needs for load control

9.6.1. *Load models*

Load modeling represents an entire domain in itself in electric system research. Load representation is particularly a problem for the stability studies [IEE 90, TAY 94]; it is thus important to find a model that is more appropriate and more adapted [ROM 02].

The major work made in the past was only devoted to the induction machine, which is critical only for a few seconds after a disturbance. But some phenomena (voltage collapse) take several minutes. The objective is thus to achieve a dynamic model covering the long and short term simultaneously. It is important to study how the instantaneous load variations affect the stability and the safety of the network. The current idea is to use load dynamic models.

Loads are difficult to model because of their diversity, their high distribution and their high numbers. Moreover, the effective presentation of the composition of the load, according to the day, hour and weather conditions, is difficult to determine and forecast. Some companies propose a load forecasting service [LEE 95], and in order to achieve this goal, new techniques have been developed for a better load representation [DOV 87]. These new techniques have reduced the error margins which positively impact the economy and the reliability of the system.

9.6.1.1. *Physical and empirical models*

A physical model is based on fundamental knowledge, on physical phenomena affecting the system. This kind of model, based on elementary law, may give accurate results in simulation, but in the case of a more complex system, as it is often difficult to obtain laws affecting the system, it is necessary to develop a model based on empirical laws, simply called an empirical model or black box. This latter model is used when there is not enough knowledge to generate a physical system or if the general operation of the system is complex. We then try to establish relations between the measured inputs and outputs.

9.6.1.2. *Data for load modeling*

Two approaches may be used to obtain data on the load characteristics. The first one is based on measurement. It uses measurements made in the substations in order to determine the load sensitivities (active and reactive powers) to the voltage and to the frequency. The measurements are obtained *in situ* during normal operation or intentional disturbances during tests. The system parameter identification is created by fitting the measured data with an estimated model. This approach is also called a "gray box" in the sense that the model is the object of assumptions. The major advantage of this empirical method is the availability of the actual data of the investigated system. Seasonal variations are followed as variations of the normal operation. On the other hand, the necessary measurement equipment for the following of the loads requires a significant economic investment.

The other approach is based on components. It requires the development of a load model from information on its constituents. Information on the mix of different load classes in a substation describes the composition percentage of the load classes in terms of consumption. The information on the composition of each class describes the relative percentages of each load type in a class in terms of active consumption. The information about the characteristics of each load is related to the physical characteristic of each load (Figure 9.2).

This method does not require on-site measurements and is easy to adapt to different systems and to different operating conditions. However, data on the load class composition vary according to the nodes and the operating conditions; it is thus often useful to update data for each system node. A better description of the load characteristic will be obtained by combining both methods.

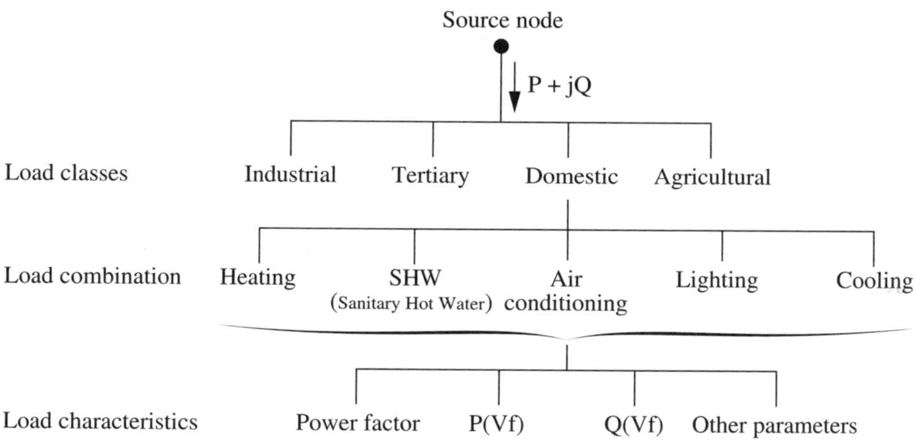

Figure 9.2. *Load modeling approach based on classes [KUN 94]*

9.6.1.3. *Measurements*

Measurements made *in situ* provide us with online knowledge of the system state. If the measurement equipment is limited to collecting data and monitoring, the operator must take the control decisions directly relative to the various risks. Other more sophisticated equipment integrates the information online, with an automatic control system and takes into account the observation of the operator. The main drawback of these measurements is the implementation and maintenance costs of the equipment.

Offline data processing enables the delivery of the system information, corresponding to periods before the measurement. This allows the analysis of different system variables, at different places, and enables us to follow their evolution. This analysis will be an important database for research objectives. The drawback is to not be able to react directly in order to observe the system reactions.

The measurement may be made during testing or during normal operation. From a technical and economical point of view, data acquisition during the normal operation is advantageous. The alternative to proceed with tests presents the drawback of a voluntary change in the load operation and thus penalizes the customers.

9.6.1.4. *Load characterization*

One aggregation contains many load types. As previously mentioned, the load composition of a specific area is characterized by data on the load classes, on the composition of each class and by the properties of each elementary load inside each class.

The load classes are generally consolidated in industrial-, residential-, commercial- and agricultural-type loads. The industrial loads are mainly linked to industrial processes corresponding in major motor parts. The residential loads correspond in large proportions to thermal comfort equipment (air conditioning or heating), and to electronic domestic apparatus. The commercial loads mainly correspond to lighting and air conditioning units. Finally, the agricultural loads are mainly induction motors for different pumping systems.

The different loads of the different classes may be loads with rapid dynamics for the electric and mechanical characteristics (induction motors), loads presenting significant discontinuities in case of voltage variations (motor protection, discharge lamps); loads whose response to voltage faults do not present significant discontinuities or delays (incandescent lamps), or loads with slow characteristics (electric heating).

Some of the major loads [TAY 94, ROM 02, KUN 94, IEE 93] are:

Induction motors: these represent an important part of electric consumption (between 60 and 70%) in the residential, commercial and industrial areas [TAY 94]. They require a constant torque, whatever the speed; they are the most constrained from a stability point of view. We find motors in pumps, ventilators, compressors for air conditioning and refrigeration.

Lighting: this is mainly made up of fluorescent mercury and sodium lamps. It represents an important part of the commercial area loads. These loads are very voltage sensitive in the sense where they switch off at 80% of the voltage. The equipment using these lamps uses the electric discharge principle. Lighting occurs when switching on and the color changes (from blue to white for mercury), with the simultaneous increase of the temperature and pressure. After the lamp is switched off, it takes a certain cooling time before switching on again. These fluorescent lamps are mainly used in offices, supermarkets and commercial districts because of their low manufacturing costs and their high light efficiency.

Thermal loads: these represent an important percentage of the loads in the residential (water heating, heating, etc.), and industrial (steel, foundry, etc.) areas: between one third and half of the energy consumption [CRI 04]. Their short-term behavior is similar to that of a constant resistance. The possible power variations at the equipment input do not immediately affect the temperature. After a few moments, if the heat generation has decreased, the ON cycle of the thermostats will continue in order to get back the required temperature. In case of low voltage, the ON cycle temperatures will increase less. In the OFF cycle, the thermostats will not be affected by the voltage changes. When the voltage is slow, most of the thermostats work in the ON cycle; the consumption is however identical as under normal operation. A sufficiently realistic model of the thermostatically controlled loads is presented in Figure 9.3.

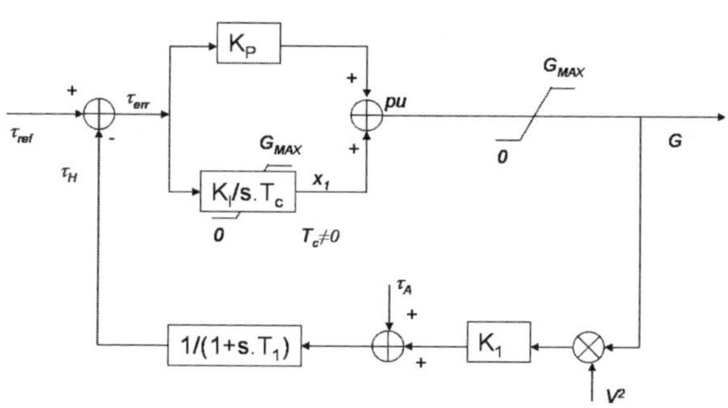

Figure 9.3. *Aggregated load realistic model, thermostatically controlled [KUN 94]*

For this model, the dynamic equation of heating equipment can be written:

$$K\frac{d\tau_H}{dt}=P_H-P_L \quad [9.1]$$

τ_H	Temperature of the heated surface	T_C	Time constant of the integrator
τ_A	Ambient temperature	τ_{ref}	Temperature of reference
P_H	Power of the heating element $\left(P_H=K_H\ G\ V^2\right)$	P_L	Heat loss by evacuation in the surrounding environment $\left(P_L=K_A\left(\tau_H-\tau_A\right)\right)()$
τ_A	Ambient temperature	T_1	Time constant of the load
G	Load conductance	K_1	Benefit associated with load model
K_P	Benefit proportional to controller	G_0	Initial value of G
K_I	Integral benefit of controller	G_{max}	Maximal value of G

Substituting the expressions of P_H and P_L, the previous equation becomes:

$$K\cdot\frac{d\tau_H}{dt}=K_H\cdot G\cdot V^2-K_A\cdot\left(\tau_H-\tau_A\right) \quad [9.2]$$

$$\frac{d\tau_H}{dt}=\frac{K_1}{T_1}\cdot G\cdot V^2+\frac{1}{T_1}\cdot\left(\tau_A-\tau_H\right) \quad [9.3]$$

with:

$$T_1=K/K_A \text{ and } K_1=K_H/K_A$$

The temperature τ_H is compared with the temperature reference and the error signal controls the load conductance via a PI controller. When all controlled loads are connected, the conductance reaches its maximum value G_{max}. The initial conditions of Figure 9.3 are written:

$$K_1=\frac{\tau_{ref}-\tau_A}{V_0^2\cdot G_0} \quad [9.4]$$

Load changers: these are not exactly load-type components. However, following a disturbance, they adjust the voltage at the distribution level, which affects the voltage sensitive loads.

The composition of the load system that is actually connected strongly depends on the day, the month, the season and the weather. The thermal comfort loads vary seasonally, but also depend on weather conditions (winter or summer, dry or humid). The weak days are strongly influenced by the commercial and industrial loads [IEE 93].

9.6.1.5. *Standard load models*

To model the loads, the mathematical equation links the voltage to the active and reactive powers used. However, because of the wide diversity and load distribution, their modeling is difficult and many alternatives have been proposed depending on the fixed goal. A classification has been made between dynamic and static models. The static model does not depend on time and describes a relationship between active and/or reactive powers, voltage and/or frequency, at any time. On the other hand, a dynamic model takes into account frequency and/or voltage according to time.

The more common static load models for active and reactive powers are expressed under an exponential or polynomial form, possibly with a term depending on frequency [IEE 93, IEE 95]. Many static models exist:

Polynomial model (ZIP): the relationship between power and voltage of a load may contain a constant power term where the relationship does not depend on the voltage variations, a constant current term, where the relation is linear, and a constant impedance term, where the relation is quadratic. The ZIP model is represented by the addition of the three terms:

$$P = P_0 \cdot \left[a_1 \cdot \left(\frac{V}{V_0} \right)^2 + a_2 \cdot \left(\frac{V}{V_0} \right) + a_3 \right] \qquad [9.5a]$$

$$Q = Q_0 \cdot \left[a_4 \cdot \left(\frac{V}{V_0} \right)^2 \right] + a_5 \cdot \left(\frac{V}{V_0} \right) + a_6 \qquad [9.5b]$$

V_0, P_0 and Q_0 are the initial condition values of the system under investigation and the coefficients a_1 to a_6 are the model parameters.

Exponential model: the powers depend on voltages according to:

$$P = P_0 \left(\frac{V}{V_0} \right)^{np} \qquad [9.6a]$$

$$Q = Q_0 \left(\frac{V}{V_0} \right)^{nq} \qquad [9.6b]$$

The parameters of this model are np, nq and the P_0 and Q_0 values of the active or reactive powers correspond to the initial conditions. Common values for the exponents of different loads [TAY 94] figure in Table 9.9. In the particular cases where np or nq equal 0.1 or 2, the load model then represents respectively a power, a current or a constant impedance.

Loads	**np**	**nq**
Air conditioned	0.5	2.5
Heat electric convector	2	0
Fluorescent lamp	1	3
Pumps, ventilators, little motors	0.08	1.6
Big industrial motors	0.05	0.5
Small industrial motors	0.1	0.6

Table 9.9. *Common values of the exponential models of the exponents [TAY 94]*

Frequency load model: the previous models may include a frequency dependence multiplying the equations by:

$$\left[1 + A \cdot \left(f - f_0\right)\right] \quad [9.9]$$

where f_0 and f are the rated and relative frequencies of the node voltage and the A parameter represents the frequency sensitivity of the model. It must be noted that although the frequency is the same at any node, a frequency variation will influence the loads differently according to their type (and the A coefficient).

Induction motor model: Figure 9.4 shows an equivalent circuit of an induction motor. Rs, Rr, Xs and Xr are respectively the stator and rotor resistances and reactances. Xm is the magnetizing reactance and s the slip of the motor. The dynamics of the stator fluxes are generally neglected in the stability analysis; it is the same for the rotor fluxes in the long-term analysis. Figure 9.5 shows the equivalent circuit for transient states in which the induction model is represented by an e.m.f E' and an impedance X'.

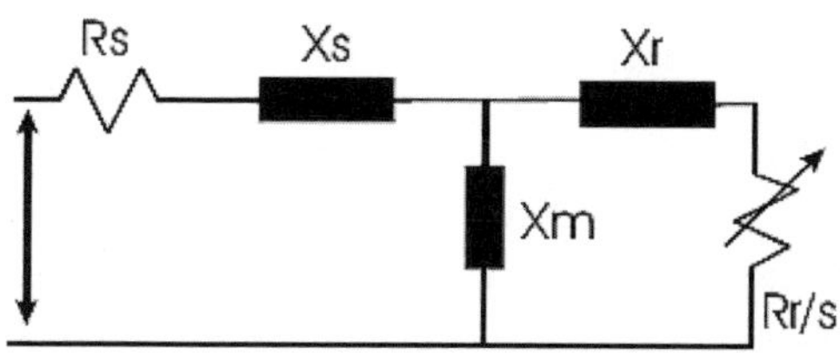

Figure 9.4. *Equivalent circuit for an induction motor in a permanent state*

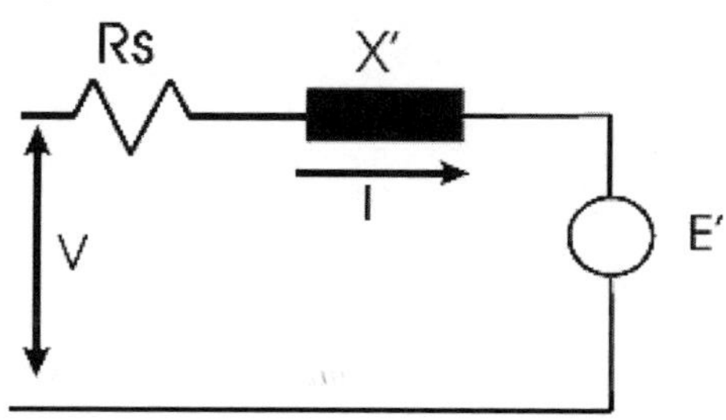

Figure 9.5. *Equivalent circuit for an induction motor in a transitory state*

9.6.1.6. *Dynamic models*

Static models are not always sufficient to represent the load behavior; dynamic models are thus necessary. These parameters can be obtained using an approach based on measurement or using an approach based on the components.

Due to the significant number of electric heating loads in Sweden, and their critical effects on the system stability, load models with an exponential representation have been proposed. These models are represented in the form of non-linear equations, where the active and reactive powers have non-linear dependency with respect to the voltage:

$$T_P \cdot \frac{dP_r}{dt} + P_r = P_0 \cdot \left(\frac{V}{V_0}\right)^{as} - P_0 \cdot \left(\frac{V}{V_0}\right)^{at} \qquad [9.10]$$

$$Pl = P_r + P_0 \cdot \left(\frac{V}{V_0}\right)^{at} \qquad [9.11]$$

where V_0 and P_0 are the voltages and powers used before the voltage variation; P_r the active recovering power, P_l the response of the total active power, T_P the recovering time constant for the active power, αt the transient dependence between the active load and the voltage and αs the permanent state dependency. Similar equations exist for the reactive power.

Loads class	Power factors	dP/dV	dQ/dV	dP/df	dQ/df
Residential					
Summer	0.9	1.2	2.9	0.8	62.2
Winter	0.99	1.5	3.2	1.0	-1.5
Commercial					
Summer	0.85	0.99	3.5	1.2	-1.6
Winter	0.9	1.3	3.1	1.5	-1.1
Industrial	0.85	0.18	6.0	2.6	1.6
Central auxiliaries	0.8	0.1	1.6	2.9	1.8

Table 9.10. *Static parameters of loads by class [KUN 94]*

Component	Power factor	dP/dV	dQ/dV	dP/df	dQ/df
Air conditioned					
Three-phase	0.9	0.088	2.5	0.98	-1.3
Mono-phase	0.96	0.202	2.3	0.89	-2.7
Window-type	0.82	0.468	2.5	0.56	-2.8
Water-heating, oven, frying	1.0	2.0	0	0	0
Dishwasher	0.99	1.8	3.6	0	-1.4
Washing machine	0.65	0.08	1.6	3.0	1.8
Tumble dryer	0.99	2.0	3.2	0	-2.5
Refrigerator	0.8	0.77	2.5	0.53	-1.5
TV	0.8	2.0	5.1	0	-2.5
Incandescent lamp	1.0	1.55	0	0	0
Fluorescent lamps	0.9	0.96	7.4	1.0	-2.8
Industrial motors	0.88	0.07	0.5	2.5	1.2
Ventilator motors	0.87	0.08	1.6	2.9	1.7
Agriculture pumps	0.85	1.4	1.4	5.0	2.0
Arc furnace	0.7	2.3	1.6	-1.0	-1.0
Unloaded transformer	0.64	3.4	11.5	0	-11.8

Table 9.11. *Static load parameters by component [KUN 94]*

Many publications and references have enabled a collection of coefficients for the static characteristics of many loads [IEE 93, IEE 95, KUN 94]. This information is summarized in Tables 9.10 and 9.11.

According to the modeling, the coefficients to be applied are different. Therefore, for a load representation depending on voltage and angular frequency, Table 9.12 presents the coefficient to be used for equations [9.12] and [9.13]:

$$P = P_0 \cdot V^{pv} \cdot \omega^{p\omega} \qquad [9.12]$$

$$Q = Q_0 \cdot V^{qv} \cdot \omega^{q\omega} \qquad [9.13]$$

Type of loads	Exponents			
	Pv	qv	pω	qω
Incandescent lamps	1.55	0	0	0
Fluorescent lamps	0.96	7.38	1	-26.6
Air conditioned	0.2	2.3	0.9	-2.67
Tumble dryer	2.04	3.27	0	-2.63
Refrigerators and freezers	0.77	2.5	0.53	-1.46
Small equipment	2	0	0	0
Pumps and ventilators	0.08	1.6	2.9	1.8
Radiators	2	0	0	0
TV, hifi or computer	2	5.2	0	-2.6

Table 9.12. *Coefficients for the models depending on voltage and pulsation [LOU 03, LOU 04]*

9.6.2. *Algorithmic aspects and loads control*

The use of load models strongly depends on the objective relative to their use. If load control is envisaged in the case of energy solutions, such as disconnection of some equipment, in the case of frequency or voltage disturbances, the load model does not need to be very complex. On the contrary, if the load control is devoted to a service aspect such as the contribution to the reserves, the model should be finer, and statistical consumption data along with probabilistic aspects relative to this consumption will be necessary. The model complexity depends on the control algorithm objective but also on the data availability (measurement, parameters, etc.).

In addition to the load disconnection aspect in case of emergencies, the load control may be included in more accurate contribution ancillary services (voltage regulation, frequency regulation). This contribution may also aim at smoothing a consumption peak, at delaying (or anticipating) consumption over a given period. Finally, the load control may be made for economic reasons, being guided by economical or tariff incentives. It is thus understandable that there is not only one control algorithm but that many algorithms may exist, adapted to different objectives: to smooth load peaks, to re-establish voltage instabilities, to re-establish frequency instabilities, to detect power oscillations, etc. [CRI 04].

9.7. Conclusion

Consideration of the problem of load control has allowed an analysis of the reasons, the means and the needs to reach some conclusions:

– the load control may have many goals: from the control of an emergency situation (voltage and frequency disturbance) to more economical aspects (energy market), in-between other applications concerning good network operation (smoothing of consumption peaks, better energy control etc.). The control objectives vary according to the actors; thus, the stakes for the operator will not be the same as for the final users;

– some programs linked to the control already exist throughout the world, especially in the USA. The load control principles and the remunerating means vary from a program to another;

– the load choice will depend on the control goal, along with the loads actually available for this control: depending on the activity areas, the days, the external conditions, the same loads will not be connected and/or controlled;

– from a material point of view, much communication, measurement and control (CMC) equipment currently exists but offers the three functions separately, which raises compatibility problems. This equipment is expensive but an extension of the loads control actions should allow manufacturing on a larger scale, which would lead to a price decrease;

– whatever the type of load control (simple disconnection, or more elaborate energy control), it will require an investment in CMC equipment. Again, this investment will depend on control goals;

– finally, load control strategy development requires load modeling and the implementation of algorithms. The model accuracy will depend on the control objective; in the same way, the algorithm structure will depend on the control type and a generic algorithm will be difficult to propose.

9.8. Bibliography

[ANG 04] ANGIOLETTI R., DESPRETZ H., "Maîtrise de l'énergie dans les bâtiments – Définitions. Usages Consommations", *Techniques de l'Ingénieur*, BE 9 020, 2004.

[BEL 05] BELHOMME R., ROBERT G., "Ancillary services provided by dispersed generation", *CIGRE 2005*, 2005.

[CAL 03] CALIFORNIA ENERGY COMMISSION, Final Report Compilation for Smart Load Control and Grid-Friendly Appliances, report prepared by Architectural Energy Corporation and Battle Memorial Institute Northwest Division, October 2003.

[CAS 95] CASTAGNONI A., "Applications Electrodomestiques – Généralités", *Techniques de l'Ingénieur*, traité Génie électrique, D 5 330, 1995.

[CER 03] "Load as a reliability resource in restructured electricity markets", *Consortium for Electric Reliability Technology Solution*, CA Energy Commission, October 2003.

[CER 04] "Review of current Southern California Edison Load Management Programs and proposal for a new market-driven, mass-market, demand-response program", *Consortium for Electric Reliability Technology Solution*, CA Energy Commission, February 2004.

[CRI 04] CRISP project, http://www.ecn.nl/crisp/, 2004.

[DOV 87] DOVAN T., DILLON T.S., BERGER C.S., FORWARD K.E., "A microcomputer based on-line identification approach to power system dynamic load modeling", *IEEE Transactions on Power System*, vol. PWRS-2, n° 3, p. 539–533, August 1987.

[ENE 04] ENERTECH ET CABINET SIDLER, http://sidler.club.fr/index.html, 2004.

[EUD 04] EUDEEP PROJECT, http://www.eudeep.com/, 2004.

[IEA 01] IEA DSB, Technologies for demand side bidding, sub-task report 4, October 2001.

[IEE 90] "Voltage stability of power systems: concepts, analytical tools and industry experience", *IEEE*, special issue: Stability, 90TH0358–2-PWR, 1990.

[IEE 93] IEEE Task Force on Load Representation for Dynamic Performance, "Load representation for dynamic performance analysis", *IEEE Transactions on Power Systems*, vol. 8, n° 2, p. 472–482, May 1993.

[IEE 95] IEEE Task Force on Load Representation for Dynamic Performance, "Standard load models for power flow and dynamic performance simulation", *IEEE Transactions on Power Systems*, vol. 10, n° 3, p. 1302–1313, August 1995.

[ISE 02] "Siebentes Kasseler Symposium Energy-systemtechnik. Energiespeicher und Energietransport", *ISET 2002*, Institute for Solar Enrgieversorgungstechnic, Kassel, 2002.

[KIR 03] KIRBY B., HIRST E., "Technical issues related to retail-load provision of ancillary services", *New England Demand Responsive Initiative*, 10 February 2003.

[KRE 05] KREMA Limited, A scoping Study: Demand Side Measures on the UK Electrical System, 29 August 2005.

[KUN 94] KUNDUR P., *Power System Stability and Control,* p. 17–40, 271- 312, 959–1000, Electric Power Research Institute, McGraw-Hill, 1992.

[LEE 95] LEE W.H., FINLEY L.A., BURI M., "Forecasting electric demand of distribution planning in rural and sparsely populated regions", *IEEE Transactions on Power Systems*, vol. 10, n° 4, p. 2008–2013, November 1995.

[LIG 03] LIGHTING INSTITUTE AT RENSSELAER POLYTECHNIC, Albany, NY, http://www.lrc.rpi.edu/researchTopics/caseStudies/index.asp, 2003.

[LOU 03] LOUIE K.W., MARTI J.R., "New aggregate load models for different load data formats", *CCECE 2003 – CCGEI 2003*, Montreal, 2003.

[LOU 04] LOUIE K.W., "A new approach to compose load devices in electric power systems", *International Journal of Applied Science and Engineering*, vol. 2, n° 2, p. 197–210, 2004.

[OH 06] OH H., MOHOLKAR A., DOUGLAS S., KLINKHACHORN P., *System Reliability and Price Responsiveness of Residential Loads*, West Virginia University Advanced Power& Electricity Research Center (APERC), January 2006.

[ORP 99] ORPHELIN M., ADNOT J., "Improvement of methods for reconstructing water heating aggregated load curves and evaluating demand-side control benefits", *IEEE Transactions on Power Systems*, vol. 14, n° 4, November 1999.

[ROM 02] ROMERO I., Dynamic load models for power systems, Thesis, Lund University, 2002.

[RTE 04] Mémento de la sûreté du système électrique, Document RTE, 2004.

[TAY 94] TAYLOR C.W., *Power System Voltage Stability*, p. 17–135, Electric Power Research Institute, McGraw-Hill, 1992.

List of Authors

Thierry ALLEAU
AFH
Meaudre
France

Pierre BAURENS
CEA
Grenoble
France

Yves BRUNET
LEG
Grenoble
France

Raphaël CAIRE
G2ELAB
ENSIEG
INP Grenoble
Saint Martin d'Hères
France

William D'HAESELEER
Energy Institute
Université catholique de Louvain
Belgium

Patrick LUICKX
Energy Institute
Université catholique de Louvain
Belgium

Jean-Philippe POIROT-CROUVEZIER
CEA
Grenoble
France

Daniel QUENARD
CSTB
Saint Martin d'Hères
France

Bertrand RAISON
G2ELAB
ENSIEG
INP Grenoble
Saint Martin d'Hères
France

Jean-Claude SABONNADIÈRE
LEG
Grenoble
France

Pierre SERRE-COMBE
CEA
Grenoble
France

Georges VAN GOETHEM
European Commission
Brussels
Belgium

Guillaume VERNEAU
IDEA
ENSIEG
Saint Martin d'Hères
France

Georges ZISSIS
CPAT
University of Toulouse 3
France

Index

F, G

H, I

L

M, N

P

R

S

T

U-Z